T0399516

Materials for Chemical Sensors

Editors

Subhendu Bhandari
Department of Plastic and Polymer Engineering
Maharashtra Institute of Technology
Aurangabad, Maharashtra
India

Arti Rushi
Electronics and Telecommunication Department
Maharashtra Institute of Technology
Aurangabad, Maharashtra
India

CRC Press
Taylor & Francis Group
Boca Raton London New York

CRC Press is an imprint of the
Taylor & Francis Group, an **informa** business

A SCIENCE PUBLISHERS BOOK

Cover illustration reproduced by kind courtesy of Dr. Subhendu Bhandari.

First edition published 2023
by CRC Press
6000 Broken Sound Parkway NW, Suite 300, Boca Raton, FL 33487-2742

and by CRC Press
4 Park Square, Milton Park, Abingdon, Oxon, OX14 4RN

Library of Congress Cataloging-in-Publication Data (applied for)

ISBN: 978-0-367-48435-4 (hbk)
ISBN: 978-1-032-45789-5 (pbk)
ISBN: 978-1-003-03977-8 (ebk)

DOI: 10.1201/9781003039778

Typeset in Times New Roman
by Radiant Productions

Preface

Sensors constitute the intelligent class that can be useful for the detection of various organic/inorganic contaminants. The sensors are the devices which belong to the category of transducers. They accept one form of signal and convert it into another form. Sensors can be classified into various sub-categories which may be either in terms of the sensing material that is being employed in designing of the sensor, or the target parameter for which the sensor is intended. However, sensors based on changes of chemical properties are always standing as monarch of research interests since in such sensors, technology component is considerably less and output signals are easy to process and analyze. Fundamentally, such sensors consider chemical reactions as the backbone of its sensing mechanism.

While fabricating any sensor, the main component which requires the highest attention is nothing but the 'sensing material'. The overall characteristics of any sensor are mostly dependent on the sensing material used in the fabrication of the sensor. 'Materials with mind' are the highest sought for resources in sensor science. Technologies in a sensor appear to be the most subsidiary part, only after the sensor material is 'intelligent' enough to extend a reliable, sensitive, selective and repetitive performance towards detection of the targeted analyte(s) as the first hand desired characteristics. Apart from very few instances, naturally occurring substances are rarely reported to be efficient sensing platforms as they fall short of meeting all requisites of an ideal sensor in a single material. Therefore, modification/modulation in naturally occurring substances or chemically/physically engineered substances, perhaps, constitutes the most significant part of consideration in sensor materials. Till date, the sensor science has witnessed the most influenced research in this particular direction.

The primary objective of this book is to discuss various materials which could be used efficiently in fabrication of chemical sensors. This book would be highly useful for the research community working in the area of sensor development. Also, for the material scientists, this book is expected to be a helping guide. The undergraduate and post graduate faculty members and students interested to work with various sensor materials may find this book as a guiding platform.

This book constitutes a spectrum of sensing materials used in the fabrication of chemical sensors. In this book, eight chapters are devoted to some of the most efficient sensing materials that have been used in the development of chemical sensors. This book begins with one of the classical materials viz. metal oxide which has attracted research attention through decades, back from 1950s. The important properties of metal oxides which are important in sensing applications are discussed in the first chapter. Also, sensing mechanism involved in metal oxide-based sensor and characteristics of this class of sensor are discussed in this chapter. The second chapter encompasses information related to one of the most promising one-dimensional nanomaterials—carbon nanotubes. Authors have provided an insight in design architecture, fabrication methods, functionalization methods, mechanism of CNT based sensors and sensing parameters. The third chapter focuses on graphene, a two-dimensional sensing material. The discussion of the physiochemical properties of this material, used in the demonstration of chemical sensor, is provided here. The fourth chapter includes review of hydrogel-based sensors. The operational principle, sensor design, sensing parameters and various sensing applications are explained in this chapter. The fifth chapter provides comprehensive idea about luminescent metal-organic frameworks (MOF) as chemical sensors. The uses of MOF materials in the detection of inorganic and organic compounds such as VOCs, ionic species, biomolecules, environmental pollutants, toxic molecules, etc., have been discussed in the chapter. The sixth chapter is assigned for biomaterials and biosensing technology. The applications of biomaterials for the detection of cancer cells, enzymes and human motions have been discussed in this chapter. The seventh chapter is based on the theme of application of textiles in chemical sensor. In this chapter, discussion is given on textile based wearable sensors for detection of sweat, wound extrudates, etc. The last chapter encompasses the information about various materials used in the chemical sensing which are not discussed in the previous chapters. The sensing capabilities of materials such as pyrylium, hydrazone, black phosphorus, diamond, electrolyte materials, quantum dots, fluorine derivatives, meta materials, ligands, crown ethers, porphyrin, etc., are discussed in this chapter. The discussion of all such 'intelligent' materials used in the chemical sensing is a herculean task; the inventory is epical that might require volumes altogether. Therefore, discussion on highly efficient materials that are used in the chemical sensing has been included in this book with a preferred orientation of different classes of materials.

The credit of initiation of this book project goes to the publishers. It is our honour to serve as the editor for this book. The contribution for the proposed chapter has been invited by the editors. Renowned scientists across the globe having interest in this particular topic have been requested to contribute in this book project. We would like to declare that all the submitted chapters are thoroughly reviewed and have undergone the revision process.

The timely submission from all the authors is appreciated in view of completion of this book project. We would like to extend our sincere thanks to all the contributing authors and co-authors for their timely submission and for their dedication in this entire journey. We owe our heartfelt regards to the publishing staff (especially, Mr. Raju Primlani) for their initiative, co-ordination, continuous support and time-to-time guidance.

Subhendu Bhandari

Arti Rushi

Contents

CHAPTER 1

Metal Oxides as Chemical Sensors

Bhagwan G. Toksha[1] and *Sagar E. Shirsath*[2,*]

Introduction

The first attempt at using metal oxide materials as chemical sensors dates seven decades back (Brattain and Bardeen 1953). The discovery revolved around the logic that semiconducting materials modify their resistance as they come in contact with some active substance depending on the atmosphere they are in contact with. This scientific exploration further expanded with continuous blood monitoring through biosensors and ion concentration monitoring through field effect based devices (Clark and Lyons 1962, Bergveld 1972). Further, the form of devices that could detect gases through heat arrangement and resistivity change could be realized (Taguchi 1972, Yagawara et al. 1990). Ideally, a sensor is a device carrying the virtues of being inexpensive, portable, accurate and instantaneous in response to some stimulus. The chemical sensors offer selective, measurable, and reproducible response to a certain target chemical substance (Fowler et al. 2009, Bandodkar et al. 2016). This 'chemical of interest' could be present in the surrounding in liquid/gaseous phase at any concentration level. These devices are fabricated with wide spectrum of material recipes to work as chemical sensors. However, with tremendous advancements in recent times, a thing such as 'ideal chemical sensor' is yet to be achieved. The chemical parameters, mostly of interest for sensing applications other than selective applications, are pH, salinity/conductivity, dissolved oxygen/carbon dioxide, etc. (Buono et al. 2021, Deji et al. 2021, Luo and Wang 2021, Sinha et al. 2021). The presence/absence of a chemical entity as a step change in any parameter, identifying a particular analyte, concentration in terms of magnitude are detected in sensing mechanisms. The metal oxide sensors reduced in size possess flexibility, simple operation, structural designs able to wrap/twist, and compatibility to integrate

---

[1] Electronics and Telecommunication Department, Maharashtra Institute of Technology, Maharashtra, India.

[2] School of Materials Science and Engineering, University of New South Wales, Sydney, NSW 2052, Australia.

* Corresponding author: s.shirsath@unsw.edu.au

with IoTs (Internet of Things), which most importantly takes the sensing applications into automated versions (Gomes et al. 2019).

The worldwide usage of chemical sensors is going to produce a business of more than 30 billion USD before 2030 (www.marketresearchfuture.com 2021). The research interest in chemical sensors is projected to witness significant increase in the forthcoming decade. The prominent reason for this growth is the increased use of chemical sensors in healthcare, oil and gas, industrial, agricultural, and biochemical sectors. The use of chemical sensors is going to be inevitable in detection of various chemical entities involved in vehicle industry, processing industry, and safety protocol and monitoring industry. In the healthcare industry in particular, the chemical sensors are of great help in treatments, portable monitors, diagnosis, and drug/alcohol abuse diagnosis. The developments in this field jointly with lower manufacturing costs involved will lead to higher clamour and mobilize the demands. Metal oxides are materials of interest for the researchers working in the field of sensors due to their affordability, adaptability in terms of properties by compositional changes, and ease in designing the connection with single recording devices. These materials also exhibit sizable step change in response parameter on exposure to the target molecule, adaptability in terms of properties by compositional changes, and ease in designing the connection with single recording devices (Sun et al. 2012). The sensors having metal oxides as active material display better results as compared to other materials in the detection of baneful pollutants, and explosive gases, and serious health-related symptoms in human exhaled breath (Fernandez et al. 2018, Yang et al. 2021).

The output of research activities in terms of publications and patents usually report sensing mechanisms limited to individual gas or chemical entities. The laboratory level experiments are found to be falling much short to real world working conditions where all other ambient conditions such as humidity, other interfering gases and temperature are present. The understanding of various mechanisms leading to sensing action could be broadly classified as 'chemiresistive' wherein the designed sensing material changes its electrical resistance in response to the chemical reaction with the target material or 'cataluminescent' where the electromagnetic radiation would be the output of a chemical reaction between the sensing material and the target material (Liu et al. 2016, Hu et al. 2019, Park et al. 2019). The structure and working mechanism of a chemical sensor is depicted in Figure 1. As the target entity to be sensed comes in contact with the sensing material, a signal as a step change in the sensing material is produced and detected through transducer action. This output from the transducer is a weak electronic pulse which is made as a measurable electrical signal in the next step. The recognition of the target entity results in some product of interaction between the sensing material and the target entity. This reaction has to be a reversible process which is indicative of the life time cycles of the sensor. The cycles of the recognition process are the results of the non-covalent chemical bonds, i.e., ionic/hydrogen bonds and van der Waals interactions. The major challenges in the effective sensing mechanism lie with the magnitude of product between sensing material and target entity. There is always a possibility of getting the same product from sensing material and some other non-intended entities in the same environment. This defeats the purpose of selectivity. The next step is to transduce them into a

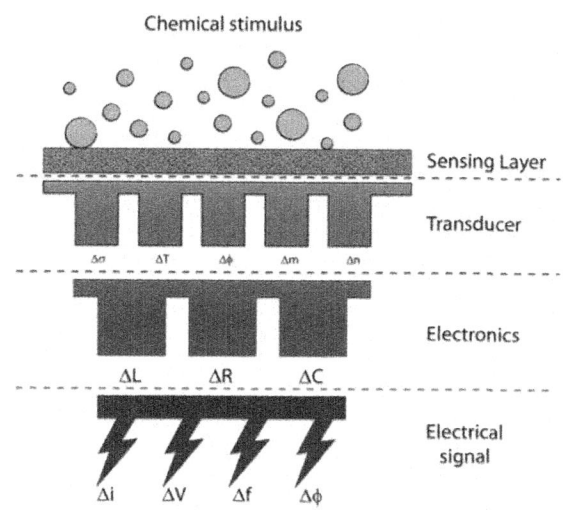

Figure 1. The layered structure and working mechanism of chemical sensors. As the target entity to be sensed comes in contact with the sensing material, a signal as a step change is produced in the sensing material and detected through transducer action with an electric signal. Reprinted with permission from Mandoj et al. (2018).

measurable electronic response. The change in the electrical property of sensing material as a function of the target entity is transduced to be a detectable signal. The change in the electrical property could be changed in resistance and/or change in the dielectric constant. The dielectric constant is measured with the use of the sensing material as a dielectric in the capacitor and the capacitance of such a capacitor is measured.

The research in the metal oxide sensing materials has benefitted from the rapid and relentless advancements due to newer nanoscale technologies. The nanoscale processing brought the possibilities of controllable manipulation of matter at the molecular level. The understanding of properties and usage of nanomaterials has led to substantial changes in sensor designs and capabilities. The reduction in size, weight, power requirements has created opportunities for previously unavailable sensitivity, and specificity. In the present chapter, the contemporary developments of various metal oxide nanostructures will be discussed in the light of the latest requirements and future directions. The present book chapter is not intended to present an exhaustive account of the metal oxide as a chemical sensor. Rather, the chapter focuses on reviewing the developments in the field of chemical sensors on the basis of various metal oxide systems. The synthesis, i.e. designing and development of the metal oxide sensors with the important sensor operating characteristics involved is also elaborated.

Metal Oxide Properties for Sensing Applications

The metal oxide nanoparticles are bestowed with properties such as high density and optimal structural features on their surfaces. Owing to their appropriate porosity, reliable electrical performance, possibility to surpass Moore's law limitation, metal oxide nano-wires have attracted research interest (Zeng et al. 2021). This class of materials have structural properties such as variation in lattice parameters and increased number of surfaces and interfaces. This leads to structural modifications which generate strain/stress and adjoining structural perturbations (Bansal et al. 2006, Shaikh et al. 2021). The nanometric phase of metal oxides exhibit specific size-dependent magnetic, chemical, and electronic properties (Liu 2006, Toksha et al. 2017). The peculiar electrical properties useful in sensing applications are strongly dependent on the particle size (Franke et al. 2006). The TiO_2 nanoparticles exhibit ionic conductivity modulating its usability in the field of sensors. These nanoparticles are demonstrated to work effectively at high temperature harsh conditions where the increase in the n-n junctions available between anatase and rutile at high temperatures contributes positively towards conductivity (Fomekong et al. 2020). The chemical/mechanical stability enables sustainability under harsh environments. The metal oxide materials possess wider band gap creating possibilities of high energy photon emission. The wide band-gap semiconducting behaviour of metal oxides makes the operation possible at higher temperatures (Pearton et al. 2010). The Ga_2O_3 thin film sensor for sensing oxygen was devised and tested for stability, sensitivity and response time at 1000°C (Baban et al. 2005). The ZnO material synthesized in one-dimensional nano-wire structure demonstrated superior potential in sensor applications with biocompatibility and improved sensing properties (Rackauskas et al. 2017). The metal oxide structural blend in case of ZnO and CuO nanowires was studied by Park et al. (Park et al. 2013). The charge carrier type junctions with other materials are designed for improvement in response and selectivity of sensors. It was reported that the ZnO nano-wire formed the hetero-junction with p-type materials, forming local p-n junctions as depicted in Figure 2.

As a well-known fact about nanomaterials, the reduction in size in nano range brings escalation in the chemical reactivity due to its larger surface area to unit mass ratio (Bhati et al. 2020). The surface area is increased with decrease in particle size and contributes towards the improvement in the performance of sensing materials. This could be further enhanced in case of 1-D nanostructure with hollow nanostructures. The study involving SnO_2 fibrous morphology with the role of sintering time reported the critical role of specific surface area and mesopore diameter (Wu et al. 2015). In another case, where WO_3 thin film morphology was synthesized by magnetron

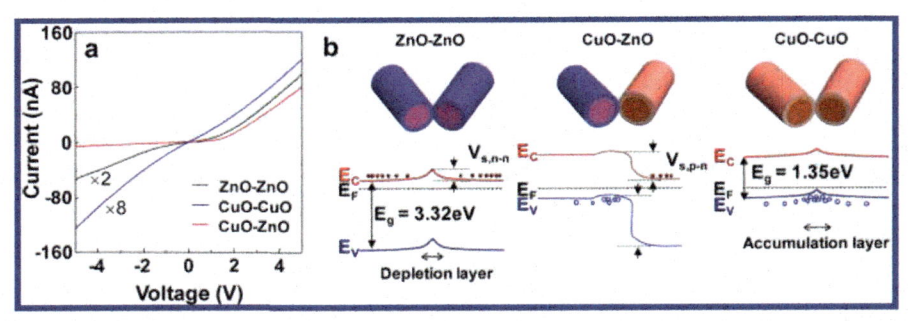

Figure 2. The contrast among the three variations of contacts between ZnO and CuO nano-wires along with current-voltage characteristics and schematic depicting energy bands. Reprinted with permission from Park et al. (2013).

sputtering, the surface areas were controlled with discharge gas pressures. The sensor thus achieved exhibited hysteresis in responding to the target gases with specific peak resistivity at a temperature difference of 100°C. The sensitivity of detecting H_2 and NO_2 gas was found to be a function of effective surface area (Shen et al. 2009).

The properties of metal oxides discussed in this section resulted from the structure, size, and shape of this class of materials. It is quite evident that there are many factors which influence the sensor related characteristics and these properties are critical in the performance of the sensing device. The combinational recipe of metal oxides needs the consideration of the p–n junction formation, evolving energy band gap structures. The selection of synthesis method and consequent processing must include optimized parameters for the best possible sensing mechanism.

Structural transformations such as grain growth, i.e., grain size modifications occurred in powder sintered tin oxide in a period over 3 years and 20 days, which was responsible for stable sensor working (Korotcenkov and Cho 2011). It is reasonable to relate the change in grain size with sensor parameters. The change in grain size could pave the way for many other physical changes such as modifications in geometric size, inter-grain contact, and pore shape/size. Further modifications in the grain size changes the electro-physical properties in terms of the band gap and concentration of point defects. It also possibly changes the crystallographic orientations to some extent faceting. The shape variation from spherulites with large number of steps for less than 10 nm size was modified to macrocrystal faceting (Korotcenkov 2005).

Sensing Mechanisms in Metal Oxides

The measurement of the resistive and capacitive components over the frequency range and the in-phase and out of phase current responses is perhaps one of the first and basic sensing mechanisms. The amperometric/potentiometric actions of sensors involve the measurements of current as a function of potential on a time scale at a constant rate. The recorded response is the magnitude of current proportional to the concentration of the target entity. The measurements could be done in many modes such as variation in electrode potential at a constant rate, testing current or only Ac component of current with varying time. A variation in this type of sensing involves transistor action with the measurement of electrical conductance of the source-drain channel being proportional to its carrier density. Table 1 presents the few cited examples involving metal oxides used as sensing elements along with the transduction type, analyzed medium and target entities sensed.

Sensor Operating Characteristics

Sensor response

The response of a sensing device is specified as the ratio of some physical parameter change recorded to the value of the parameter before the exposure of the target entity. A non-exhaustive list of physical parameters includes current, resistance, capacitance, optical absorbance, and emission intensity before and after the interaction with target entity. The rate at which this change occurs may remain constant (linear) or vary (nonlinear) in an experiment. In case of metal oxide chemical sensors the

Table 1. Various metal oxides used as sensing elements along with the transduction type, analyzed medium and target entities sensed.

Metal oxide involved as sensing element	Transduction type	Analyzed medium	Target entity	References
Ag	Potentiometric Transistor action	Liquid	K^+, Na^+, Ca^{2+}, and H^+	Moser et al. 2020
In_2O_3	Potentiometric Transistor action	Liquid	pH, $C_6H_{12}O_6$	Chen et al. 2017
In_2O_3, ZnO, Co_3O_4	Potentiometric Transistor action	Vapour	CO,H_2	Goto et al. 2015
α-Fe_2O_3 polymer composite	Amperometric	Vapour	N_2H_4	Harraz et al. 2016
Reduced graphene oxide and ZnO	Amperometric	Vapour	C_2H_5OH, and CH_3COCH_3	Galstyan et al. 2016
Reduced graphene oxide and TiO_2	Amperometric	Vapour	$C_6H_8O_6$	Harraz et al. 2019
γ-Al_2O_3	Impedimetric	Vapour	Moisture	Islam et al. 2015
$InSnO_3$	Impedimetric	Vapour	Humidity	McGhee et al. 2020
TiO_2, SiO_2, and Al_2O_3	Impedimetric	Vapour	Humidity	Steele et al. 2008
SnO_2	Optical	Vapour	NH_3	Poole et al. 2014
CuO and ZnO	Optical	Liquid	Li^+ and Ag^+ ions	Maruthupandy et al. 2017
In_2O_3, SnO_2	Optical	Vapour	H_2, CH_4 and CO	Jee et al. 2018

chemical constituents, surface-modification and structures of sensing layers, and ambient conditions are critical in determining the sensitivity (Wang et al. 2010). The improvement in the sensor response of metal oxide sensors could be achieved by various ways. The mixing and coating of Ti and Au, respectively, reportedly enhanced the sensor response of ZnO as well as mixed sensor. The alloying effect and metal catalytic effect were related with the improvement in the sensor response (Wongrat et al. 2012). In another experiment, the sensor response along with all other figures of merit was improved with nanoparticles of gold decorated over SWCNT. The only challenge this arrangement reportedly faced was effect of relative humidity reducing the sensor response (Lone et al. 2018). The improvement in sensing response with the similar lines of decorating host matrix of zinc oxide leading to expansion of depletion layers with the tungsten trioxide were also reported (Park 2019). The synthesis schematic of gold nanoparticles with citrate capping used to improve upon the sensor response in indium oxide nanowires is depicted in Figure 3 (Singh et al. 2011).

Stability and repeatability

The sensor virtue to be able to produce the same output signal in a given experiment over a period of time is its stability and the magnitude with which the independent measurements taken under the identical conditions coincide is its repeatability. The

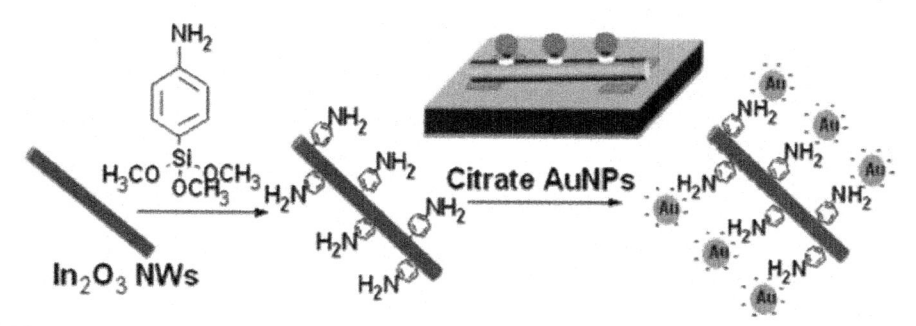

Figure 3. The improvement in sensor response for detecting low concentrations of carbon monoxide gas at room-temperature using gold nanoparticle functionalized indium oxide nanowire. Reprinted with permission from Singh et al. (2011).

fluctuations or change in magnitude of signal from sensor is undesirable and at the same time unavoidable in most of the cases. The signal from the sensor needs to be stable and repeatable which measures the possibility of multiple recoveries of sensing parameters after many cycles. The sensors with working life stability and reproducibility not less than 2 years and working of at least 17000 hours are considered to be standard. There is a lack in the standard data points arising from the practical limit, with laboratory studies designed in a limited time frame and manufacturing data being proprietary and mostly protected. The review in this direction by Korotcenkov and Cho concluded that all the tasks and allied parameters in designing of sensors are of utmost important (Korotcenkov and Cho 2011). The early detection of poisoning before it hampers the sensor functioning is required (Schuler et al. 2015). One of the hurdles to overcome for achieving higher stability is virulent molecules reducing the possibilities of adsorption sites and availability of oxygen species (Karpova et al. 2014). These being the channels of sensing reaction there, reduction leads to slowing down the rate of surface reactions. The virulent could be in the chemical form such as deposition of sulphur, carbon in the form of coke, or in the physical form such as fluctuation in temperature and water as humidity. These physical/chemical virulent reduces the stability of sensing devices. The substances, however, are not limited to sulphur, carbon, or humidity as any entity having a reacting electronic configuration or excitation of electrons will be potential candidates in the category of virulent molecules. The bi-layer design covering the sensing layer with hydrophobic material not disturbing the functioning regarding the selectivity is reported by Jeong et al. (Jeong et al. 2021).

Limit of detection

In the real-world harsh conditions, the target entity will of course will not be a pre-defined parameter in the context of sensor working. It is, therefore, obvious that the sensor must be enabled to detect very low concentrations of the target entity. This is quantified as 'the limit of detection (LOD)' which is the smallest possible limit of the target entity, which can be detected by the sensor against the reference null sample with reasonable reliability. The determination of limits of detection of sensors depends on several performance criteria. Moreover, these criteria are interrelated;

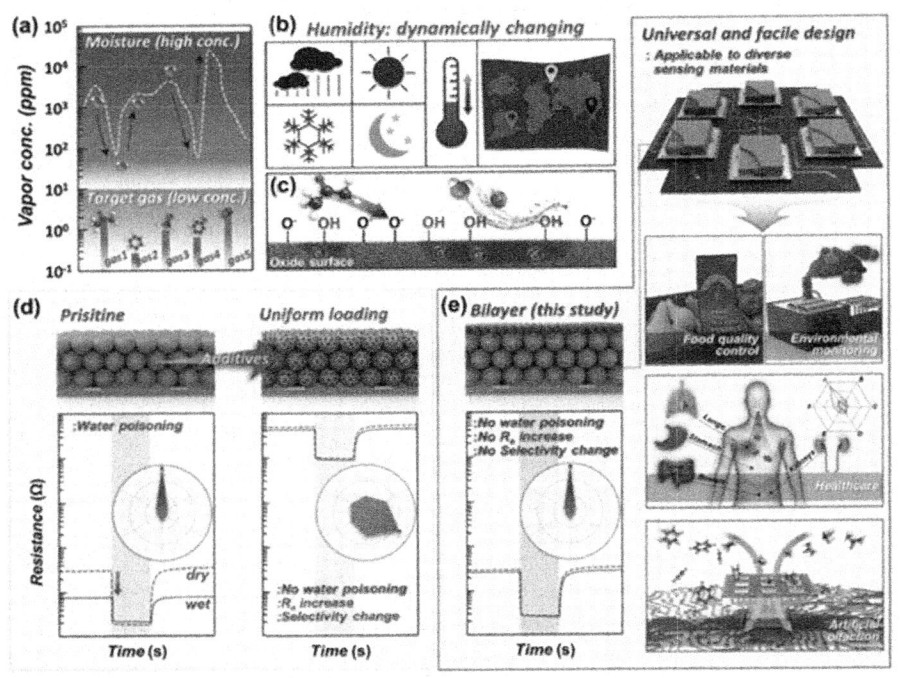

Figure 4. (a) and (b) The fluctuations in the moisture with the change in the environmental conditions (c) the mechanism of water poisoning on the metal oxide sensor (d) the changes brought in from the untreated sample to the uniform loading though increasing the resistivity, undesirable change in selectivity is accompanied (e) addressing the challenge of alteration in other parameters with bi-layer approach, corresponding curves and applications in various contexts. Reprinted with permission from Jeong et al. (2021).

for example, temperature fluctuations could significantly affect the limit of detection (Figure 4). The metal oxide sensors suffer non-linear responses, cross-sensitivities and temporal stability problems which put challenges in the measurement of limit of detection (Burgués et al. 2018). The approach involving the standard deviations at low concentration and banking on the calibration curve are the methods to determine the limit of detection (Loock and Wentzell 2012). The approach involves repeated measurement of the sensor's response with blank reference and target entity at a concentration close to the LOD. The other method of determining the limit of detection involves the use of a linear calibration curve. The limit of detection in the range of 3 to 100 ppm with various metal oxide nanopowders and nanofibres in case of CO_2 gas could be located in literature (Gautam et al. 2021). The limit of detection often confused with sensitivity or sensibility is quite different from both the terms. This statistical term is affected by large random variation and involves type I and type II errors (false positives, false negatives) both (Bernal 2014). The limit of detection could be reduced to the calculation step as 'multiply rms noise by 3 and divide it by slope', where rms noise represents the standard deviation of the sensor signal, and the slope is response first derivative versus the target entity concentration (Mirzaei et al. 2019).

Selectivity

The selective response and measurement of the target entity in presence of other non-intended entities under the same experimental conditions is the selectivity of the sensor. The sensing material's properties, surface microstructure, porosity, grain size, grain boundaries, the catalyst incorporation and the sensing temperature are the decisive parameters towards selective response (Liewhiran and Phanichphant 2007, Fine et al. 2010). The approaches involving various sensing mechanisms combining different response spectra in one sensor and the other one involving the electronic nose were explored in a specific experiment to improve upon gas sensor selectivity and were reported by Ponzoni et al. This study concluded the electronic nose approach being more compact and consistent in terms of selectivity (Ponzoni et al. 2017). The systematic implementation of steps in developing network of nanowires in an integrated platform of various metal oxides, i.e., ZnO, Co_3O_4, In_2O_3 and SnO_2 as shown in Figure 3, was reported by Yi et al. (Yi et al. 2013, 2014). The processing involved the combined effect of sequential screen printing and micro-injecting followed by calcination. This resulted in a sensor of 4 metal oxide nano networks on an alumina substrate and studied for improvement in the selectivity of C_7H_8, NH_3, HCHO or CH_3COCH_3 gases. The work presented a compact, multi-gas sensor device with an aim to assess high selectivity. The adsorption/desorption variation in the metal oxide surface with the target entities, in this case the gases, the variation in surface structure due to nature of metal oxides, specific surface area, increase in active sites for the chemisorption target entities, grain boundary barrier mechanism and nanowire morphology has contributed towards the selectivity of the sensor while the lower operating temperatures were reported to be a requirement of the effective working for all four metal oxides involved (Yi et al. 2014).

The functionality of selective detection towards any target entity for a wide range is one of the final frontiers in the evolution of sensors. A design with bi-layer structure with oxide layer working for chemiresistor sensing and gold over-layer for catalytic action providing high selectivity of methylbenzene was proposed by Moon et al. (Moon et al. 2019). It was recorded that the selectivity was modulated with the gold over-layer morphology, amount, and density. The use of a micro-porous filter in a metal oxide sensor simultaneously addressing the problem of improving selectivity and stability was reported by Hwang et al. (Hwang et al. 2020). The reduced interference of moisture was addressed by using a micro sized structure used for developing the metal organic framework. The filtering approach contributed towards the selectivity criteria of carbon monoxide.

Response and recovery time

The sensing signals as a physical step change of sensors do not reach to their intended maximum value instantaneously with the exposure of the target entity. The signal grows and attends to the maxima in a certain time interval labelled as response time. This time contributes towards the cycles of sensing events. Moreover, the sensors are required to recover to the base line as soon as the intended maxima are achieved. Generally, this time is known as the recovery time that the sensor takes to return to be ready for sensing the next event on the removal of the target entity. Not all the

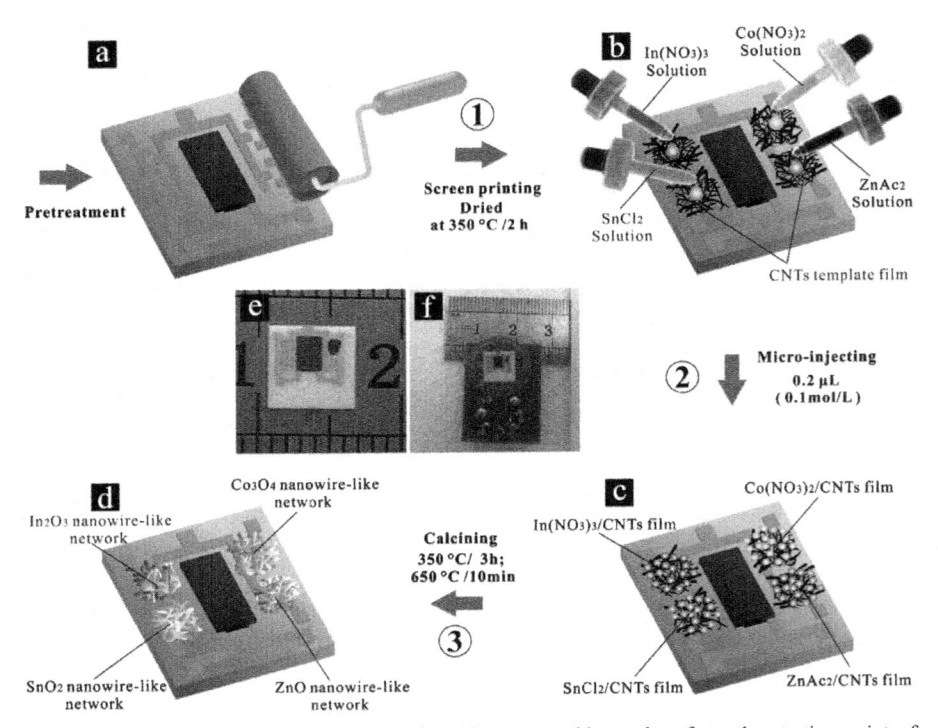

Figure 5. The scheme presenting the evolution of sensor usable product from the starting point of substrate. (Reprinted with permission from Yi et al. 2014).

sensors are always designed or have the functionality of being reversible (Figure 5). Once the sensor responds to the target entity with a permanent change, it leads in an irreversible action. The requirements for a better sensing action are higher response and shorter response time. The higher porosity of nano-crystalline microstructure was found to be a more dominant factor as compared to smaller size. This was related to the fact that the smaller size of the nanoparticles need to overcome agglomeration tendencies these particles possess (Liu et al. 2009). One of the studies involving robots in sensing operations identified the longer recovery time after once the sensor is subjected to the target entity as a major flaw. This extended recovery time causes robot guided sensor system failure in intended chemical entity sensing operation due to delay in responses (Eu and Yap 2020). The response and recovery of sensors could be improved with the design being a more porous structure while sensing vapour and liquid target entities. The porous structure is related with abundant channels for the diffusion of target entities (Basu et al. 2019).

Target Entities

The modern era sensors would be specifically aimed at many target entities. The newer technologies and use of newer material recipe is consequently going to modify the requirements of the sensors to monitor substances. This section is devoted to non exhaustive discussion of a few prominent target entities that could be effectively sensed, and monitored by metal oxide-based sensors.

Uric acid

Uric acid is a blood waste in humans excreted via urine or serum by the kidney. The purine nucleotide metabolism leads to its secretion in the body. The accumulation in the body could lead to medical complications in heart and kidneys. The reduced levels of uric acid are one of the causes of Parkinson's disease or multiple sclerosis. The severity of this chemical makes the sensing and monitoring of uric acid an all-important task. The metal oxide based sensors used to assess the concentration of uric acid were found beneficial as compared to diagnostic testing which are not cost effective, involve more time in testing and are generally complicated (Ponnaiah et al. 2018).

Oxygen

Oxygen is a naturally occurring gaseous element which forms the basis of life. The sensing of this element could possibly be the most important and crucial in numerous fields. Its absence and excess both are critical. The outburst of Covid-19 pandemic in 2019 reasserted the importance of controlled availability of oxygen. The medical conditions requiring oxygen supply as a life support, anaesthesia, and patients' recovery from various medical conditions need the external oxygen supply. The scarcity of oxygen gas in enclosures could put human life at potential risk of asphyxiation, leading to an unconscious state, or even to death (Sieber et al. 2012). There is a need for appropriate concentration of oxygen to get the intended outcome in the form of chemical and combustion processes. The power generating engines rely on the correct mixture to achieve the expected performances (Sari et al. 2017). On the other hand, though oxygen itself does not burn, it significantly increases the chances of something catching the fire or spreading of the fire (Huang and Gao 2021).

Blood sugar/glucose

The blood sugar levels are very crucial in medical conditions of diabetes. This disease is a double trouble as it is one of the most commonly diagnosed diseases and deadly as well. If not leading to death, it causes serious discomfort and significantly disturbs the living styles. The complications are due to insulin deficiency in the body or inability of the body in utilizing the available insulin. The sensing in terms of diagnosis and continuous monitoring of blood sugar levels becomes a very important aspect in any of the medical procedures to be conducted on patients. Thus, sensors which are cost effective, fast and spontaneous can help patients live a quality life and medical practitioners to treat patients in better ways. Metal oxide based sensors could be effective in developing glucose levels (Rahman et al. 2010, Abunahla et al. 2019, Dong et al. 2021).

Carbon dioxide

Carbon dioxide is a gas released through various oxidation and combustion processes occurring everywhere. This gas, famously known as greenhouse gas, is released through respiration and consumed by plants through photosynthesis. The gas is colourless, odourless and remains inflammable at normal temperature and pressure. The dynamics of concentration of this gas has individual to mass implications. The mass change in carbon dioxide levels is a serious environmental concern while its accumulation in an enclosure could lead to death due to suffocations (Gupta Chatterjee et al. 2015). The other closely related chemical is carbon monoxide which results from the incomplete combustion due to lower concentrations of oxygen or insufficient temperature. The major threat with this gas is that it has more affinity to haemoglobin than oxygen. This leads to serious medical complications of hypoxia as it makes the protein unable to carry the second gas to the body cells (Hampson 2015).

pH

One of the most common terms omnipresent in science measurements is the pH. It is the hydrogen potential of a substance, i.e. the concentration of hydrogen ions. The scale on which pH is measured is logarithmic, indicating that small changes in pH value have very high magnitudes. This makes the accurate sensing of pH much more important. The pH scale is a maintaining factor of reaction rates and quality everywhere, from potable water to wastewater (Siregar et al. 2017), agriculture (Satoh and Kakiuchi 2021), food and beverages (Vivaldi et al. 2020), pulp and paper industries (Manjakkal et al. 2020), and oil and gas industry (Sliem et al. 2021).

Refrigerant gases

The control over temperature is required in the food industry, hospitals and faster computational facilities. The temperature control systems rely on the working substance to adjust the required temperature. The continuous monitoring is required as the leakage of refrigerant gases could lead to destruction of the system and surroundings, increased atmospheric pollution, the greenhouse effect and ozone layer depletion.

Volatile organic compounds

These are the toxic vapours of organic compounds, released due to the man-made sources or natural combustion of fossil fuels. The slow poison of volatile organic matter has detrimental, carcinogenic and close to fatal effects on humans. They are ubiquitous in paint removers, polishing liquids of furniture, mosquito repellents, stain remover used in floor, glass, carpet cleaning, air fresheners, adhesives, nail polish, thinner, etc. These are serious threats to human health as they cause damage to respiratory system and various other medical complications (Fermo et al. 2021). The sensors based on metal-oxide nano-materials have exhibited promising results in sensing volatile organic compounds for various figures of merits (John and Ruban Kumar 2021).

LPG

LPG is an odourless and colourless gas formation of propane and butane. It is an indispensable commodity due to its widespread use in heater, cooking, and automotive vehicles. The LPG is flammable and explosive causing serious concerns over its hazardous consequences to the humans as well as environment. The early detection and monitoring could avoid the damage caused by explosions (Park et al. 2006, Turgut et al. 2013). Nickel oxide nano-powder produced via sol–gel route with an aim to elaborate the role of annealing temperatures, reduced graphene-gold thin film structure, cerium oxide-iron oxide nano-composite thin films are few recent studies in the detection of LPG gas (Taheri et al. 2018, Gupta et al. 2021, Kabure et al. 2021).

Ammonia

Ammonia is a part of environmental gas composition and also released through fertilizers in agricultural activity. It gets released as gas in atmosphere and also runs off in aquatic ecosystems. It is a colourless gas causing irritation and has pungent smell. Ammonia coming in contact with eyes causes blindness; when inhaled, it causes lung damage and can even be fatal. The dissolved ammonia is a severe environmental concern as it causes physical/physiological damage, behavioural alteration, and is fatal to aquatic life. A range of metal oxide Zn, In, Ga, etc., based sensors is synthesized for ammonia detection (Priya et al. 2021).

Hydrogen sulphide

Hydrogen sulphide smells like pungent rotten egg odour. It is a colourless, flammable, corrosive, and highly poisonous gas. The source of releasing this gas can be either natural, decomposition of organic compounds, or from sewage. Many other activities such as refining of crude oil, sulphur springs, coal pits, coal gasification plants, decomposition of sulphur containing organic matter, food processing industry, and many analytical/catalytic activities are also responsible for the release of this gas. Higher concentration can cause unconsciousness, personal distress, deactivation of the olfactory system, organ damage, and can be deadly to human beings. The monitoring of hydrogen sulphide in the human body is also critical because it is an indicator of many clinical conditions such as carcinogenic, cardiovascular, and diabetic complications. The brain, heart, and kidneys' well-being and proper functioning also can be assessed by monitoring this gas (Kumar et al. 2021).

Conclusion

The systematic review of the recent literature in the metal oxide sensor brings us to important fact that there is still a requirement of design of experiments to be as close as possible to the real-world conditions. There is a significant improvement in the state-of-the-art sensor devices with the help of various sophisticated designs of nanometric dimensional approaches. The use of metal oxide nanomaterials is a concurrent attractive and active research area in the field of chemical sensor science and most of the conventional methods and materials will become absolute by these developments. The recent outbreak of Covid-19 pandemic situation warrants the need of research on the biochemical sensing devices. The sensor operating characteristics discussed herein indicates that the promising features of metal-oxide materials are the ease in constructive principle, facile measurement procedure, superior strength, sensor response, response and recovery time. These features are going to contribute useful developments in the field of chemical sensors. The selectivity and limit of detection presents the scope of further research combined with other operating characteristics of these materials.

References

Abunahla, H., Mohammad, B., Alazzam, A., Jaoude, M. A., Al-Qutayri, M., Abdul Hadi, S., and Al-Sarawi, S. F. 2019. MOMSense: Metal-oxide-metal elementary glucose sensor. Sci. Rep. 9: 1–10.

Baban, C., Toyoda, Y., and Ogita, M. 2005. Oxygen sensing at high temperatures using Ga_2O_3 films. Thin Solid Films 484: 369–373.

Bandodkar, A. J., Jeerapan, I., and Wang, J. 2016. Wearable chemical sensors: Present challenges and future prospects. ACS Sens. 1: 464–482.

Bansal, V., Poddar, P., Ahmad, A., and Sastry, M. 2006. Room-Temperature biosynthesis of ferroelectric barium titanate nanoparticles. Journal of the Am. Chem. Soc. 128: 11958–11963.

Basu, A. K., Chauhan, P. S., Awasthi, M., and Bhattacharya, S. 2019. α-Fe_2O_3 loaded rGO nanosheets based fast response/recovery CO gas sensor at room temperature. Appl. Surf. Sci. 465: 56–66.

Bergveld, P. 1972. Development, Operation, and application of the ion-sensitive field-effect transistor as a tool for electrophysiology. IEEE Trans. Biomed. Eng. 5: 342–351.

Bernal, E. 2014. Limit of detection and limit of quantification determination in gas chromatography. Advances in Gas Chromatography. IntechOpen. 57–81.

Bhati, V. S., Hojamberdiev, M., and Kumar, M. 2020. Enhanced sensing performance of ZnO nanostructures-based gas sensors: A review. Energy Rep. 6: 46–62.

Brattain, W. H., and Bardeen, J. 1953. Surface properties of germanium. Bell Syst. Tech. J. 32: 1–41.

Buono, C., Mirabella, D. A., Desimone, P. M., and Aldao, C. M. 2021. Effects of Schottky barrier height fluctuations on conductivity: Consequences on power-law response in tin oxide gas sensors. Solid State Ion. 369: 115725.

Burgués, J., Jiménez-Soto, J. M., and Marco, S. 2018. Estimation of the limit of detection in semiconductor gas sensors through linearized calibration models. Anal. Chim. Acta. 1013: 13–25.

Chen, H., Rim, Y. S., Wang, I. C., Li, C., Zhu, B., Sun, M., Goorsky, M. S., He, X., and Yang, Y. 2017. Quasi-two-dimensional metal oxide semiconductors based ultrasensitive potentiometric biosensors. ACS Nano. 11: 4710–4718.

Clark, L. C. and Lyons, C. 1962. Electrode systems for continuous monitoring in cardiovascular surgery. Ann. N. Y. Acad. Sci. 102: 29–45.

Deji, Kaur, N., Choudhary, B. C., and Sharma, R. K. 2021. Carbon-dioxide gas sensor using co-doped graphene nanoribbon: A first principle DFT study. Mater. Today: Proc. 45: 5023–5028.

Dong, Q., Ryu, H., and Lei, Y. 2021. Metal oxide based non-enzymatic electrochemical sensors for glucose detection. Electrochim. Acta. 370: 137744.

Eu, K. S. and Yap, K. M. 2020. Overcoming long recovery time of metal-oxide gas sensor with certainty factor sensing algorithm. Int. J. Smart Sens. Intell. Syst. 7.

Fermo, P., Artíñano, B., De Gennaro, G., Pantaleo, A. M., Parente, A., Battaglia, F., Colicino, E., Di Tanna, G., Goncalves da Silva Junior, A., Pereira, I. G., Garcia, G. S., Garcia Goncalves, L. M., Comite, V., and Miani, A. 2021. Improving indoor air quality through an air purifier able to reduce aerosol particulate matter (PM) and volatile organic compounds (VOCs): Experimental results. Environ. Res. 197: 111131.

Fernandez, A. C., Sakthivel, P., and Jesudurai, J. 2018. Semiconducting metal oxides for gas sensor applications. J. Mater. Sci. Mater. Electron. 29: 357–364.

Fine, G. F., Cavanagh, L. M., Afonja, A., and Binions, R. 2010. Metal oxide semi-conductor gas sensors in environmental monitoring. Sensors. 10: 5469–5502.

Fomekong, R. L., Kelm, K., and Saruhan, B. 2020. High-temperature hydrogen sensing performance of Ni-Doped TiO2 prepared by co-precipitation method. Sensors. 20: 5992.

Fowler, J. D., Allen, M. J., Tung, V. C., Yang, Y., Kaner, R. B., and Weiller, B. H. 2009. Practical chemical sensors from chemically derived graphene. ACS Nano. 3: 301–306.

Franke, M. E., Koplin, T. J., and Simon, U. 2006. Metal and metal oxide nanoparticles in chemiresistors: Does the nanoscale matter? Small. 2: 36–50.

Galstyan, V., Comini, E., Kholmanov, I., Ponzoni, A., Sberveglieri, V., Poli, N., Faglia, G., and Sberveglieri, G. 2016. A composite structure based on reduced graphene oxide and metal oxide nanomaterials for chemical sensors. Beilstein J. Nanotechnol. 7: 1421–1427.

Gautam, Y. K., Sharma, K., Tyagi, S., Ambedkar, A. K., Chaudhary, M., and Pal Singh, B. 2021. Nanostructured metal oxide semiconductor-based sensors for greenhouse gas detection: progress and challenges. R. Soc. Open Sci. 8: 201324.

Gomes, J. B. A., Rodrigues, J. J. P. C., Rabêlo, R. A. L., Kumar, N., and Kozlov, S. 2019. IoT-enabled gas sensors: Technologies, applications, and opportunities. J. Sens. Actuator Netw. 8: 57.

Goto, T., Hyodo, T., Ueda, T., Kamada, K., Kaneyasu, K., and Shimizu, Y. 2015. CO-sensing properties of potentiometric gas sensors using an anion-conducting polymer electrolyte and Au-loaded metal oxide electrodes. Electrochim. Acta. 166: 232–243.

Gupta Chatterjee, S., Chatterjee, S., Ray, A. K., and Chakraborty, A. K. 2015. Graphene–metal oxide nanohybrids for toxic gas sensor: A review. Sens. Actuators B Chem. 221: 1170–1181.

Gupta, P., Kumar, K., Pandey, N. K., Yadav, B. C., and Saeed, S. H. 2021. Effect of annealing temperature on a highly sensitive nickel oxide-based LPG sensor operated at room temperature. Appl. Phys. A. 127: 289.

Hampson, N. B. 2015. Cost of accidental carbon monoxide poisoning: A preventable expense. Prev. Med. Rep. 3: 21–24.

Harraz, F. A., Ismail, A. A., Al-Sayari, S. A., Al-Hajry, A., and Al-Assiri, M. S. 2016. Highly sensitive amperometric hydrazine sensor based on novel α-Fe$_2$O$_3$/crosslinked polyaniline nanocomposite modified glassy carbon electrode. Sens. Actuators B Chem. 234: 573–582.

Harraz, F. A., Faisal, M., Ismail, A. A., Al-Sayari, S. A., Al-Salami, A. E., Al-Hajry, A., and Al-Assiri, M. S. 2019. TiO$_2$/reduced graphene oxide nanocomposite as efficient ascorbic acid amperometric sensor. Journal of Electroanal. Chem. 832: 225–232.

Hu, J., Zhang, L., and Lv, Y. 2019. Recent advances in cataluminescence gas sensor: Materials and methodologies. Appl. Spectrosc. Rev. 54: 306–324.

Huang, X., and Gao, J. 2021. A review of near-limit opposed fire spread. Fire Saf. J. 120: 103141.

Hwang, K., Ahn, J., Cho, I., Kang, K., Kim, K., Choi, J., Polychronopoulou, K., and Park, I. 2020. Microporous elastomer filter coated with metal organic frameworks for improved selectivity and stability of metal oxide gas sensors. ACS Appl. Mater. Interfaces. 12: 13338–13347.

Islam, T., Nimal, A. T., Mittal, U., and Sharma, M. U. 2015. A micro interdigitated thin film metal oxide capacitive sensor for measuring moisture in the range of 175–625ppm. Sens. Actuators B Chem. 221: 357–364.

Jee, Y., Yu, Y., Abernathy, H. W., Lee, S., Kalapos, T. L., Hackett, G. A., and Ohodnicki, P. R. 2018. Plasmonic conducting metal oxide-based optical fiber sensors for chemical and intermediate temperature-sensing applications. ACS Appl. Mater. Interfaces. 10: 42552–42563.

Jeong, S. -Y., Moon, Y. K., Kim, J. K., Park, S. -W., Jo, Y. K., Kang, Y. C., and Lee, J. -H. 2021. A general solution to mitigate water poisoning of oxide chemiresistors: Bilayer sensors with Tb$_4$O$_7$ overlayer. Adv. Funct. Mater. 31: 2007895.

John, R. A. B., and Ruban Kumar, A. 2021. A review on resistive-based gas sensors for the detection of volatile organic compounds using metal-oxide nanostructures. Inorg. Chem. Commun. 133: 108893.

Kabure, A. A., Shirke, B. S., Mane, S. R., Garadkar, K. M., Sargar, B. M., and Pakhare, K. S. 2021. LPG gas sensor activities of CeO$_2$-Fe$_2$O$_3$ nanocomposite thin film at optimum temperature. Appl. Phys. A. 127: 711.

Karpova, E., Mironov, S., Suchkov, A., Karelin, A., Karpov, E. E., and Karpov, E. F. 2014. Increase of catalytic sensors stability. Sens. Actuators B Chem. 197: 358–363.

Korotcenkov, G. 2005. Gas response control through structural and chemical modification of metal oxide films: state of the art and approaches. Sens. Actuators B Chem. 107: 209–232.

Korotcenkov, G., and Cho, B. K. 2011. Instability of metal oxide-based conductometric gas sensors and approaches to stability improvement (short survey). Sens. Actuators B Chem. 156: 527–538.

Kumar, V., Majhi, S. M., Kim, K. -H., Kim, H. W., and Kwon, E. E. 2021. Advances in In_2O_3-based materials for the development of hydrogen sulfide sensors. Chem. Eng. J. 404: 126472.

Liewhiran, C., and Phanichphant, S. 2007. Effects of palladium loading on the response of thick film flame-made ZnO gas sensor for detection of ethanol vapor. Sensors. 7: 1159–1184.

Liu, H., Gong, S. P., Hu, Y. X., Liu, J. Q., and Zhou, D. X. 2009. Properties and mechanism study of SnO_2 nanocrystals for H_2S thick-film sensors. Sens. Actuat. B Chem. 140: 190–195.

Liu, J., Han, T., Sun, B., Kong, L., Jin, Z., Huang, X., Liu, J., and Meng, F. 2016. Catalysis-based cataluminescent and conductometric gas sensors: Sensing Nanomaterials, mechanism, applications and perspectives. Catalysts. 6: 210.

Liu, W. -T. 2006. Nanoparticles and their biological and environmental applications. J. Biosci. Bioeng. 102: 1–7.

Lone, M. Y., Kumar, A., Ansari, N., Husain, S., Zulfequar, M., Singh, R. C., and Husain, M. 2018. Enhancement of sensor response of as fabricated SWCNT sensor with gold decorated nanoparticles. Sens. Actuators. A. 274: 85–93.

Loock, H. -P., and Wentzell, P. D. 2012. Detection limits of chemical sensors: Applications and misapplications. Sens. Actuators B Chem. 173: 157–163.

Luo, M., and Wang, Q. 2021. A reflective optical fiber SPR sensor with surface modified hemoglobin for dissolved oxygen detection. Alex. Eng. J. 60: 4115–4120.

Mandoj, F., Nardis, S., Di Natale, C., and Paolesse, R. 2018. Porphyrinoid thin films for chemical sensing. In: K. Wandelt, ed. Encyclopedia of Interfacial Chemistry. Oxford: Elsevier. 422–443.

Manjakkal, L., Szwagierczak, D., and Dahiya, R. 2020. Metal oxides based electrochemical pH sensors: Current progress and future perspectives. Prog. Mater. Sci. 109: 100635.

Maruthupandy, M., Zuo, Y., Chen, J. -S., Song, J. -M., Niu, H. -L., Mao, C. -J., Zhang, S. -Y., and Shen, Y. -H. 2017. Synthesis of metal oxide nanoparticles (CuO and ZnO NPs) via biological template and their optical sensor applications. Appl. Surf. Sci. 397: 167–174.

McGhee, J. R., Sagu, J. S., Southee, D. J., Evans, Peter. S. A., and Wijayantha, K. G. U. 2020. Printed, fully metal oxide, capacitive humidity sensors using conductive indium tin oxide inks. ACS Appl. Electron. Mater. 2: 3593–3600.

Mirzaei, A., Lee, J. -H., Majhi, S. M., Weber, M., Bechelany, M., Kim, H. W., and Kim, S. S. 2019. Resistive gas sensors based on metal-oxide nanowires. Int. J. Appl. Phys. 126: 241102.

Moon, Y. K., Jeong, S. -Y., Kang, Y. C., and Lee, J. -H. 2019. Metal oxide gas sensors with au nanocluster catalytic overlayer: Toward tuning gas selectivity and response using a novel bilayer sensor design. ACS Appl. Mater. Interfaces. 11: 32169–32177.

Moser, N., Leong, C. L., Hu, Y., Cicatiello, C., Gowers, S., Boutelle, M., and Georgiou, P. 2020. Complementary metal–oxide–semiconductor potentiometric field-effect transistor array platform using sensor learning for multi-ion imaging. Anal. Chem. 92: 5276–5285.

Park, K., Mannan, M. S., Jo, Y. -D., Kim, J. -Y., Keren, N., and Wang, Y. 2006. Incident analysis of Bucheon LPG filling station pool fire and BLEVE. J. Hazard. Mater. 137: 62–67.

Park, S. 2019. Enhancement of hydrogen sensing response of ZnO nanowires for the decoration of WO_3 nanoparticles. Mater. Lett. 234: 315–318.

Park, S. Y., Kim, Y., Kim, T., Eom, T. H., Kim, S. Y., and Jang, H. W. 2019. Chemoresistive materials for electronic nose: Progress, perspectives, and challenges. InfoMat. 1: 289–316.

Park, W. J., Choi, K. J., Kim, M. H., Koo, B. H., Lee, J. -L., and Baik, J. M. 2013. Self-assembled and highly selective sensors based on air-bridge-structured nanowire junction arrays. ACS Appl. Mater. Interfaces. 5: 6802–6807.

Pearton, S. J., Ren, F., Wang, Y. -L., Chu, B. H., Chen, K. H., Chang, C. Y., Lim, W., Lin, J., and Norton, D. P. 2010. Recent advances in wide bandgap semiconductor biological and gas sensors. Prog. Mater. Sci. 55: 1–59.

Ponnaiah, S. K., Periakaruppan, P., and Vellaichamy, B. 2018. New electrochemical sensor based on a silver-doped iron oxide nanocomposite coupled with polyaniline and its sensing application for picomolar-level detection of uric acid in human blood and urine samples. J. Phys. Chem. B. 122: 3037–3046.

Ponzoni, A., Baratto, C., Cattabiani, N., Falasconi, M., Galstyan, V., Nunez-Carmona, E., Rigoni, F., Sberveglieri, V., Zambotti, G., and Zappa, D. 2017. Metal Oxide gas sensors, a survey of selectivity issues addressed at the SENSOR Lab, Brescia (Italy). Sensors. 17: 714.

Poole, Z. L., Ohodnicki, P., Chen, R., Lin, Y., and Chen, K. P. 2014. Engineering metal oxide nanostructures for the fiber optic sensor platform. Opt. Express. 22: 2665–2674.

Priya, G., Savita, M., Narendra, K. P., and Vernica, V. 2021. Metal-oxide based ammonia gas sensors: A Review. Nanosci. Nanotechnol. - Asia. 11: 270–289.

Rackauskas, S., Barbero, N., Barolo, C., and Viscardi, G. 2017. ZnO nanowire application in chemiresistive sensing: A Review. Nanomaterials. 7: 381.

Rahman, M. M., Ahammad, A. J. S., Jin, J. -H., Ahn, S. J., and Lee, J. -J. 2010. A comprehensive review of glucose biosensors based on nanostructured metal-oxides. Sensors. 10: 4855–4886.

Sari, W., Smith, P., Leigh, S., and Covington, J. 2017. Oxygen Sensors based on screen printed platinum and palladium doped indium oxides. Proceedings. 1: 401.

Satoh, Y., and Kakiuchi, H. 2021. Calibration method to address influences of temperature and electrical conductivity for a low-cost soil water content sensor in the agricultural field. Agric. Water Manag. 255: 107015.

Schuler, M., Sauerwald, T., and Schütze, A. 2015. A novel approach for detecting HMDSO poisoning of metal oxide gas sensors and improving their stability by temperature cycled operation. J. Sens. Sens. Syst. 4: 305–311.

Shaikh, B. B. R., Toksha, B. G., Shirsath, S. E., Chatterjee, A., Tonde, S., and Chishty, S. Q. 2021. Microstructure, magnetic, and dielectric interplay in NiCuZn ferrite with rare earth doping for magneto-dielectric applications. J. Magn. Magn. Mater. 537: 168229.

Shen, Y., Yamazaki, T., Liu, Z., Meng, D., Kikuta, T., and Nakatani, N. 2009. Influence of effective surface area on gas sensing properties of WO_3 sputtered thin films. Thin Solid Films. 517: 2069–2072.

Sieber, A., Enoksson, P., and Krozer, A. 2012. Smart electrochemical oxygen sensor for personal protective equipment. IEEE Sens. J. 12: 1846–1852.

Singh, N., Gupta, R. K., and Lee, P. S. 2011. Gold-nanoparticle-functionalized In_2O_3 nanowires as CO gas sensors with a significant enhancement in response. ACS Appl. Mater. Interfaces. 3: 2246–2252.

Sinha, S., Pal, T., Kumar, D., Sharma, R., Kharbanda, D., Khanna, P. K., and Mukhiya, R. 2021. Design, fabrication and characterization of TiN sensing film-based ISFET pH sensor. Mater. Lett. 304: 130556.

Siregar, B., Menen, K., Efendi, S., Andayani, U., and Fahmi, F. 2017. Monitoring quality standard of waste water using wireless sensor network technology for smart environment. *In*: 2017 International Conference on ICT For Smart Society (ICISS). Presented at the 2017 International Conference on ICT For Smart Society (ICISS). 1–6.

Sliem, M. H., Fayyad, E. M., Abdullah, A. M., Younan, N. A., Al-Qahtani, N., Nabhan, F. F., Ramesh, A., Laycock, N., Ryan, M. P., Maqbool, M., and Arora, D. 2021. Monitoring of under deposit corrosion for the oil and gas industry: A review. J. Pet. Sci. Eng. 204: 108752.

Steele, J. J., Taschuk, M. T., and Brett, M. J. 2008. Nanostructured metal oxide thin films for humidity sensors. IEEE Sens. J. 8: 1422–1429.

Sun, Y. -F., Liu, S. -B., Meng, F. -L., Liu, J. -Y., Jin, Z., Kong, L. -T., and Liu, J. -H. 2012. Metal oxide nanostructures and their gas sensing properties: A Review. Sensors (Basel, Switzerland). 12: 2610–2631.

Taguchi, N. 1972. Gas detecting device, 3695848. United States Patent Office. https://patents.google.com/patent/US3695848A/en.

Taheri, M., Feizabadi, Z., Jafari, S., and Mansour, N. 2018. Fast response and high sensitivity of reduced graphene oxide/gold nano-hybrid for LPG sensors at room temperature. J. Electron. Mater. 47: 7232–7239.

Toksha, B. G., Shirsath, S. E., Mane, M. L., and Jadhav, K. M. 2017. Auto-ignition synthesis of $CoFe_2O_4$ with Al^{3+} substitution for high frequency applications. Ceram. Int. 43: 14347–14353.

Turgut, P., Arif Gurel, M., and Kadir Pekgokgoz, R. 2013. LPG explosion damage of a reinforced concrete building: A case study in Sanliurfa, Turkey. Eng. Fail. Anal. 32: 220–235.

Vivaldi, F., Santalucia, D., Poma, N., Bonini, A., Salvo, P., Del Noce, L., Melai, B., Kirchhain, A., Kolivoška, V., Sokolová, R., Hromadová, M., and Di Francesco, F. 2020. A voltammetric pH sensor for food and biological matrices. Sens. Actuators B Chem. 322: 128650.

Wang, C., Yin, L., Zhang, L., Xiang, D., and Gao, R. 2010. Metal oxide gas sensors: Sensitivity and influencing factors. Sensors (Basel, Switzerland). 10: 2088–2106.

Wongrat, E., Hongsith, N., Wongratanaphisan, D., Gardchareon, A., and Choopun, S. 2012. Control of depletion layer width via amount of AuNPs for sensor response enhancement in ZnO nanostructure sensor. Sens. Actuators B Chem. 171–172: 230–237.

Wu, J., Zeng, D., Tian, S., Xu, K., Li, D., and Xie, C. 2015. Competitive influence of surface area and mesopore size on gas-sensing properties of SnO_2 hollow fibers. J. Mater. Sci. 50: 7725–7734.

www. marketresearchfuture. com, 2021. Chemical sensors market growth, size, share and forecast 2027 | MRFR.

Yagawara, S., Manaka, J., and Ohta, W. 1990. Gas detecting device.

Yang, D., Gopal, R. A., Lkhagvaa, T., and Choi, D. 2021. Metal-oxide gas sensors for exhaled-breath analysis: A review. Meas. Sci. Technol. 32: 102004.

Yi, S., Tian, S., Zeng, D., Xu, K., Zhang, S., and Xie, C. 2013. An In_2O_3 nanowire-like network fabricated on coplanar sensor surface by sacrificial CNTs for enhanced gas sensing performance. Sens. Actuators B Chem. 185: 345–353.

Yi, S., Tian, S., Zeng, D., Xu, K., Peng, X., Wang, H., Zhang, S., and Xie, C. 2014. A novel approach to fabricate metal oxide nanowire-like networks based coplanar gas sensors array for enhanced selectivity. Sens. Actuators B Chem. 204: 351–359.

Zeng, H., Zhang, G., Nagashima, K., Takahashi, T., Hosomi, T., and Yanagida, T. 2021. Metal–oxide nanowire molecular sensors and their promises. Chemosensors. 9: 41.

CHAPTER 2

Carbon Nanotube-Based Chemical Sensors

Sovan Lal Banerjee[1,2] *and Matthew V. Tirrell*[1,2,*]

Introduction

Nanomaterials are defined as the materials with a dimension in between 1–100 nm (Shul'ga et al. 2012). Among different types of nanomaterials, interest in the CNT-based materials is growing gradually because of the combinations of some interesting properties that cannot be attained in a single and polycrystalline structures (Sinha et al. 2006, Hierold et al. 2008). Due to worldwide commercial requirements, CNT production has jumped to several tons in annual capacity (De Volder et al. 2013). Fascinating properties of CNT attract attention for wide arrays of applications such as electrochemical biosensors (Wang 2005), medicinal therapy (Dinh et al. 2019), microelectronics (Wong et al. 2003), computing (Georgantzinos and Anifantis 2010), absorption and screening of electromagnetic waves (Das et al. 2009), hydrogen storage (Liu et al. 2010), nanoprobes (Song et al. 2010), composite materials (coatings or fillers) (Dai et al. 2016), supercapacitors (Wu et al. 2021), etc. Different advanced procedures are used to synthesize the single-walled carbon nanotubes (SWCNTs) and multiwalled carbon nanotubes (MWCNT), respectively. Although there are some reported developments in the synthesis process of SWCNTs having metallic and semiconducting properties with a selectivity of 90–95% (Qin et al. 2014), the adopted techniques still suffer from generating semiconducting CNTs economically. Bottom-up synthesis of CNT has shown a promising way to scale up the SWCNTs production. Still, this process is limited to produce SWCNTs with high purity, unimodal distribution of length, diameter, and chirality (Jasti and Bertozzi

---

[1] Material Science Division and Center for Molecular Engineering, Argonne National Laboratory, Lemont, 60439.
[2] Pritzker School of Molecular Engineering, University of Chicago, Chicago. IL, 60637.
* Corresponding author: mtirrell@uchicago.edu

2010). There are some processes available, such as separation by conjugated polymer wrapping (Wang et al. 2014), gel chromatography (Tanaka et al. 2009), density-gradient centrifugation with selective surfactants (Antaris et al. 2010), but these are unable to give a scale up production of SWCNTs in its purest form. A similar limitation is also present in case of MWCNTs production. Although a high volume of MWCNTs can be produced using chemical vapor deposition (CVD) process (Kunadian et al. 2009), structural deviations and contaminations in the produced MWCNTs are the unavoidable properties of this process. Therefore, we can say that all these advanced techniques are complex and still need to be mature enough to prepare the cleanest forms of CNTs. But apart from this, unconventional electrical, mechanical, magnetic properties of the CNTs can be the starting point of many breakthrough researches especially in nanoelectronics (Ruoff and Lorents 1995).

Experimentally, it has been observed that CNTs consist of π-electron wires having quantized electronic states and coherence length. This length is much longer than the length possible for the conducting polymers, which certainly makes CNT a superior candidate for preparing the nanoelectronic devices. CNTs can also be oriented in a nano-wire network with addition of suitable modifying surface agents for efficient detection of analytes (Ackermann et al. 2016). Due to the presence of a large surface to volume ratio and long percolating paths in nanotubes, from early 1990s, it has been believed that the nanotubular structure is the best possible architecture for the sensing applications (Swager 1998). Following the principle, for the first time, Kong and co-workers (Kong et al. 2000) prepared SWCNTs-based chemical sensors and they studied the change in the electrical conductivity of the CNT in contact with the reductive n-doping ammonia (NH_3) and oxidative p-doping nitrogen dioxide (NO_2) gases, respectively. Due to the tubular structure of CNTs, the entire weight is concentrated in the surface layer of CNTs, accelerating the absorption of the analyte molecules and subsequent sensing properties of CNTs (Eletskii 2004). Because of these excellent properties, CNTs-based materials are the promising candidates for designing the superminiaturized bio and chemical sensors. It has been observed that most of the CNT sensors are SWCNTs-based sensors because of their doping and carrier pinning viabilities (Barone et al. 2005). The operational principle of these sensors is based on the change in the *I-V* curve upon adsorption of the specific molecules on the CNT surface. Inspired from the biological olfactory design, CNTs-based array-type sensors have also been designed for the effective sensing of proximal molecules. This kind of chemical nose/tongue-type design is suitable to fabricate "universal sensors", where channels are designed in such a way that different kinds of molecules can be sensed individually (Fitzgerald and Fenniri 2017). But as stated earlier, this kind of sensor still suffers from very precise selectivity of the analytes, which restricts the wide commercialization of the CNT-based sensors.

In this chapter, we intend to provide a comprehensive perspective of the progress and contributions of the modern engineering, technologies, and basic sciences in designing of CNTs-based sensors, which mainly includes the surface functionalization of CNTs with specialized functional polymers and different inorganic nanoparticles (NPs). The mechanism of chemical sensing by CNTs can be divided into two categories such as "inter-CNT" and "intra-CNT". Except these, the other interfaces such as CNT-electrodes and CNT-dielectric are also responsible for the sensing by CNTs. It has been observed that changes in the electronic properties of CNTs are not always indications of their sensing properties but significant response by the collective system/systems is also a manifestation of the sensing property by the CNTs, which might be controlled by the external thermal, mechanical, and chemical interactions. Chen et al., in their study, reported the non-covalent interactions of proteins with CNTs that gave excellent alternations in the electrical properties of CNTs leading to the superior bio-sensing characteristics (Chen et al. 2001). However, in their latter study, they also mentioned that interactions between the electrodes and CNTs perturb the actual sensing response, which eventually limited the sensing selectivity by the CNTs (Chen et al. 2004). It was observed that SWCNTs are superior in sensing compared to the MWCNTs (Wepasnick et al. 2010). Due to the single-layer structure and high electrical transport properties, SWCNTs are much more accessible for the approaching molecules and so SWCNTs have high sensing properties. All these characteristics, modulation in sensing properties by different functionalized CNTs have been elaborated in the following sections, which we believe makes this chapter interesting for both the new and experienced researchers and readers of this field.

Structural Profile of Carbon Nanotubes (CNTs)

Nanotubes consist of several hexagonal graphite planes rolled up in elongated cylindrical structures having a diameter of 1 nm to more than dozens of nanometers and with lengths up to several microns. It is fascinating that the properties of the CNTs solely depend on their shape (Zaporotskova et al. 2016). S. Injima from Japan first discovered carbon nanotubes in 1991. Surface of the CNT consists of regular hexagonal carbon cycles (Dresselhaus et al. 1996, 1998, Smalley 2003). The structure of the tubulenes is conventionally delineated as an interconnected carbon atoms orienting in a single network hexagonal cells, commonly known as sp^2-network (D'jačkov 2011, Martin and Nierengarten 2012). Depending on the synthesis procedures, CNT with versatile structures such as one or multi-layered tubulenes with open or close terminals can be obtained. The important structural parameter, i.e., chirality of the CNT depends on the longitudinal axis of a nanotube and the conjoint orientation of the hexagonal network. Chirality of the CNT can be determined by the angle Θ known as the chiral angle or the orientation angle, which is formed by the direction of the common edge of two adjacent hexagons and rolling direction of the nanotube as shown in Figure 1a (Zaporotskova et al. 2016). The chirality of the nanotubes can also be depicted by two integers such as "n and m", which locates the hexagon of the network which will coincide after the rolling of the nanotube. Although multiple rolling options are available for the nanotube formation, the preferred ones are those which retain the structure of the hexagonal network unperturbed such as $\Theta = 0$ and 30 arc deg related to the (n, 0) and (n, n) chiralities, respectively. The conductivity (metallic or semiconducting) of the CNT depends on the rolling or orientation angle of the CNT. However, it is evidenced that the band gap present in the CNT is 0.1 to 0.2 eV, which makes them semiconductor.

Nevertheless, one can fabricate various CNT-based electronic devices just by altering the band gap using relevant modifications.

Depending on the symmetry, CNT can be subdivided in two categories, namely, chiral, having screw symmetry and, achiral, with cylindrical symmetry. Achiral CNT can be further classified in two categories: zig-zag nanotubes, where two edges of each hexagon are parallel to the cylindrical axis (Figure 1b), and arm-chair nanotubes, where two edges of hexagon are oriented perpendicular to the cylindrical axis (Figure 1c) (Zaporotskova et al. 2016).

Typically, CNT can be characterized by specifying some parameters such as

Chiral vector ($\mathbf{C}_h$), that can be represented as

$$C_h = na_1 + ma_2 \tag{1}$$

where, $\mathbf{a}_1$ and $\mathbf{a}_2$ are the unit vectors.

The chiral angle (Θ), specified by the following expression

$$\Theta = tan^{-1}\left(\frac{\sqrt{3}\,m}{m + 2n}\right) \tag{2}$$

The diameter (d_t) of the tubulene can be written as

$$d_t = \frac{C_h}{\pi} = \frac{\sqrt{3}\,a_{c-c}(m^2 + mn + n^2)^{1/2}}{\pi} \tag{3}$$

Where C_h is the length of the chiral vector $\mathbf{C}_h$, and a_{c-c} represents the difference between the nearest carbon atoms (0.1421 nm for graphite), respectively.

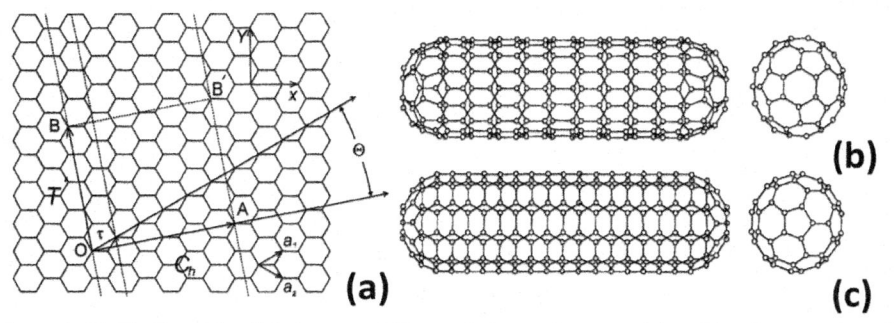

Figure 1. Idealized models of (a) zig-zag and (b) arm-chair monolayer nanotubes; (c) main parameters of nanotube lattice: OA = Ch = na$_1$ + ma$_2$ is the chiral vector specified by the unit vectors a$_1$ and a$_2$; Θ is the chiral angle, OB = T is the lattice unit cell vector and τ is the translation vector. Reprinted with permission from Zaporotskova et al. (2016).

Finally, the lattice vector **T**, is a very important parameter for one-dimensional (1D) CNT. This can be expressed as follows

$$T = \frac{(2m + n)a_1 - (2n + m)a_2}{d_k} \qquad (4)$$

where d_k can be written as
d_k = d if n-m is not a multiple of 3d or 3d if n = m is a multiple of 3d
where "*d*" is the greatest common divisor of (n, m).

Chemical Sensing Mechanism

From the experimental findings, it has been observed that chemical sensing mechanism of CNTs is different from those in conducting polymers, whose sensing techniques can be illustrated by considering the molecular mechanism, whereas presence of the extended π electron system in CNTs imparts complications in describing the sensing process (Hangarter et al. 2013, Braik et al. 2016). Their sensing mechanism can be well understood by describing the band structures rather than the molecular orbitals (Boyd et al. 2014).

The sensing techniques by CNTs can be categorized as intra-CNT (sensing effects due to the interactions within the tube), inter-CNT (sensing due to the contact between the CNTs), and Schottky barrier modulation (sensing due to the interaction between the CNTs and electrodes) as shown in Figure 2a.

Intra-CNT sensing techniques involve the interaction of CNT nanotube bundles or nanotubes with the analyte, where formation of the defects on the

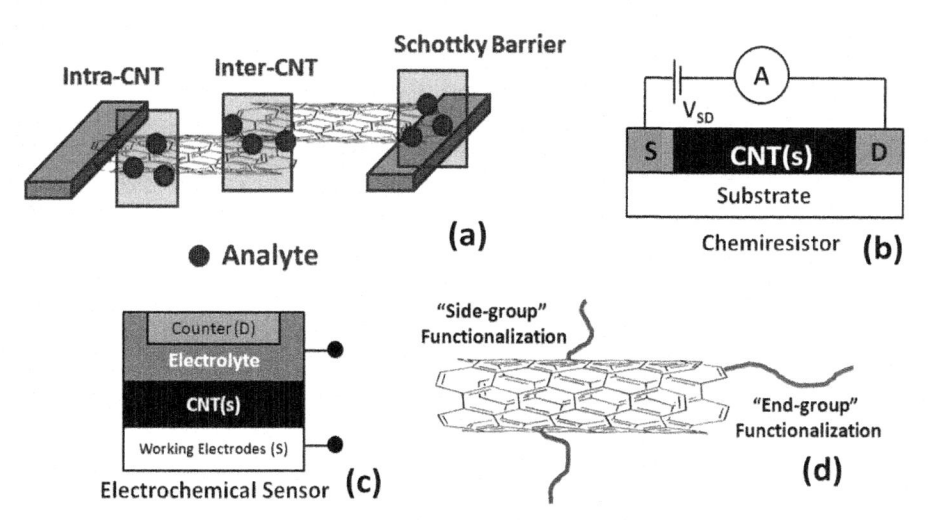

Figure 2. (a) Schematic of the sensing mechanisms observed in the CNTs-based sensors such as intra-CNT, inter-CNT, and Schottky Barrier Modulation; schematic representation of the (b) chemiresistor and (c) electrochemical sensor, respectively; and (d) schematic of the surface functionalization of CNTs including both "side-group" functionalization and "end-group" functionalization.

wall of nanotubes or alternations in the mobility of the charge carriers happens. Immobilizations of the analyte on the surface of CNTs induce decrease or increase in the electron conductance by the CNTs that can be monitored using different instruments. For individual SWCNTs, charge transfer between the tubes and the analytes can be observed experimentally using current-voltage (I-V) characteristics, Raman spectroscopy, and photoemission spectroscopy (PES) (Collins et al. 2000). It has been observed that under ambient condition, CNTs used to be p-doped due to the physisorption of oxygen molecules on the surface of CNTs (Kong et al. 2000). Interactions with the p-dopants will obviously enhance the concentrations of holes that eventually decrease the resistance. A reverse phenomenon occurs in case of interactions with n-dopants. For probing the sensing mechanism of CNTs, field effect transistors (FET) are the important tool where I-V characteristics of CNTs have been monitored upon interactions with analytes (Kong et al. 2000).

The charge transfer to CNTs during the doping can also be monitored from shifts in the Raman spectroscopy. In graphitic materials, a shift in the G-band-stretching related to the sp^2 C-C bond to the higher wave numbers indicates an electron accepting analyte and vice-versa (Rao et al. 1997, Voggu et al. 2008). Carrier pinning to the CNTs can enhance the D/G ratio due to the strong localized interactions between the tubes and analytes.

In case of the inter-CNT sensing mechanism, interactions between the nanotubes in a nanotube network can contribute to the electronic properties of the whole network system. A small alternation in the inter-tube distances significantly modulates the contact resistance due to the fluctuation in the charge tunneling as a result of the inter-tube interactions (Li et al. 2007). The inter-tube charge tunneling can be controlled by couple of ways such as swelling of CNTs by the wrapper or the supporting matrix and second one is introduction of the analytes into the interstitial spaces of CNTs. A swelling of the CNT-network by the analyte can decrease the bulk conductance due to the increase in the distance between the tubes. For example, Lobez and Swager (Lobez and Swager 2010) and Zeininger et al. (Zeininger et al. 2018) reported the poly(olefin sulfone) (POS)-wrapped CNTs for detecting the ionizing radiation. In this case, radiation can cause depolymerization of metastable POS, and an enhancement of interactions between tubes, causing an increase in the electron conductivity.

In some cases, neither intra nor inter-CNTs interactions play the role behind the sensing mechanism. The modulation in the conductivity has been controlled by the phenomenon that occurs at the surface of the electrode and CNTs (Liu et al. 2005). This is called Schottky Barrier (SB) modulation. Many research groups have noticed this behavior by observing the property of CNTs in the presence or absence of the passivating agents at the electrode/CNTs interfaces. Both nature of CNTs and electrodes can contribute to this SB modulation. Schematic of the different sensing mechanisms has been shown in Figure 2a.

Theoretical Models of Sensing

In this section, we have discussed the insights of the theoretical modeling in design and sensing mechanism of CNTs-based sensors. Theoretical modeling always plays an important role and provides great support to understand the experimentally obtained results. In CNTs-based sensors, theoretical studies also significantly contribute to gain the idea regarding electronic properties of CNTs at its pristine state or after binding with the analytes, elasticity and flexibility of the CNTs' structure, thermal conductivity, mechanical properties, etc. (Saito et al. 1992, Lu 1997, Bernholc et al. 1998, Berber et al. 2000). Cho and co-workers (Peng and Cho 2000) demonstrated the theoretical models related to the adsorption of different gases to the CNTs-based sensors using density functional theory (DFT) calculations. There, they have investigated changes in the electronic and energy states of the CNTs after adsorption-induced doping by the gas molecules such as NO_2, CO_2, O_2, NH_3, and H_2O, respectively. Alternation of the electronic energy states in CNTs was demonstrated as a result of the electron donations from the analytes such as NH_3 to the CNTs, and electron acceptance by O_2 and NO_2 from CNTs. Several articles are available based on the theoretical modeling of the adsorption of NO_2 on the CNTs-based sensors. It has been stated that changes in the sensing properties are governed by many factors such as the electron localization effects, modulations in the density of states (DOS), alternations in the dipole moment of NO_2, and changes in the electronic states of CNTs due to the chemisorptions of the NO_2 and subsequent formation of the nitrite or nitro groups (Adjizian et al. 2014, Cui et al. 2012, Li et al. 2013). The computational studies generally use periodic boundary conditions to prepare the model for the quasi one-dimensional CNTs having a typical length to diameter ratio of about 1000:1 (Bettinger 2004). However, the limitation of this process is highlighted when the truncated CNTs-with saturated end states with hydrogen have been used. Due to the saturation of the end states and very small length of CNTs, calculation of the change in the electronic state of CNTs becomes difficult. It has been observed that the accuracy of the theoretical modeling can be improved by considering the Clar sextet rule, which was postulated in 1964 (Clar and Schoental 1964). Details of the implementation of the Clar sextet rule in CNTs have been elaborated by Ormsby and King (Ormsby and King 2004), Matsuo et al. (Matsuo et al. 2003), and Baldoni et al. (Baldoni et al. 2009). Readers can go through these references for more detailed study.

Although several studies on acting mechanisms of CNTs-based sensors have been supported and justified by the theoretical studies, a number of complexities and limitations have been revealed in case of the computational modeling of CNTs, such as: (a) sometimes the interaction between the sensor and the analyte stands to be very weak and it becomes difficult to accurately calculate the physisorbed states of the analyte-adsorbed CNTs by DFT analysis, (b) presence of the high strained bonds in CNTs due to the existence of the conjugated π-electrons is mostly cumbersome to represent computationally, and (c) the electronic states of the CNTs have been strongly affected by its chirality, which hinders the corroboration of the theoretical modeling and experimental findings (Seo et al. 2005, El-Barbary et al. 2014).

Often, it has been observed that intrinsic doping of CNTs with heteroatoms enhances the selectivity and sensitivity of the sensors toward the electron donating and accepting analytes. For example, p-type dopants (Hizhnyi et al. 2015, Rocha et al. 2008) like aluminum and boron enhance the bonding of electron-donating analytes like cyanides, hydrogen halides, ammonia, formaldehyde etc., whereas n-type dopants such as phosphorous, nitrogen, and sulfur increase the binding energies as well as the charge transfer between the CNTs-based sensors and the electron-accepting analyte like NO_2.

Presence of stronger charge transfer and high binding energies between the metal and analytes in case of the metal nanoparticle-ornamented CNTs having single metal or cluster of metals sometimes enhances the selectivity and sensitivity of the sensors. There are a number of reports available where CNTs decorated with metal nanoparticles have been used for sensing and was successfully supplemented by the computational studies (Cui et al. 2012, Singh et al. 2013). Ellis et al. (Ellis et al. 2015) designed the indium oxide (In_2O_3)-decorated SWCNTs-based sensor to selectively identify the presence of acetone and ethanol, and the study was supported by the DFT analysis to elucidate the mechanism of charge transfer during the interaction between sensor and analyte. Therefore, it can be said that the computational study is an important contribution to understand the sensing mechanism of CNTs-based sensors along with the experimental outputs.

Sensor Device Architecture and Fabrication Processes

CNTs-based sensors can be fabricated in different architectures using different fabrication techniques. The sensitivity and selectivity of a sensor mainly depends on the chemistry of pristine and functionalized CNTs. However, alignment of CNTs, use of single CNT or network of CNTs, and interactions between the CNTs (intra, inter, and SB) also affects the sensing properties. It has been observed that single-CNT based sensors are more accurate in sensing having lower limit-of-detection (LOD), lower baseline noise, and these are able to mollify the complexity in sensing as observed in case of the multiple CNTs by reducing the drift (Lee et al. 2010, Feldman et al. 2008). Although single-CNT based sensors are superior in detection, fabrication of the CNT-network is much easier and less time consuming as well as easy to characterize. Along with this, CNT-network-based sensors can provide more repeatability than single-CNT-based sensors. For effective sensing, CNTs should be debundled to allow the analyte to interact with the CNTs. Another important factor for superior sensing is the nanotube density (Ishikawa et al. 2010); a lower density with higher surface-to-volume ratio can enhance the sensitivity and improve the lower limit of detection.

Two basic fabrication techniques exist for the incorporation of CNTs into electric devices: fabrication of device using as-grown CNT films (Kong et al. 1998), and another method is deposition of the purified CNTs (Komatsu and Wang 2010). In the former method, high temperature chemical vapor deposition process has been adopted to design the CNTs-based sensor devices having no/minimum bundling effect or inter-CNT effects. However, controlling of the alignment angles of as-grown CNTs is a bit difficult in this process, especially for the larger channel length design. In the solution casting method of the purified CNTs, it is easy to maintain the chiralities and specific diameter of CNTs (Komatsu and Wang 2010). Different techniques have been adopted to fabricate the CNTs-based sensors via solution–based methods such as spin-coating, patterned lithography, use of the surface template, drop casting, spray-coating, inkjet printing, doctor's blade coating, etc., where rapid evaporation of the solvent supports easy binding of the CNTs with the substrate (electrode). Alternating current dielectrophoresis (DEP) (Suehiro et al. 2003) and layer-by-layer (Qu et al. 2005) approaches are some of the other relevant solvent-based methods to fabricate CNTs-based sensor devices.

Examples of solvent-free purified CNTs deposition methods are also available. For example, Mirica and co-workers (Frazier et al. 2014) fabricated sensor devices via abrasive deposition techniques where selectors were ball milled with pallets of CNTs. Some commercial chemi-resistive sensor devices have been fabricated by adhesive deposition of "bucky gels" (Sekitani et al. 2008), consisting of CNT-containing nonvolatile, viscous mixture with ionic liquids and deep eutectic liquids.

CNT-sensors can be designed in the form of different architectures such as transistors (Kong et al. 1998), two-electrode solid-state sensors (Azzarelli et al. 2014), electrochemical sensors (Zhu 2017, Li et al. 2014), and array-based sensors (Ishihara et al. 2017). Schematic diagrams of chemi-resistors and electrochemical sensors have been shown in Figure 2b and Figure 2c, respectively.

Selective Functionalization of CNT for Designing Chemical Sensors

Experimentally, it has been observed that pristine CNTs are less prone to detect different kinds of analytes and cannot act as a sensor in most cases. Selective functionalization of the CNTs with different components such as metal nanoparticles, polymers, small molecules can impart the desired sensitivity and selectivity (Mallakpour and Soltanian 2016). Different biomolecules, macromolecules, and anchored chemical groups that are used for the functionalization and selective identification of the analyte/analytes are known as selectors. There are two methods of functionalization of CNTs such as non-covalent functionalization and covalent fictionalization (Mallakpour and Soltanian 2016). In case of the covalent bonding in between the reacting molecule and the surface of CNTs, there is a change in the hybridization state from sp^2 to sp^3 has been observed at the reacting sites. This eventually affects the electronic and optical properties of CNTs and perturb the sensing mechanism, which is not the case during the non-covalent functionalization. Generally, non-covalent functionalization involves the anchoring of metal nanoparticles or small organic molecules via hydrogen bonding or π-π interactions leading to less interference into the electronic conducting properties of CNTs, but non-covalent functionalization is less stable compared to the robust covalent functionalized CNTs (Mallakpour and Soltanian 2016). Several approaches have been reported to functionalize CNTs either by non-covalent or covalent approaches. Here, we will discuss some of those examples to highlight the contribution of the surface functionalization of CNT toward its sensing properties.

Non-covalent functionalization of CNTs

Metal Nanoparticles-decorated CNTs

Metal nanoparticles decorated CNTs are the important class of sensors for detecting many gas molecules. Experimentally, it has been observed that incorporation of selective metal nanoparticles can enhance the selectivity and sensitivity of CNTs-based sensors. For example, palladium (Pd)-modified single CNT or CNT networks can efficiently sense the presence of hydrogen (H_2) gas at a concentration level of 4–400 ppm in air at room temperature. Sun et al. reported the fabrication of flexible chemiresistive CNTs-based sensors having Pd NPs that can withstand 1000 cycles of bending and relaxing without much deviation in the sensing property (Sun and Wang 2007). Lu et al. fabricated a tin oxide (SnO_2) NPs (~ 2–3 nm) decorated MWCNTs for the detection of H_2, NH_3, and CO gasses (Lu et al. 2009).

Functionalization with small molecules

CNTs can be functionalized with small organic molecules such as surfactants or aromatic molecules (Frazier and Swager 2013). Selective functionalization of CNTs with selectors can identify the targeted analyte/analytes via formation of halogen bonding (Weis et al. 2016), π-π interactions (Im et al. 2016), hydrogen bonding (Zhu et al. 2016), and host-guest interactions (Wang et al. 2008a).

Non-covalent functionalization of CNTs with polymer

Wrapping of CNTs with selective polymers improves the dispersibility of CNTs in the solvents (both polar and non-polar solvents). It has been observed that CNTs can be effectively dispersed in water in the presence of single-strand DNA (ssDNA) under sonication (Zheng et al. 2003). Different kinds of saccharides and polysaccharides are also used for the non-covalent modifications of CNTs. Conjugated polymers are sometimes employed for the modifications of CNTs. Swager and co-workers (Wang et al. 2008a, Fennell et al. 2017) reported the modification of CNTs with conjugated polymers for the detection of specific analyte. Polymeric surfactants such as Tween 20 have also been used for the non-covalent modifications of CNTs (Chen et al. 2003).

Covalent functionalization of CNTs

As discussed earlier, covalent functionalization of CNTs are a superior alternative over the non-covalent functionalization due to the robustness of the structure that provides the stability to the composite system even in the *in vivo* condition. Selectors can be precisely attached either on the surface of the CNTs or at the termini (Mann et al. 2011, Wang and Swager 2011).

Modification of the CNTs with alkynes or allyl groups can facilitates the functionalization of CNTs with required functional groups via cycloaddition reaction, azide-alkyne reaction, thiol-ene reactions etc. Using these processes, carboxylic groups, thioalkyl chain, allyl and propergyl groups, crown ether groups can be introduced in the surface of CNTs. Paoletti et al. fabricated pentafluorobenzoate and pentaflurobenzoic acid functionalized SWCNTs via aziridination for the ppm level detection of NH_3 and triethylamine at room temperature (Paoletti et al. 2018). Liu and co-workers (Kong et al. 2010) developed hexafluoroisopropanol (HFIP) groups modified SWCNTs for the detection of nerve agent mimicking compound dimethyl methylphophonate (DMMP). In this case, functionalization was carried out using diazonium ions, which is a very convenient process for the modification of carbon nanomaterials. Liang et al. (Liang et al. 2004) fabricated tin oxide (SnO_2) nanocrytals modified MWCNTs for the ppm level detection of NO_2, ethanol, NO, acetylene gases at 300°C. Haddon et al. (Bekyarova et al. 2010) used pyridines (basic groups) modified SWCNTs for the detection of hydrogen chloride. Indium oxide modified SWCNTs were used for the sensing of ethanol and acetone. Biomolecules such as DNA can be introduced at the surface or termini of the CNTs for sensing applications (Weizmann et al. 2010).

Different approaches for functionalization of CNTs have been shown in Figure 2d.

Sensor Performance Parameters

Sensor performance can be monitored by using some parameters such as selectivity, sensitivity, specificity, stability, and lastly the response and recovery times. Selectivity of a sensor can be defined as the capability to differentiate the targeted analyte, besides the other contaminants present in a sample. Apart from this, low drift, and reproducibility of the data are also determining factors for efficient sensors. Chemical sensors are devices that are capable of recognizing and transducing the chemical information obtained from a sample. Ideally, a sensor should be selective, sensitive, and stable under the applied conditions (Usher and Keating 1996). In this respect, CNTs-based sensors are the perfect candidates for replacing the existing commercial sensors in the analytical lab due to its portability, robustness under field condition, and inexpensive nature, which are the implicit characteristics of a suitable sensor device (Wang and Musameh 2004a). This section will explain the required performance parameters of a CNTs-based chemical sensor in terms of the chemical interactions of the sensor with the analyte. A sensor-response curve can describe how the sensor behaves in contact with the analyte as a function of time. Relative changes in the measured current (I), resistance (R), conductance (G) ($G = I/V$), power gain, capacitance (C), and resonant frequency (f_0) w.r.t. to time are the major measuring parameters by a sensor to detect an approachable analyte. For example, upon interaction of an analyte with the sensor, the change in the conductivity ($-\Delta G/G_0$) in the device can be represented as-

$$-\frac{\Delta G}{G_o} = \left[\frac{I_o - I}{I_o}\right] \times 100$$

where I_0 and I represent changes in the current that has been shown by the sensor before and after exposure to the analyte.

Along with changes of these parameters, another important criterion to be a good sensor is its rapidness of sensing viz. the sensitivity. This can be defined as how prompt a sensor can response to the analyte and how it can discriminate the small change in mass/concentration of the targeted molecules. This can be practically determined by measuring the slope of a calibration curve (signal vs concentration), plotted using a standard solution.

In discussing sensitivity, the term Limit Of Detection (LOD) (Li et al. 2003) should be mentioned. This can be defined as the minimum amount of the analyte present in a sample that can be sensed with reasonable accuracy. LOD can be measured using the following equation

$$LOD = 3 \times \frac{rms_{noise}}{slope}$$

where rms_{noise} is the root mean-square noise of the sensor, determined as the deviation of the sensing response curve from the suitable polynomial fit, and the slope can be determined from the calibration curve (linear regression fit of the sensor response vs. concentration of the analyte).

It has been observed that sensors with low LOD often deviate from linearity at very high concentrations of the analyte due to the progression toward the saturation point. That is why the LOD of a sensor has been completely determined based on the application-dependent requirements. Apart from sensitivity, specificity is an important factor for the sensor as well. In a direct interpretation, specificity also represents the selectivity of a sensor that can be elucidated as the ability to solely detect the targeted molecules and no other contaminants. This is true for an ideal sensor but it is very hard to achieve because of the cross-reactivity, also known as cross-signaling, generated due to the formation of new compounds via the reaction between the sensor and the different analytes. To avoid this cross-signaling and obtain error-free results, two methods have been adopted. In the first approach, the targeted analyte has been mixed with other contaminants to reproduce the real-world environment and the selectivity of the sensor has been monitored by determining the differences in the signal of the analyte in the presence and absence of the interfering agents or contaminants. In the second approach, the signals of the targeted sample and the interfering molecules are measured separately, considering there are no cross-compounds generates due to the reactions between sensor and analytes. By comparing the calibration curve of these two and the subsequent ratio of the obtained signals from individual analyte and the interfering compounds, one can determine the selectivity of the particular sensor. This process is relatively easy to adopt and generates data with quantifiable selectivity.

Stability of the sensor is also a prime factor that determines the repeatability of the sensing process in an identical environment for a long period of time or for subsequent steps of sensing. This can be determined by exposing the sensor to the analyte for multiple cycles of measurements. Stability of the sensor is mainly affected by three parameters such as drift, hysteresis, and irreversibility. Drift can be defined as the change in the output signal over time irrespective of external stimuli (Ellis and Star 2016). This can be aroused as a result of small displacement in the CNT-CNT junctions, any physical changes in the analytes or selectors, electromobility of the ions under very small applied potential, disturbance in the sensing layer due to the presence of gas flow, thermal changes, etc. The change in the output signal by a sensor as a result of increasing or decreasing the concentration of the approached molecules can be defined as a hysteresis of a sensor. The hysteresis value also depends on the irreversibility of the sensor, defined as the background signal of the sensor that appears before and after the exposure to the targeted analyte/analytes.

Finally, response and recovery times are very important factors for a sensor. Response time is the minimum time required for a sensor to achieve 90% of its steady-state or maximum value upon coming in contact with the analyte, whereas recovery time is how fast the sensor can get back its peak value, which is typically 10%. Reversibility of a sensor can also be delineated as the recovery time that means how rapid the output of a sensor can be restored after the exposure to the analyte/analytes.

Various Application Fields of CNTs-based Sensors

In this context, we have described the contributions of CNTs-based sensors in versatile fields of applications. The discussions have encompassed some of the important sections such as utilization of CNTs-based sensors in biomedical application, monitoring environment parameters, national security, and finally in food and agriculture application.

CNT-based biological sensors

Identifications of biomolecules and health monitoring

In this section, we have summarized the contribution of CNT-based sensors in the detection of the biomolecules and health monitoring system, which is a crucial factor for meticulous diagnosis of different diseases and real-time monitoring of patients' health. Nowadays, medical costs are increasing enormously due to the use of expensive equipment, laboratory testing, and diagnosis costs. Fostering the developments of CNTs-based sensors can produce more cost-effective treatments and point-of-care diagnostics.

CNT-based DNA Sensors. In this section, we will discuss about the contribution of CNTs-based sensors in the selective identification of DNA, which is an important factor to recognize different diseases related to genetic disorders and their way of treatment, drug discovery, prevent the spread of infectious diseases, and handling the bio-warfare agents (Wang et al. 2004b). The mechanism behind the sensing of DNA by CNTs-based sensor is changes in the electrical current as a result of the immobilization of single-stand DNA on the electrodes and subsequent hybridization of the corresponding sequence (Balasubramanian and Burghard 2006). For example, He and Dai (He and Dai 2004) reported the fabrication of ssDNA functionalized CNT electrodes for the detection of target sequence of DNA and complementary DNA. Initially, they synthesized carboxylic acid modified CNTs and then post-functionalized them with ssDNA. Star et al. fabricated a CNTs-based FET device that can selectively sense the immobilization and subsequent hybridization of DNA (Star et al. 2006), which was observed as a result of reduction in the conductance at a particular gate voltage upon interaction of targeted DNA sequence. Recently, Khan et al. (Khan et al. 2018) fabricated a low cost CNT/ssDNA composite ink and coated it on a screen-printed carbon electrode using inkjet-printing method for the selective detection of protein biomarker at a detection limit of 90 ng/mL.

CNT-based Glucose Sensors. Determination of the blood glucose level is the commonly performed medical test to monitor the patients with diabetes. For the CNTs-based glucose sensors, both chemFET and electrochemical sensors have been reported using glucose oxidase (GOx) as selectors (Balasubramanian and Burghard 2006). During the sensing process, direct transfer of electrons from GOx to CNT electrodes and formation of the hydrogen peroxide (H_2O_2) during the reduction process was enormously helpful to make GOx/CNT composite as an efficient electrochemical sensor for detection of glucose (Balasubramanian and Burghard 2006). But, the problems associated with the noncovalently attached GOx/CNT composite system are leaching of enzymes from the composite system as well as the denaturation of the enzyme (Balasubramanian and Burghard 2006). These issues can be resolved just by introducing metal/metal oxide NPs on the CNT system, which helps to bind the GOx enzyme over the surface of CNT. Functionalization of CNT with different polymer materials also enhance the binding of GOx on the surface of CNTs. Chen et al. (Chen et al. 2012) fabricated Pt-Pd bimetallic NPs decorated MWCNTs for the detection of glucose, and have found 15% decrease in the sensitivity after 28 days of study, whereas Wang et al. (Wang et al. 2009) reported a similar method to detect glucose in which MWCNTs were decorated with zinc oxide (ZnO) NPs. They reported a 10% loss in the detection current after 160 days of study. In both the cases, the metal/metal oxide-ornamented MWCNTs were wrapped with polymer film to avoid the certain inference by the compounds like fructose, ascorbic acid, and uric acid. Tang et al. (Tang et al. 2004) designed the platinum (Pt)-decorated CNT electrodes for efficient detection of glucose, where they found retention of 73.5% of sensitivity after 22 days of study. The designed sensor has an initial level of sensitivity of 14 μA/mM.

Apart from the addition of metal/metal oxide NPs, introduction of the conducting polymer such as poly(aniline) (PANI) and poly(pyrrole) (PPy) also show contributing sensing properties toward glucose. Pilan and Raicopol (Pilan and Raicopol 2014) designed a PANI-functionalized SWCNT/Prussian blue composite for electrochemical detection of glucose. This showed a selective glucose sensing over the rational interfering agents such as acetaminophen, lactate, ascorbic acid, and uric acid. Gao et al. (Gao et al. 2003) electropolymerized polypyrrole (PPy) on Fe particles-decorated aligned MWCNTs and attached GOx for detection of glucose. They observed a large linear range of 2.5–20 mM, and they concluded that the combination of Fe particles and aligned CNTs contributed to the sensing property. In a report, Lee and Cui (Lee and Cui 2010) reported design of flexible sensor via a layer-by layer approach where SWCNTs, polystyrenesulfonate (PSS), poly(diallyldimethylammonium chloride) (PDDA), and GOx were arranged in the mentioned technique on polyethylene terephthalate (PET) surface. This sensor showed a significant linear range of 0.5 to 25 mM and a glucose sensing ability at a minimum glucose concentration of 0.5 mM. Cella et al. (Cella et al. 2010) reported a CNTs-based glucose sensor having a displacement mechanism where the SWCNTs were initially functionalized with hydrophobic dextran derivative (DexP), which later formed a complex with concanavlin A (ConA). When this complex came in contact with glucose, ConA was replaced with glucose resulting in noticeable change in the resistance. Some of the examples of nonenzymatic CNTs-based glucose sensors are also reported. For example, Baghayeri et al. (Baghayeri et al. 2016) demonstrated the fabrication of MWCNTs decorated with Ag NPs for the efficient detection of glucose where they found a low LOD of 0.3 nM. The basic working principle of these kinds of sensors relies on the catalytic properties of the metal nanoparticles.

CNT-based breath analyzer

Volatile molecules exhaled from a human breath is an important link that can be used to identify different diseases such as gastric, colon, breast, lung, prostate cancers, multiple sclerosis, Parkinsons's and Alzhimer's diseases, hyperglycemia, diabetes, etc. This is a portable and noninvasive approach to determine the physical condition of a subject. Typically, gas chromatography-mass spectrometry (GC-MS) (Konvalina and Haick 2014) technique has been used to precisely quantify the desired molecules. Group of Haick et al. (Peng et al. 2008) has extensively worked on the breath analysis technique to identify a person's health condition. From several experimental outputs, CNTs-based materials are found to be an ideal alternative in this respect due to their chemical versatility and easy to fabrication technique. Inorganic gases and the volatile organic compounds (VOCs) are the two important factors that can be discussed to embellish this section.

Experimentally, it has been established that inorganic gases are the important biomarkers for determining the patient's health condition. The commonly used method to monitor the partial pressure or the concentration of the carbon dioxide (CO_2) is known as "capnography", which actually indicates the patients' obstructive conditions such as cardiac arrest, asthma, bronchitis etc. as well as their metabolic conditions too (Richardson et al. 2016). Rise in the ammonia (NH_3) concentration from the ppb to ppm level indicates renal disorder or failure. This can be used as a biomarker for the sensing application. Multiple research groups have developed different CNTs-based sensors for monitoring the permissible NH_3 level. Penza et al. (Penza et al. 2007) developed the Fe-grown MWCNTs/Au/Pt nanocomposite sensor via sputter deposition of Au and Pt over Fe-grown MWCNTs to improve the NH_3 sensing ability. In addition, halitosis and diabetes can be monitored by observing the hydrogen sulfide (H_2S) level. In human body, nitric oxide (NO) plays an important role for many physiological processes such as neural communication, regulation of blood pressure, etc. However, detection of the NO is a bit tricky process because of its very short half-life in an aerobic atmosphere. Star et al. (Star et al. 2004) reported the FET device containing PEI-coated CNTs for the detection of NO. Because of the limited sensitivity of CNTs towards NO, they oxidized the NO to NO_2 using chromium oxide (CrO_3) prior to sensing.

There are some excellent reports available demonstrating the superior capability of CNT-based sensors towards CO_2 sensing. Although many of the literature focused on the CNTs-based inorganic-organic composites sensors, polymer-wrapped CNTs sensors are now predominantly accepted. Star et al. reported the CNT-based CO_2 sensors having starch and PEI-modified CNTs incorporated in the FET device and found a wide range of sensing from 500 ppm to 10% of CO_2 (Star et al. 2004). Varghese et al. (Varghese et al. 2001) described the silica-modified MWCNTs (MWCNTs-SiO_2)-based composite as a CO_2 sensor but they suffered from the cross-sensitivity with undesired gases and humidity. Olney et al. (Olney et al. 2014) developed a PEDOT:PSS-modified SWCNT composite and showed a CO_2 sensitivity upto 10 ppm, which was due to the conformational change and phase separation of the PEDOT:PSS upon adsorption of CO_2. In another report, Li et al. (Li et al. 2012) reported poly(ionic liquid) (PIL)-modified CNT-based chemiresistive CO_2 sensor, where PIL and SWCNT were deposited onto integrated microelectrodes. The interactions between the adsorbed CO_2 and $[BF_4^-]$ anion were the governing factor to show a dynamic CO_2 sensitivity between 500 ppt to 10 ppm.

In addition to the inorganic gases, volatile organic compounds (VOCs), which are the product or byproducts generated due to the cellular metabolism or oxidative reactions in the presence of the reactive oxidative species (ROS), are an imperative human health monitoring biomarker (Hakim et al. 2012). The typical components of VOCs that are generated during cellular activities and exhaled in the form of excreted fluids are hydrocarbons, aldehydes, and ketones (Hakim et al. 2012). Some of the other compounds are ester, nitriles, and aromatic compounds. Concentrations of these components solely rely on the activities of the human body such as change in the diet plan, alternations in the living environments, and presence of any disease(s), etc. For example, free radical oxidations of lipids in the presence of ROSs can produce compounds like pentane, aldehydes, and ethane. Halitosis generates high level of dimethyl sulfide, methyl mercaptan, and hydrogen sulfide.

In case of the detection of VOCs, which is combinations of different gases, use of the array-based sensors is the most convenient and adopted process. This kind of sensors contains multiple units having selective "fingerprint" regions for different compounds, which can be used to easily identify the mixture of analytes. Sometimes array-based sensors are connected parallel with the sensors that are able to selectively identify individual analyte. Wang et al. (Wang et al. 2016) reported vertically aligned conducting polymer coated CNT to detect n-pentane with a significant LOD of 50 ppm and reasonably good selectivity over toluene and methanol. CNT was sequentially coated with a conducting polymer, poly(3,4-ethylenedioxythiophene) (PEDOT) followed by a non-conducting polymer poly(styrene) (PS). Here, interaction of PEDOT with the pentane changes the conductivity whereas PS repels the polar components to provide the selectivity. The working principle of this kind of array-based sensors lies in two subsequent steps: sensing of the analytes by the "fingerprints" region of the sensor followed by the analyzing of the sample by computational methods viz. principal component analysis (PCA) and linear discriminant analysis (LDA). Haick et al. (Peng et al. 2008) were the first to employ the CNT-based sensors for the detection of the VOCs to diagnose the patients suffering from renal failure and lung cancer. In their follow up work, they designed FET-based sensors, which are devoid of the effects of humidity that actually affect the chemiresistors-based sensors. They succeeded in determining 17 different diseases from 1404 subjects with incredible 86% accuracy, using their SWCNTs-based sensors coated with the molecularly modified gold (Au) nanoparticles. Shirsat et al. (Shirsat et al. 2012) reported the metalloporphyrene-functionalized SWCNTs-based (noncovalent interactions) sensors arrays for detection of different VOCs such as hydrocarbons, ketones, amines, aromatics, and alcohols. Sensors based on the interactions between the MWCNTs and different surfactants such as sodium dodecylbenzenesulfonate (SDBS), sodium deoxycholate (DOC), triton x-405 (TX405), 1-hexadecyl trimethylammonium bromide (CTAB), etc., were reported by Chatterjee et al. (Chatterjee et al. 2015), where they spray-coated CNT films onto integrated electrodes using layer-by-layer concept. They were succeeded to separately identify between ethanol, water, methanol, acetone, chloroform, and toluene. Apart from the polymer-CNT composite, addition of the inorganic metal oxide nanoparticles improves the sensing property. Ding et al. (Ding et al. 2013) fabricated chemFET sensors consisting of SWCNTs and TiO_2 and they succeeded in achieving a good sensing property towards acetone over H_2, CO, NH_3, and NO upto a concentration of 20 ppm even in the presence of air.

CNT-based environment monitoring sensors

Alternations in the environmental parameters in significant amounts have intense effects over the climate change. In this section, we will discuss about the contributions of CNTs-based sensors on monitoring some of the sections that are related to the environment such as change in concentration of the metal-ions in water, pH of the aqueous system, and presence of the toxic gases in air. Contemplated research by various groups over these topics will be highlighted during the discussions.

CNT-based metal-ions sensors

In terms of toxicity, some metal ions are toxic at very high level of concentrations such as cobalt, iron, zinc, and copper but they are the necessary ingredients for many signaling and metabolic processes of the living being at low level of concentrations, whereas some of the metal ions like arsenic lead, mercury, chromium, and cadmium are very toxic even at their low concentrations (Gumpu et al. 2015). The permissible levels of these metal ions in water have been set by the US Environmental Protection Agency (EPA) and the World Health Organization (WHO), because presence of these toxic metal ions in water above the tolerance limit could result in severe health and environmental hazards (Gumpu et al. 2015). As per the documented safe vulnerable limit of these metal ions, lead, arsenic, cadmium, mercury, and chromium can be present in aqueous solution at a concentration of 1–10 ppb, while the upper limit for ions like, iron, zinc, and copper are in the order of 1–10 ppm (Gumpu et al. 2015).

From several interesting reports, it has been observed that CNTs-based chemical sensors can be a very promising and fruitful candidate for detection of these metal ions in water because of CNTs' strong selectivity towards metal ions, small size, large surface area, excellent electrical conductivity, ease of functionalization, etc. (Musameh et al. 2011). The technique acting behind the electrochemical sensing of CNTs-based electrodes is anodic stripping voltammetry (ASV). In this method, metal/metals (analyte) is/are deposited on the CNT-electrode under constant potential and after that they are stripping out from the surface of the electrode for analysis. Changes in the redox potential are the driving force for the deposition of these metal ions on the electrodes, but for selective and fast deposition of the ions at the solution/electrode interface, sometimes CNTs' surface is functionalized with specific molecules or compounds. In addition to the electrochemical sensing, FET-based sensors and chemiresistive sensors are also used because of their fast response and these processes are devoid of the stripping process, which is involved in the ASV technique.

Here, we will discuss about some of the well-known examples of CNTs-based chemical sensors. In a report, Forzani et al. (Forzani et al. 2006) described different peptide sequence coated SWCNTs-based FET sensors for detection of metal ions. They used Gly-Gly-His for detecting Cu^{2+} ions and His_6 for the recognition of Ni^{2+}. Star and co-workers (Gou et al. 2013) developed SWCNTs-based metal ions sensor having noncovalently functionalized polyazomethine (PAM) to detect CO^{2+} ions where, upon interaction of the analyte, a large conformational change was observed in PAM due to the complexation with CO^{2+} ion. Wanekaya and co-workers (Morton et al. 2009) fabricated a cysteine-modified MWCNTs-based sensor for sensing Cu^{2+} and Pb^{2+} with a LOD of 15 ppb and 1 ppb, respectively. Due to the presence of the amine and carboxylic acid groups in cysteine, it is very active towards metal ions. Wang et al. (Wang et al. 2012) reported an interesting observation where they functionalized MWCNTs with thiacalexarene, which is very prone to selecting Pb^{2+} ion over Zn^{2+}, Cd^{2+}, and Ni^{2+} ions.

CNT-based pH sensors

Fluctuations in the pH level can impart significant effects in many chemical processes involved in the industrial, environmental, and biological applications. Rapid and accurate sensing of the pH drift is a possible solution to retain the appropriate balance in each of the mentioned applications. From several studies, it has been observed that CNTs-based sensors can be a supple alternative compared to the conventional pH sensors, which are sometimes very costly, complicated, and not as sensitive as required (Jeon et al. 2020). Both functionalized and non-functionalized CNTs-based sensors have been reported as sensors for real-time monitoring of alternations in pH (Mzoughi et al. 2012). Some of the examples are discussed below.

Functionalization of CNTs with pH responsive polymers is a frequently adopted method to identify the pH change in the solution (Kaempgen and Roth 2006). For CNTs-based sensors, we can readily avoid the reference electrodes, glass membranes, and the requirements of high power supply, which are the basic requirements of conventional pH sensors. Use of different types of pH responsive and conducting polymers such as poly(aniline) (PANI) (Kaempgen and Roth 2006), and poly(pyrolle) (PPy) (Ferrer-Anglada et al. 2006) have been reported to be used by several research groups. For example, Roth and co-workers fabricated PANI/SWCNTs (Kaempgen and Roth 2006) and PPy/SWCNTs-coated (Ferrer-Anglada et al. 2006) transparent and flexible electrodes for efficient detection of the pH change, whereas Liao et al. (Liao et al. 2011) reported PANI/SWCNTs composite-based sensors having tunable conductivity with change in pH in the aqueous solution. The change in the degree of protonation of PANI is the responsible factor for pH-sensing. Although these sensors are very rapid to detect pH changes, they suffered from the stability issues. To solve this problem and increase the selectivity, Gou et al. (Gou et al. 2014) reported the fabrication process of poly(1-amino anthracene) (PAA) functionalized oxidized SWCNTs that are highly electroconductive as well as thermally stable along with being very prompt to sense the pH between 2–12 over

120 days. Electrpolymerization method was adopted to polymerize 1-amino anthracene. The best aspect of this process is that PAA is very much sensitive towards hydronium ions over the interfering sodium ion (Na^+) and calcium ions (Ca^{2+}), respectively. There are some other reports available where pristine CNTs-based sensors have been used as a pH sensor. Li et al. (Li et al. 2011) fabricated the microelectrodes consisting of carboxyl groups-ornamented CNTs using dielectrophoresis and studied their potential in pH sensing. As per their observations, the sensor was able to work in the pH range of 5–9 with a shelf-life of 10 days. Takeda et al. (Takeda et al. 2007) designed FET devices having carboxy-functionalized CNTs, and they found the designed sensor was very stable in the liquid medium and sensitive towards pH change too. Compared to the polymer-wrapped CNT sensors, carboxy-treated CNT sensors are less stable and sensitive.

CNT-based gas sensors

In this section, we have discussed about the gas sensing behavior of CNTs where recognition of different toxic and hazardous gases such as ammonia (NH_3), nitrogen dioxide (NO_2), hydrogen (H_2), and greenhouse gases such as carbon dioxide (CO_2), methane (CH_4) has been particularly highlighted. Apart from this, we have also added some interesting references for the detection of benzene, toluene, and xylene (BTX), hydrogen sulfide (H_2S), and sulfur dioxide (SO_2).

NH_3 and NO_2 sensors. Gas sensors are one of the most important devices that can detect the presence and concentration of various hazardous and noxious gases and vapors. Ammonia, an inorganic compound, naturally found in air, water, and soil has severe effects in its high concentration on the human health such as irritation in nose and throat. Pulmonary edema may occur at a minimum NH_3 vapor concentration of 500 ppm (Timmer et al. 2005). Lower concentration of ammonia also impacts skin, eye, and respiratory track (Michaels 1999). Similarly, NO_2 also has adverse effects on human body and environment. In the atmosphere, NO_2 acts as a pollutant and it causes acid rain and photochemical smog. Percentage of NH_3 and NO_2 increases due to the combustion process of chemical plants and motor vehicles. OSHA (Occupational Safety Health Administration) has defined the TWA (Total Weight-Average) of permissible NH_3 and NO_2 exposure limit over 8h to be between 25 to 50ppm and 5 ppm, respectively (Code, n.d.).

CNT based sensors for NH_3 and NO_2 are almost same. Responsiveness towards the pristine CNTs and its effect on the functionalization with nanoparticles (NPs), polymers, and small molecules has been reported in these references (Penza et al. 2007, Mirica et al. 2012).

Mirica et al. (Mirica et al. 2012) demonstrated a solvent free method to deposit pristine SWCNTs by the abrasion process of compressed powder onto the surface of sensing materials in cellulose fibers. Detection limit of this sensor is upto 0.5 ppm. Rigoni et al. (Rigoni et al. 2013) showed the effect on chemiresistive sensors composing pristine SWCNTs with the level of 20 ppb sensitivity to NH_3 and a LOD of 3 ppb. This sensor was developed based on a 'drop casting' method. It was shown that SWCNTs are more aligned for the dielectrophoresis method rather than the disordered SWCNTs.

Penza et al. reported the sensitivity of the Fe-grown MWCNTs films to NH_3 and NO_2 (Penza et al. 2007). He et al. (He et al. 2009) fabricated polyaniline (PANI)-coated MWCNTs through the *in situ* polymerization for the detection of ammonia level. At room temperature, it showed good response and reproducibility. At the concentration range of 0.2 ppm to 15 ppm of NH_3, a linear response was obtained. Liu et al. (Liu et al. 2015) reported a functionalized SWCNTs chemi-resistive sensor to detect amines (including NH_3) with the cobalt-meso-aryl-porphyrin complex ($Co(tpp)ClO_4$). High sensitivity and high selectivity towards amine (including NH_3) among different analytes like hexane, CO, and isoprene was shown by this sensor. Various biogenic amines (e.g., Cadaverine, putrescine) and the spoilage in raw meat and fish products (i.e., chicken, salmon, cod) over time was also observed in this case.

Zhang et al. developed electrochemical functionalized SWCNTs with polyaniline (PANI). This PANI-SWCNTs-based sensor is capable of the on-line detection of NH_3. The functionalized SWCNTs showed a level of 2.44% AR/R per ppm of NH_3 which is 60 times more than that of pristine SWCNTs (Lee et al. 2013). Sharma et al. constructed $CuPcOC_8$/MWCNTs-COOH hybrid material to determine various gases like NO_2, NH_3, and Cl_2 at 150°C (Sharma et al. 2016).

H_2 sensor. Hydrogen is the lightest element, and it can be a reason of explosion above particular concentration. Due to the weak interaction between CNTs and H_2, pristine CNTs manifest some advantages in the detection of hydrogen. Therefore, functionalization is always required when CNTs are used to detect H_2. Kong et al. reported a Pd-decorated SWCNTs that act as H_2 sensor (Kong et al. 2001). In this sensor, thin layer of Pd was accumulated by electron beam evaporation on an individual SWCNTs and after the detection of ppm level of H_2, previously measured conductance was found to be decreased (Kong et al. 2001).

Another H_2 sensor is Schottky-contact based sensors for the development of the contact electronic barrier height between the electron and the CNTs on the exposure of H_2 gas. Javey et al. (Javey et al. 2003) reported this mechanism in a Pd-contacted SWCNTs field-effect transistor. Sayago et al. (Sayago et al. 2007) demonstrated a Pd-functionalized SWCNT gas sensor for the detection of hydrogen. This sensor showed an increased resistance at the exposure of 0.1%–2% hydrogen. Pd-MWCNTs were also exhibited as sensing material to construct strong and flexible chemiresistor for hydrogen detection (Randeniya et al. 2012).

Randeniya et al. used PANI-coated MWCNTs as sensing materials for the detection of H_2 with high sensitivity (Randeniya et al. 2012). Sensor properties are enhanced when the interaction between hydrogen and -OH, $-NH_2$ groups occur. A carbon nanostructured graphene based MWCNTs was used to detect hydrogen (Li et al. 2010). This sensor shows an increased sensitivity of 17% when the concentration level of hydrogen is as low as 4% volume.

Greenhouse Gases Chemiresistive Sensors. Global warming is caused by increased concentration of greenhouse gases (e.g., water vapor, CO_2, methane, freon, etc.) in the atmosphere. CO_2 and methane are the main components of greenhouse gases. CO_2 is toxic when it reaches above 5%. Therefore, it is necessary to detect CO_2 for both human health and environment. CH_4 is more dangerous and powerful than CO_2 in terms of greenhouse effect. Although pristine SWCNTs have no response towards methane, functionalized CNTs could be a better option as methane sensors (Lu et al. 2004).

Li et al. (Li et al. 2012) constructed a highly sensitive and selective chemi-resistive CO_2 sensor with a LOD of 500 ppt. A uniform PIL/SWCNTs film is fabricated using poly(ionic liquid) (PIL) covered SWCNTs in this type of sensor. It showed some features such as selectivity, sensitivity, reproducibility, and resistance to relative humidity.

Sainato et al. (Sainato et al. 2016) developed a MWCNTs/ZnO composite based chemical sensor for the detection of methane. By atomic layer deposition (ALD), ZnO was applied as the functionalizing material to the previously treated surface of MWCNTs. Optimization of ALD temperature leads to the improvement of crystalline quality and the LOD is 10 ppm.

Three main characteristics were considered for the detection of methane: (a) significant change of relative resistance of ZnO NPs to the low concentration of CH_4; (b) energetically feasible electron transport at MWCNTs/ZnO junction, and (c) strong electrical current modulation potential because of the ballistic transport of electrons through MWCNTs. This mechanism implied the adsorption of CH_4 molecule on the metal oxide NPs resulting in the increase in relative resistance of the sensor.

Apart from these examples, CNTs-based sensors have also been used for detection of SO_2 (Zhang et al. 2013), H_2S (Zhang et al. 2014), BTX (Rushi et al. 2014), etc.

CNT-based sensors for national security

CNTs-based sensors can play a major role in maintaining the national security. Nowadays, explosives and chemical warfare agents (CWAs) are the blooming threats for the society. Growing level of terror attacks has not only shaken the national security of a country, but it includes loss of innocent civilians' lives too (Liu et al. 2016). Considering this intimidation to the society, rapid detection of these agents and subsequent neutralization is the best possible way to come out from this situation. Blood agents, respiratory agents, nerve agents, and pulmonary agents are the major sections under the chemical warfare agents (CWAs). Apart from these uses, different kinds of explosives are a menace to the society.

Since World War 1, different compositions of the toxic gases such as phosgene, mustard gas, and chlorine have been used as chemical warfare agents (CWAs) resulting in approximately 1.3 million causalities (Noort et al. 2002). These are the mass destructing agents used during the military conflicts. In this section, we will discuss about the detection of different CWAs using CNTs-based sensors by considering the biochemical interactions of the sensor with blood agents, nerve agents, vesicant agents, pulmonary agents, etc. Current methods of detecting CWAs include mass spectrometry (Kolakowski and Mester 2007), surface acoustic wave (SAW) (Joo et al. 2007), ion mobility spectrometry (Kolakowski and Mester 2007), infrared spectroscopy (Braue and Pannella 1988), electrochemical sensors (Liu and

Lin 2005), colorimetric sensors (Weis and Swager 2015), and fluorescence sensors (Yang and Swager 1998). Although different types of methods are unique in their sensing and measurement techniques, they suffer from different limitations such as sensitivity, selectivity, portability, use of very sophisticated circuits, etc. CNTs-based sensors are promising candidates due to their portable size, and other required properties such as rapid changes in the electrochemical behaviors, chemi-resistive nature, etc.

Among different CWAs, explosive-based weapons are the frequently adopted way for the mass destructions due to their easy preparation methods (Shvetsov et al. 2017). Rapid and exact detection of the explosive's chemical compounds can be the best way to prevent intimidating casualties. From the extensive researches, it has been proved that pristine CNTs are very much sensitive toward the nitroaromatic compounds because of the effective π-π stacking as well as charge transfer properties (Woods et al. 2007). Nitroaromatic compounds are the base material for designing the well-known explosive 2,4,6-trinitrotoluene (TNT), which is a high-boiling explosive. Designing of CNTs-based sensors for the detection of TNT is found to be the best way to sense this explosive. For example, Ruan and co-workers (Ruan et al. 2013) fabricated a piezoelectrical microcantilever coated with pristine CNTs to detect TNT. The basic work principle lying behind this sensor is heating of the microcantilever, which accelerates the decomposition of the nitroaromatic compounds via an exothermic reaction that eventually increases the bending of the microcantilever, and can be detected via piezoelectrical system. Kumar et al. (Kumar et al. 2018) developed a chemiresistive sensor having pristine CNTs-based film to detect DNT at a concentration of 0.2–2 ppm of level. In another report, Kim et al. (Kim et al. 2011a) reported a synthesis scheme of tripeptide receptor (tryptophan-histidine-tryptophan) anchored polydiacetylene polymer and subsequent modification of SWCNTs with the obtained polymer to selectively detect TNT.

Blood agents are also a part of CWAs. The acting mechanism of blood agents involves the disruption of the oxygen flow in the blood stream by making complex with hemoglobin. Carbon monoxide (CO) and cyanide are the main active molecules of blood agents. Poisoning with carbon monoxide and cyanide generally causes dizziness, nausea, headache, but extreme intake of these components can cause unconsciousness and even death. CO is used to bind with the heme groups, which eventually interrupts the oxygen flow from lungs to tissue (Thom 2002), whereas cyanide binds with the enzyme cytochrome c oxidase of the heme center, resulting in the prevention of utilization of the oxygen in the body (Baud 2007). There are few examples available of the CNTs-based sensors for the detection of blood agents. Yari et al. (Yari and Sepahvand 2011) fabricated MWCNTs filled with silver nitrate ($AgNO_3$) for the chemical identification of cyanide, and the important aspect of this work is that the sensor is selective to cyanide over iodide, thiocyanate, hydroxide, and acetate.

Pulmonary agents, which are also a part of CWAs, include phosgene, chlorine, and thionyl chloride ($SOCl_2$) that can cause coughing, dyspnea, lung damage etc. (Guo et al. 1990). Different groups have developed CNTs-based sensors for the detection of these pulmonary agents. For example, Li et al. (Li et al. 2006) reported design of functionalized CNTs-based sensors having hydroxypropyl cellulose and chlorosulfonated polyethylene for the selective response to HCl and Cl_2. Gohier et al. (Gohier et al. 2011) fabricated nitrogen-doped, and polyethyleneimine (PEI)-functionalized CNTs for the selective sensing of Cl_2. Lee and Strano (Lee and Strano 2008) reported the sodium dodecyl sulfate (SDS)-wrapped CNTs for the chemiresistive sensing of $SOCl_2$. Apart from some modifications of CNTs with organic molecules, inorganic nanoparticles modified CNTs have also been used for the detection of pulmonary agents. Popa et al. (Popa et al. 2013) reported the blend of hollow Pt-nanocubes with CNTs for the detection of Cl_2 via catalytic interactions of Cl_2 with metals. It showed 6-folds enhancement in the sensitivity toward Cl_2 over pristine CNTs.

Nerve agents and the vesicant agents are also the members of CWAs. Nerve agents basically affect the nervous system of the patient by interrupting the activity of the enzyme acetylcholineesterase (AChE), which is responsible for the breaking down of the neurotransmitter acetylcholine (ACh) (Szinicz 2005). Inhibition of activities of the AChE shows symptoms such as convulsions, confusions, muscle weakness, respiratory failure, and death. Even at minimum concentrations of 10–100 ppb, typical organophosphorous nerve agents can be lethal (Wismer 2009).

Vesicant agents are responsible for eye injuries, respiratory disorders, and skin blisters. Typical vesicant agents are nitrogen mustard and sulfur mustard. These were first used during the World War I (Noort et al. 2002). Novak et al. (Novak et al. 2003) reported for the first time the use of CNTs-based sensor for the detection of the simulant dimethyl methylphosphonate (DMMP) in air at a ppb level of concentration. They observed a reduction in the conductance of the p-doped CNT along with the shift in the threshold voltage of FET by –2 V as a result of the electron charge donation from the adsorbent to the CNTs. Pristine CNTs were also used as a chemocapacitors for the detection of DMMP at a concentration level of ppb. Names of some of the nerve agents mimicking compounds and decomposition products are dimethyl methylphosphonate (DMMP), pinacolylmthylphophonate (PMP), methyl phophonic acid (MPA), paraoxon, tabun, sarin, VX, etc.

Many research groups have frequently used polymer modified CNTs for the detection of nerve agents. For example, Chuang et al. (Chuang et al. 2013) reported the fabrication of sensor arrays consisting of 30 channels having 15 different polymer/CNT compositions to identify several CWAs and organic solvents. Wang et al. fabricated hexafluroisopropanol (HFIP)-substituted polythiophene (Wang et al. 2008b) and poly(3,4-ethylenedioxythiophene) (Fennell et al. 2017) functionalized CNTs-based chemiresistive sensors for the selective detection of DMMP. Kumar et al. (Kumar et al. 2016) reported design of noncovalent functionalization of CNTs with *p*-hexafluroisopropanol aniline for the selective detection of DMMP with enhanced response time (13-folds). Recently, Johnson and co-workers (Staii et al. 2005) fabricated the single-stand DNA (ssDNA) modified CNTs-based sensor to detect pinene, homologous carboxylic acids, and enantiomers of limonene.

CNT-based sensors in food and agriculture applications

The food industry and agriculture sectors are crucial platforms, contributing to the economies of every developing and developed nation. Agricultural commodities play an important role in fulfilling the increasing global food demand. Any type of adulteration, contamination, and spoilage, food ripening or pathogenic invasion can likely reduce the nutritive values of food. Different processes in this aspect can be regulated and developed by sensing the chemical analyte. 1.3 billion tons of food is wasted because of the inappropriate and real time measurements of the conditions of the easily spoilt products along with the supply chain (Timmermans et al. 2014). Therefore, the development of economically sensitive detection techniques can modify the co-operation of harvest and transportation system by controlling food and agricultural analytes using faster and simpler methods.

CNT based sensors provide some advantages like low-power consumption, small size, and simplicity, which are necessary in food application. Complex analyte detection is possible by allowing the production of economically viable sensors that are appropriate for food chain and supply management. These sensors have been integrated into circuitry, which precedes the real-time information on the condition of food using different connected devices (e.g., smart phones) for use in the detection of fruit ripening (Esser et al. 2012), smart packaging (Zhu et al. 2017), pesticide detection (Chen et al. 2008), and food spoilage (Liu et al. 2015).

Food quality

Food Taste and Smell. CNT sensors can detect the taste and smell besides the regulation of fruit ripening. Sometimes, sensors in this regard are known as electronic tongues and noses (e-tongues and e-noses). Humans have almost 400 different olfactory receptors that also sense smell simultaneously (Zozulya et al. 2001). If olfactory system can sense a smell once (specially binding to a receptor with odorant), it can respond to the neural signal also.

Researchers have integrated olfactory receptors in CNTs-based sensors to recognize several odorants by single carbon center. CNTs-based sensors have been constructed and modified by using olfactory receptors from humans (Jin et al. 2012, Kim et al. 2009), insects (Lee et al. 2015), rodents (Yoon et al. 2009), and canines (Park et al. 2012). The use of human olfactory protein was first reported by Park and co-workers in 2009 (Kim et al. 2009, Yoon et al. 2009). Detection of odorant was supported by the feasibility of coating the SWCNT networks also. The olfactory receptors shift from the inactive, neural state to the active negatively charged state attached to the odorant and decreasing the mobile charge-carrier density onto its network. A few carbon atoms at picomolar concentrations can also differentiate between different butyrate molecules. Cell-signaling pathways in natural olfactory system were also demonstrated by mimicking the sensitivity of the sensors. Besides this, CNT based sensors are also included in different olfactory receptors for the quantification of sour, salt, sweet, bitter, and umami tastes, etc. (Kim et al. 2011b).

Fruit Ripeness. The climacteric is a period of fruit ripening associated with increased ethylene production and a rise in the cellular respiration. Generally, fruit ripening defines the change of fruit pigment and sugar release. Ethylene is one of the smallest biologically active gaseous plant hormone that plays a crucial role in inducing the ripening process of fruits with other hormones and signals (Burg and Burg 1965).

Therefore, the process of ripening and senescing of the climacteric fruits (e.g., apple, banana, melons, apricots, etc.) can be managed through the concentration of ethylene in the atmosphere.

In storage and transportation, optimization of harvesting time and preservation of freshness can be sorted out by regulating the ethylene concentration. CNT based sensors were used for this detection of ethylene, both experimentally and computationally. Esser et al. (Esser et al. 2012) showed an ethylene selective sensor by combining a fluorinated tris(pyrazolyl)-borate copper (II) complex to SWCNTs that was able to form air-stable complexes with ethylene. At room temperature, CNTs do not show sensitivity towards ethylene gas. In this respect, Legrhib et al. reported ethylene concentration detection at room temperature along with MWCNTs grafted with tin oxide (SnO_2) nanoparticles (Leghrib et al. 2009). It has been observed that climacteric fruits show a decrease in the ethylene concentration over time once they have reached at the peak of ripeness, whereas the non-climacteric fruits show a significant low production of ethylene.

Food Spoilage. Food spoilage and food microbial safety are complex factors, improving the risk for effective control without increasing or affecting food quality. Real-time prevention of food spoilage can decrease unnecessary wastage of food. Different CNT based sensors have been integrated to detect the spoilage of food. These sensors can identify volatile biomarkers indicating biogenic ammines and ammonia for spoilage in meat and fish products (Ruiz-Capillas and Jiménez-Colmenero 2005), 1-octen-3-ol for fungal infections of grain (Sinha et al. 1988), and hexanal for spoiled milk (Mehta and Bassette 1978). Liu et al. developed a chemi-resistive sensor to monitor the spoilage of meat and seafood over days, differentiating the changes between the sample stored at room temperature and the sample stored at 4°C in the refrigerator (Liu et al. 2013). It gives similar result to the other products like oysters (Chiu et al. 2014, Hussain and Dawson 2013, Lee et al. 2015), meats (Yoon et al. 2009), and a wide range of seafood (Lim et al. 2013). Human olfactory protein or CNTs-based devices have also been demonstrated for the spoiled milk. It depends on the hexanal, changing with the amount of the headspace over the liquid.

Foodborne Pathogens. Foodborne pathogens like viruses, bacteria, and parasites are biological agents that can result in food-borne illness event. Food-borne pathogen detection is a direct process, apart from the detection of food spoilage associated with the biomarkers.

Recently, different methods for the detection of these pathogens comprise of optical based sensors, mass sensors, electrochemical sensors, colony counting, immunology-based methods and polymerase chain reaction (Velusamy et al. 2010, Mortari and Lorenzelli 2014). Different groups reported functionalization of CNTs with antibodies (Ab) to the specific interaction for the target bacteria or toxin to ease the detection of microorganisms in complex media. Jun and co-workers reported the rapid detection of food pathogens using functionalized nanojunctions (Frenzen et al. 2005). Zhao et al. demonstrated electrochemical sensors for *Shigella flexneri* for Ab-functionalized MWCNT/SA composite (Zhao et al. 2011).

Pesticide Contamination. Pesticides in food show severe health implications. Excessive use of pesticide and contamination of food affects human health and biodiversity (Liu et al. 2012). For agricultural benefits, millions of tons of pesticides are used per year. It also links to severe physical condition like cancer (breast, pancreatic, mom-Hodgkin lymphoma, brain, prostate, and kidney) (Bassil et al. 2007).

CNTs-based sensors have been developed for the measurement of pesticide concentration in complex food grains like soybean, cabbage, celery samples, beer, juice, milk (Yu et al. 2015, Bhatt et al. 2017, Gan et al. 2016). Functionalized CNT based electrodes are useful for a large number of sensors related to the selective oxidation or reduction of the pesticide (Yu et al. 2015, Bhatt et al. 2017, Gan et al. 2016).

Food Safety. Food safety is used as a scientific method handling the preparation and proper storage of food in various ways that prevent food-borne illness. Food related illness is a concerned factor which was marked as a large economic loss due to premature death, medical care, and loss of productivity (Hussain and Dawson 2013). To improve the quality control and meet the need of food qualities, effective technologies are useful.

As per the literature, CNTs-based sensors are employed to detect the spoilage by indicating the nutritional status through the biomarker along with the quality of packaging.

Integrity of Packaging and Oxygen Sensors. Quality of the content of the food depends on the packaging system. Sensitivity of biogenic decomposition of the products can be detected via sensing methods. Oxidative degeneration of food and pharmaceutical products can be prevented from the packaging in inert atmosphere (Brody et al. 2001, Mahajan et al. 2005). To measure the intactness of the food, oxygen detection inside the packaging is a difficult method. On the other hand, oxygen sensors have many other biological and medical applications like BOD (Biological Oxygen Demand) characterization in wastewater, partial pressure in arterial blood for the diagnosis of the patient's condition, etc.

Zettl and co-workers reported the physical characteristics of CNTs including local density of electrical state and electrical resistance which significantly trace the presence of oxygen (Bradley et al. 2000, Collins et al. 2000). The doping of O_2 on the electronic structure of CNT was considered both experimentally and computationally (Kang et al. 2005, Rajavel et al. 2015).

Modified Atmosphere Packaging (MAP) is important for the oxygen sensitive food and drugs because of the harmful effect resulting in the cumulative oxygen level (Zhu et al. 2017). In this regard, Zhu et al. developed a wireless oxygen dosimeter using Fe^{II}-poly(4-vinylpyridine)-SWCNT complexes with passive RFID tags. Fe^{II} binds with the pyridyl ligands using SWCNTs through the poly(4-vinylpyridine) (P4VP). As Fe^{II} is reducing agent, it transfers electron density to the carbon nanotube and increase the resistivity removing the hole-carriers. Exposure of oxygen variable concentration can be detected via passive RFID tags that can permit the wireless oxygen detection inside the food packaging (Zhu et al. 2017). A thin layer of PDMS is given on the device as it acts as an oxygen permeable moisture barrier which limits the effects of humidity to avert the contact of the food. Kauffman et al. fabricated a chemiresistive device having SWCNTs ornamented with oxygen-sensitive Eu^{3+} dendrimer complex (Eu_8) that can be set to its original activity just by applying ultraviolet illumination (UV, 365 nm) after contact with oxygen.

Summary, Future Outlook, and Commercial Viability of CNTs-based Sensors

Since the discovery of CNTs, extensive innovations have taken place in different sectors of applications; CNTs-based chemical sensors are a part of this scientific revolution. This chapter explains the basic mechanisms associated with the CNTs-based sensors, which are mainly dependant on the alternation in resistance, dielectric constant, electrical conductivity, etc., of CNTs upon interaction of analyte/analytes with the CNT-based electrode. To increase the selectivity, response time, and sensitivity, pristine CNTs are frequently modified with different approaches like doping, decoration of the CNTs' surface with metal/metal oxide nanoparticles, functionalization with different functional groups or polymers, etc. These pristine or modified CNTs are then used as a chemiresistive or CNT-FET sensor to detect different analytes including gas, metal ions, biomolecules, WCAs compounds, food and agriculture products, etc.

In this chapter, we have included the mechanism of sensing by CNTs-based sensors, their design architecture and fabrication methods, functionalization of CNTs along with the performance parameters. Theoretical modeling of the sensing by CNTs has also been included in this chapter. Literature studies on CNTs-based sensors show that, compared to industrially used conventional and accepted sensing techniques, CNT sensors are smaller in size, more rapid to sensing analyte, low cost, and simple. CNTs-based sensors are sometimes very prone to be affected by the external environments like chemical environment (interfering analytes including ions, gases, etc.), and physical environment (temperature, humidity), resulting in loss in their selectivity, stability, and self-life. It is discussed in this chapter that cross-sensitivity in the CNTs-based sensors happens due to several factors like generation of defects due to the oxidation, electron transfer from the contaminants, change in orientation of the aligned CNTs, etc. Most of the time, interference to the sensor comes from the very close neighbors of the targeted analyte. For example, during the detection of ethane, frequently methane and propane act as interfering agents. To mitigate these issues, efforts have been made in preparing the array-based sensors having different channels with selective chemical agents. This architecture also suffers from limitations in the presence of very complex environment (occurrence of many interfering analytes).

Despite all these interesting properties, CNTs-based sensors are still suffering from accuracy, which limits their commercial implementations. Another important aspect that restricts the commercialization of CNTs-based sensor is the material quality of the commercially available nanotubes. It has been observed that properties of CNTs, procured from the same company, can vary with different batches because of the existence of some defects in CNTs, purification protocols, etc., which eventually affect the sensitivity, selectivity, and response time of the exactly designed CNTs-based sensor towards detection of the same analyte/analytes. Sometimes minor defects exist in the CNT-based sensors due to the undesired variation in the microfabrication process. This also affects the sensing process. Therefore, attention should be paid to standardize the involving processes such as the synthesis of CNTs and fabrication of sensors.

In spite of all these limitations, CNTs-based sensors are the promising candidates for monitoring the occurrence and limits of different molecules/compounds even at a concentration level of ppb. We already have observed this phenomenon from the literature, as discussed above. A successful knowledge and technology transfer from lab scale studies to the industrial scale up process is the best possible way to commercialize the CNTs-based sensor, which is really under preliminary stage right now. Culmination of the unresolved and targeted areas where CNTs-based sensors can impart significant contribution should be the parallel wing of the continuing research. We anticipate that CNTs-based sensors will enter the commercial arena very soon and will replace the existing commercially used sensors in the foreseeable future.

References

Ackermann, T., Neuhaus, R., and Roth, S. 2016. The effect of rod orientation on electrical anisotropy in silver nanowire networks for ultra-transparent electrodes. Sci. Rep. 6: 1–9.

Adjizian, J. J., Leghrib, R., Koos, A. A., Suarez-Martinez, I., Crossley, A., Wagner, P., Grobert, N. et al. 2014. Boron-and nitrogen-doped multi-wall carbon nanotubes for gas detection. Carbon. 66: 662–673.

Antaris, A. L., Seo, J. W. T., Green, A. A., and Hersam, M. C. 2010. Sorting single-walled carbon nanotubes by electronic type using nonionic, biocompatible block copolymers. ACS Nano. 4: 4725–4732.

Azzarelli, J. M., Mirica, K. A., Ravnsbæk, J. B., and Swager, T. M. 2014. Wireless gas detection with a smartphone via rf communication. Proc. Natl. Acad. Sci. 111: 18162–18166.

Baghayeri, M., Amiri, A., and Farhadi, S. 2016. Development of non-enzymatic glucose sensor based on efficient loading Ag nanoparticles on functionalized carbon nanotubes. Sens. Actuators B: Chem. 225: 354–362.

Balasubramanian, K., and Burghard, M. 2006. Biosensors based on carbon nanotubes. Anal. Bioanal. Chem. 385: 452–468.

Baldoni, M., Selli, D., Sgamellotti, A., and Mercuri, F. 2009. Unraveling the reactivity of semiconducting chiral carbon nanotubes through finite-length models based on Clar sextet theory. J. Phys. Chem. C. 113: 862–866.

Barone, P. W., Baik, S., Heller, D. A., and Strano, M. S. 2005. Near-infrared optical sensors based on single-walled carbon nanotubes. Nat. Mater. 4: 86–92.

Bassil, K. L., Vakil, C., Sanborn, M., Cole, D. C., Kaur, J. S., and Kerr, K. J. 2007. Cancer health effects of pesticides: systematic review. Can. Fam. Physician. 53: 1704–1711.

Baud, F. J. 2007. Cyanide: critical issues in diagnosis and treatment. Hum. Exp. Toxicol. 26: 191–201.

Bekyarova, E., Kalinina, I., Sun, X., Shastry, T., Worsley, K., Chi, X. et al. 2010. Chemically engineered single-walled carbon nanotube materials for the electronic detection of hydrogen chloride. Adv. Mater. 22: 848–852.

Berber, S., Kwon, Y. K., and Tománek, D. 2000. Unusually high thermal conductivity of carbon nanotubes. Phys. Rev. Lett. 84: 4613.

Bernholc, J., Brabec, C., Nardelli, M. B., Maiti, A., Roland, C., and Yakobson, B. I. 1998. Theory of growth and mechanical properties of nanotubes. Appl. Phys. A. 67: 39–46.

Bettinger, H. F. 2004. Effects of finite carbon nanotube length on sidewall addition of fluorine atom and methylene. Org. Lett. 6: 731–734.

Bhatt, V. D., Joshi, S., Becherer, M., and Lugli, P. 2017. Flexible, low-cost sensor based on electrolyte gated carbon nanotube field effect transistor for organo-phosphate detection. Sens. 17: 1147.

Boyd, A., Dube, I., Fedorov, G., Paranjape, M., and Barbara, P. 2014. Gas sensing mechanism of carbon nanotubes: From single tubes to high-density networks. Carbon. 69: 417–423.

Bradley, K., Jhi, S. H., Collins, P. G., Hone, J., Cohen, M. L., Louie, S. G. et al. 2000. Is the intrinsic thermoelectric power of carbon nanotubes positive? Phys. Rev. Lett. 85: 4361.

Braik, M., Barsan, M. M., Dridi, C., Ali, M. C., Ben, M. A., and Brett. 2016. Highly sensitive amperometric enzyme biosensor for detection of superoxide based on conducting polymer/CNT modified electrodes and superoxide dismutase. Sens. Actuators B: Chem. 236: 574–582.

Braue, E. H., and Pannella, M. G. 1988. FT-IR analysis of chemical warfare agents. Microchim. Acta. 94: 11–16.

Brody, A. L., Strupinsky, E. P., and Kline, L. R. 2001. Active packaging for food applications. CRC press.

Burg, S. P., and Burg, E. A.1965. Ethylene action and the ripening of fruits: Ethylene influences the growth and development of plants and is the hormone which initiates fruit ripening. Science. 148: 1190–1196.

Cella, L. N., Chen, W., Myung, N. V., and Mulchandani, A. 2010. Single-walled carbon nanotube-based chemiresistive affinity biosensors for small molecules: ultrasensitive glucose detection. J. Am. Chem. Soc. 132: 5024–5026.

Chatterjee, S., Castro, M., and Feller, J. F. 2015. Tailoring selectivity of sprayed carbon nanotube sensors (CNT) towards volatile organic compounds (VOC) with surfactants. Sens. Actuators B: Chem. 220: 840–849.

Chen, H., Zuo, X., Su, S., Tang, Z., Wu, A., Song, S. et al. 2008. An electrochemical sensor for pesticide assays based on carbon nanotube-enhanced acetycholinesterase activity. Analyst. 133: 1182–1186.

Chen, K. J., Lee, C. F., Rick, J., Wang, S. H., Liu, C. C., and Hwang, B. J. 2012. Fabrication and application of amperometric glucose biosensor based on a novel PtPd bimetallic nanoparticle decorated multi-walled carbon nanotube catalyst. Biosens. Bioelectron. 33: 75–81.

Chen, R. J., Zhang, Y., Wang, D., and Dai, H. 2001. Noncovalent sidewall functionalization of single-walled carbon nanotubes for protein immobilization. J. Am. Chem. Soc. 123: 3838–3839.

Chen, R. J., Bangsaruntip, S., Drouvalakis, K. A., Kam, N. W. S., Shim, M., Li, Y. et al. 2003. Noncovalent functionalization of carbon nanotubes for highly specific electronic biosensors. Proc. Natl. Acad. Sci. 100: 4984–4989.

Chen, R. J., Choi, H. C., Bangsaruntip, S., Yenilmez, E., Tang, X., Wang, Q. et al. 2004. An investigation of the mechanisms of electronic sensing of protein adsorption on carbon nanotube devices. J. Am. Chem. Soc. 126: 1563–1568.

Chiu, S. W., Wu, H. C., Chou, T. I., Chen, H., and Tang. K. T. 2014. A miniature electronic nose system based on an MWNT–polymer microsensor array and a low-power signal-processing chip. Anal. Bioanal. Chem. 406: 3985–3994.

Chuang, P. K., Wang, L. C., and Kuo, C. T. 2013. Development of a high performance integrated sensor chip with a multi-walled carbon nanotube assisted sensing array. Thin Solid Films. 529: 205–208.

Clar, E., and Schoental, R. 1964. Polycyclic hydrocarbons (Vol. 2). Springer.

Code, U. S. (n.d.). Table Z-1-Limits for Air Contaminants. Cellulose, 9004: 34–36.

Collins, P. G., Bradley, K., Ishigami, M., and Zettl, D. A. 2000. Extreme oxygen sensitivity of electronic properties of carbon nanotubes. Science. 287: 1801–1804.

Cui, S., Pu, H., Mattson, E. C., Lu, G., Mao, S., Weinert, M. et al. 2012. Ag nanocrystal as a promoter for carbon nanotube-based room-temperature gas sensors. Nanoscale. 4: 5887–5894.

D'jačkov, P. N. 2011. Élektronnye svojstva i primenenie nanotrubok. Binom. Laboratorija Znanij.

Dai, H., Gallo, G. J., Schumacher, T., and Thostenson, E. T. 2016. A novel methodology for spatial damage detection and imaging using a distributed carbon nanotube-based composite sensor combined with electrical impedance tomography. J. Nondestr. Eval. 35: 1–15.

Das, N. C., Liu, Y., Yang, K., Peng, W., Maiti, S., and Wang, H. 2009. Single-walled carbon nanotube/ poly (methyl methacrylate) composites for electromagnetic interference shielding. Polym. Eng. Sci. 49: 1627–1634.

De Volder, M. F. L., Tawfick, S. H., Baughman, R. H., and Hart, A. J. 2013. Carbon nanotubes: present and future commercial applications. Science. 339: 535–539.

Ding, M., Sorescu, D. C., and Star, A. 2013. Photoinduced charge transfer and acetone sensitivity of single-walled carbon nanotube–titanium dioxide hybrids. J. Am. Chem. Soc. 135: 9015–9022.

Dinh, T., Dau, V., Tran, C., Nguyen, T., Phan, H., Nguyen, N., and Dao, D. V. 2019. Polyacrylonitrile-carbon Nanotube-polyacrylonitrile: A Versatile Robust Platform for Flexible Multifunctional Electronic Devices in Medical Applications. Macromol. Mater. Eng. 304: 1900014.

Dresselhaus, G., Dresselhaus, M. S., and Saito, R. 1998. Physical properties of carbon nanotubes. World scientific.

Dresselhaus, M. S., Dresselhaus, G., and Eklund, P. C. 1996. Science of fullerenes and carbon nanotubes: their properties and applications. Elsevier.

El-Barbary, A. A., Eid, K. M., Kamel, M. A., Osman, H. M., and Ismail, G. 2014. Effect of Tubular Chiralities and Diameters of Single Carbon Nanotubes on Gas Sensing Behavior: A DFT Analysis. J. Surf. Eng. Mater. Adv. Technol. 4: 66–74.

Eletskii, A. V. 2004. Sorption properties of carbon nanostructures. Phys.-Usp. 47: 1119.

Ellis, J. E., Green, U., Sorescu, D. C., Zhao, Y., and Star. A. 2015. Indium Oxide- Single-Walled Carbon Nanotube Composite for Ethanol Sensing at Room Temperature. J. Phys. Chem. Lett. 6: 712–717.

Ellis, J. E., and Star, A. 2016. Carbon nanotube based gas sensors toward breath analysis. Chem. Plus. Chem. 81: 1248.

Esser, B., Schnorr, J. M., and Swager, T. M. 2012. Selective detection of ethylene gas using carbon nanotube-based devices: utility in determination of fruit ripeness. Angew. Chem. Int. Ed. 51: 5752–5756.

Feldman, A. K., Steigerwald, M. L., Guo, X., and Nuckolls, C. 2008. Molecular electronic devices based on single-walled carbon nanotube electrodes. Acc. Chem. Res. 41: 1731–1741.

Fennell, J. F., Hamaguchi, H., Yoon, B., and Swager, T. M. 2017. Chemiresistor devices for chemical warfare agent detection based on polymer wrapped single-walled carbon nanotubes. Sensors. 17: 982.

Ferrer-Anglada, N., Kaempgen, M., and Roth, S. 2006. Transparent and flexible carbon nanotube/polypyrrole and carbon nanotube/polyaniline pH sensors. Phys. Status Solidi (B). 243: 3519–3523.

Fitzgerald, J., and Fenniri, H. 2017. Cutting edge methods for non-invasive disease diagnosis using e-tongue and e-nose devices. Biosensors. 7: 59.

Forzani, E. S., Li, X., Zhang, P., Tao, N., Zhang, R., Amlani, I. R. et al. 2006. Tuning the chemical selectivity of SWNT-FETs for detection of heavy-metal ions. Small. 2: 1283–1291.

Frazier, K. M., Mirica, K. A., Walish, J. J., and Swager, T. M. 2014. Fully-drawn carbon-based chemical sensors on organic and inorganic surfaces. Lab Chip. 14: 4059–4066.

Frazier, K. M., and Swager, T. M. 2013. Robust cyclohexanone selective chemiresistors based on single-walled carbon nanotubes. Anal. Chem. 85: 7154–7158.

Frenzen, P. D., Drake, A., Angulo, F. J., and E. I. P. F. W. Group. 2005. Economic cost of illness due to Escherichia coli O157 infections in the United States. J. Food Prot. 68: 2623–2630.

Gan, T., Lv, Z., Sun, Y., Shi, Z., Sun, J., and Zhao, A. 2016. Highly sensitive and molecular selective electrochemical sensing of 6-benzylaminopurine with multiwall carbon nanotube@ SnS 2-assisted signal amplification. J. Appl. Electrochem. 46: 389–401.

Gao, M., Dai, L., and Wallace, G. G. 2003. Biosensors based on aligned carbon nanotubes coated with inherently conducting polymers. Electroanalysis: An International Journal Devoted to Fundamental and Practical Aspects of Electroanalysis. 15: 1089–1094.

Georgantzinos, S. K., and Anifantis, N. K. 2010. Carbon nanotube-based resonant nanomechanical sensors: a computational investigation of their behavior. Phys. E: Low-Dimens. Syst. Nanostructures. 42: 1795–1801.

Gohier, A., Chancolon, J., Chenevier, P., Porterat, D., Mayne-L'Hermite, M., and Reynaud, C. 2011. Optimized network of multi-walled carbon nanotubes for chemical sensing. Nanotechnology. 22: 105501.

Gou, P., Kraut, N. D., Feigel, I. M., Bai, H., Morgan, G. J., Chen, Y., Tang, Y. et al. 2014. Carbon nanotube chemiresistor for wireless pH sensing. Sci. Rep. 4: 1–6.

Gou, P., Kraut, N. D., Feigel, I. M., and Star, A. 2013. Rigid versus flexible ligands on carbon nanotubes for the enhanced sensitivity of cobalt ions. Macromolecules. 46: 1376–1383.

Gumpu, M. B., Sethuraman, S., Krishnan, U. M., and Rayappan, J. B. B. 2015. A review on detection of heavy metal ions in water–an electrochemical approach. Sens. Actuators B: Chem. 213: 515–533.

Guo, Y. L., Kennedy, T. P., Michael, J. R., Sciuto, A. M., Ghio, A. J., Adkinson, N. F. Jr. et al. 1990. Mechanism of phosgene-induced lung toxicity: role of arachidonate mediators. J. Appl. Physiol. 69: 1615–1622.

Hakim, M., Broza, Y. Y., Barash, O., Peled, N., Phillips, M., Amann, A. et al. 2012. Volatile organic compounds of lung cancer and possible biochemical pathways. Chem. Rev. 112: 5949–5966.

Hangarter, C. M., Chartuprayoon, N., Hernández, S. C., Choa, Y., and Myung, N. V. 2013. Hybridized conducting polymer chemiresistive nano-sensors. Nano Today. 8: 39–55.

He, L., Jia, Y., Meng, F., Li, M., and Liu, J. 2009. Gas sensors for ammonia detection based on polyaniline-coated multi-wall carbon nanotubes. Mater. Sci. Eng. B. 163: 76–81.

He, P., and Dai, L. 2004. Aligned carbon nanotube–DNA electrochemical sensors. Chem. Comm. 3: 348–349.

Hierold, C., Helbling, T., Roman, C., Durrer, L., Jungen, A., and Stampfer, C. 2008. CNT based sensors. Adv. Sci. Technol. 54: 343–349.

Hizhnyi, Y., Nedilko, S. G., Borysiuk, V., and Gubanov, V. A. 2015. Computational studies of boron-and nitrogen-doped single-walled carbon nanotubes as potential sensor materials of hydrogen halide molecules HX (X= F, C l, B r). Int. J. Quantum Chem. 115: 1475–1482.

Hussain, M. A., and Dawson, C. O. 2013. Economic impact of food safety outbreaks on food businesses. Foods. 2: 585–589.

Im, J., Sterner, E. S., and Swager, T. M. 2016. Integrated gas sensing system of swcnt and cellulose polymer concentrator for benzene, toluene, and xylenes. Sensors. 16: 183.

Ishihara, S., Labuta, J., Nakanishi, T., Tanaka, T., and Kataura, H. 2017. Amperometric detection of sub-ppm formaldehyde using single-walled carbon nanotubes and hydroxylamines: A referenced chemiresistive system. ACS Sens. 2: 1405–1409.

Ishikawa, F. N., Curreli, M., Olson, C. A., Liao, H. I., Sun, R., Roberts, R. W. et al. 2010. Importance of controlling nanotube density for highly sensitive and reliable biosensors functional in physiological conditions. ACS Nano. 4: 6914–6922.

Jasti, R., and Bertozzi, C. R. 2010. Progress and challenges for the bottom-up synthesis of carbon nanotubes with discrete chirality. Chem. Phys. Lett. 494: 1–7.

Javey, A., Guo, J., Wang, Q., Lundstrom, M., and Dai, H. 2003. Ballistic carbon nanotube field-effect transistors. Nature. 424: 654–657.

Jeon, J. Y., Kang, B. C., and Ha, T. J. 2020. Flexible pH sensors based on printed nanocomposites of single-wall carbon nanotubes and Nafion. Appl. Surf. Sci. 514: 145956.

Jin, H. J., Lee, S. H., Kim, T. H., Park, J., Song, H. S., Park, T. H. et al. 2012. Nanovesicle-based bioelectronic nose platform mimicking human olfactory signal transduction. Biosens. Bioelectron. 35: 335–341.

Joo, B. S., Huh, J. S., and Lee, D. D. 2007. Fabrication of polymer SAW sensor array to classify chemical warfare agents. Sens. Actuators B: Chem. 121: 47–53.

Kaempgen, M., and Roth, S. 2006. Transparent and flexible carbon nanotube/polyaniline pH sensors. J. Electroanal. Chem. 586: 72–76.

Kang, D., Park, N., Ko, J., Bae, E., and Park, W. 2005. Oxygen-induced p-type doping of a long individual single-walled carbon nanotube. Nanotechnology. 16: 1048.

Khan, N. I., Maddaus, A. G., and Song, E. 2018. A low-cost inkjet-printed aptamer-based electrochemical biosensor for the selective detection of lysozyme. Biosensors. 8: 7.

Kim, T. H., Lee, B. Y., Jaworski, J., Yokoyama, K., Chung, W. J., Wang, E. et al. 2011a. Selective and sensitive TNT sensors using biomimetic polydiacetylene-coated CNT-FETs. ACS Nano. 5: 2824–2830.

Kim, T. H., Lee, S. H., Lee, J., Song, H. S., Oh, E. H., Park, T. H. et al. 2009. Single-carbon-atomic-resolution detection of odorant molecules using a human olfactory receptor-based bioelectronic nose. Adv. Mater. 21: 91–94.

Kim, T. H., Song, H. S., Jin, H. J., Lee, S. H., Namgung, S., Kim, U. et al. 2011b. "Bioelectronic super-taster" device based on taste receptor-carbon nanotube hybrid structures. Lab on a Chip. 11: 2262–2267.

Kolakowski, B. M., and Mester, Z. 2007. Review of applications of high-field asymmetric waveform ion mobility spectrometry (FAIMS) and differential mobility spectrometry (DMS). Analyst. 132: 842–864.

Komatsu, N., and Wang, F. 2010. A comprehensive review on separation methods and techniques for single-walled carbon nanotubes. Materials. 3: 3818–3844.

Kong, J., Soh, H. T., Cassell, A. M., Quate, C. F., and Dai, H. 1998. Synthesis of individual single-walled carbon nanotubes on patterned silicon wafers. Nature. 395: 878–881.

Kong, J., Franklin, N. R., Zhou, C., Chapline, M. G., Peng, S., Cho, K. et al. 2000. Nanotube molecular wires as chemical sensors. Science. 287: 622–625.

Kong, J., Chapline, M. G., and Dai, H. 2001. Functionalized carbon nanotubes for molecular hydrogen sensors. Adv. Mater. 13: 1384–1386.

Kong, L., Wang, J., Fu, X., Zhong, Y., Meng, F., Luo, T. et al. 2010. p-Hexafluoroisopropanol phenyl covalently functionalized single-walled carbon nanotubes for detection of nerve agents. Carbon. 48: 1262–1270.

Konvalina, G., and Haick, H. 2014. Sensors for breath testing: from nanomaterials to comprehensive disease detection. Acc. Chem. Res. 47: 66–76.

Kumar, D., Jha, P., Chouksey, A., Rawat, J., Tandon, R. P., and Chaudhury, P. K. 2016. 4-(Hexafluoro-2-hydroxy isopropyl) aniline functionalized highly sensitive flexible SWCNT sensor for detection of nerve agent simulant dimethyl methylphosphonate. Mater. Chem. Phys. 181: 487–494.

Kumar, D., Jha, P., Chouksey, A., Tandon, R. P., Chaudhury, P. K., and Rawat, J. S. 2018. Flexible single walled nanotube based chemical sensor for 2, 4-dinitrotoluene sensing. J. Mater. Sci. Mater. Electron. 29: 6200–6205.

Kunadian, I., Andrews, R., Qian, D., and Mengüç, M. P. 2009. Growth kinetics of MWCNTs synthesized by a continuous-feed CVD method. Carbon. 47: 384–395.

Lee, C. Y., and Strano, M. S. 2008. Amine basicity (p K b) Controls the Analyte Binding Energy on Single Walled Carbon Nanotube Electronic Sensor Arrays. J. Am. Chem. Soc. 130: 1766–1773.

Lee, C. Y., Choi, W., Han, J. H., and Strano, M. S. 2010. Coherence resonance in a single-walled carbon nanotube ion channel. Science. 329: 1320–1324.

Lee, D., and Cui, T. 2010. Low-cost, transparent, and flexible single-walled carbon nanotube nanocomposite based ion-sensitive field-effect transistors for pH/glucose sensing. Biosens. Bioelectron. 25: 2259–2264.

Lee, K., Scardaci, V., Kim, H. Y., Hallam, T., Nolan, H., Bolf, B. E. et al. 2013. Highly sensitive, transparent, and flexible gas sensors based on gold nanoparticle decorated carbon nanotubes. Sens. Actuators B: Chem. 188: 571–575.

Lee, M., Jung, J. W., Kim, D., Ahn, Y. J., Hong, S., and Kwon, H. W. 2015. Discrimination of umami tastants using floating electrode-based bioelectronic tongue mimicking insect taste systems. Acs Nano. 9: 11728–11736.

Lee, S. H., Lim, J. H., Park, J., Hong, S., and Park, T. H. 2015. Bioelectronic nose combined with a microfluidic system for the detection of gaseous trimethylamine. Biosens. Bioelectron. 71: 179–185.

Leghrib, R., Llobet, E., Pavelko, R., Vasiliev, A. A., Felten, A., and Pireaux, J. J. 2009. Gas sensing properties of MWCNTs decorated with gold or tin oxide nanoparticles. Procedia Chem. 1: 168–171.

Li, C. A., Han, K. N., Pham, X. H., and Seong, G. H. 2014. A single-walled carbon nanotube thin film-based pH-sensing microfluidic chip. Analyst. 139: 2011–2015.

Li, C., Thostenson, E. T., and Chou, T. W. 2007. Dominant role of tunneling resistance in the electrical conductivity of carbon nanotube–based composites. Appl. Phys. Lett. 91: 223114.

Li, J., Lu, Y., Ye, Q., Cinke, M., Han, J., and Meyyappan, M. 2003. Carbon nanotube sensors for gas and organic vapor detection. Nano Lett. 3: 929–933.

Li, J., Lu, Y., and Meyyappanm, M. 2006. Nano chemical sensors with polymer-coated carbon nanotubes. IEEE Sens. J. 6: 1047–1051.

Li, L., Zhang, G., Chen, L., Bi, H. M., and Shi, K. Y. 2013. Ni (NiO)/single-walled carbon nanotubes composite: Synthesis of electro-deposition, gas sensing property for NO gas and density functional theory calculation. Mater. Res. Bull. 48: 504–511.

Li, P., Martin, C. M., Yeung, K. K., and Xue, W. 2011. Dielectrophoresis aligned single-walled carbon nanotubes as pH sensors. Biosensors. 1: 23–35.

Li, W., Hoa, N. D., and Kim, D. 2010. High-performance carbon nanotube hydrogen sensor. Sens. and Actuators B: Chem. 149: 184–188.

Li, Y., Li, G., Wang, X., Zhu, Z., Ma, H., Zhang, T., and Jin, J. 2012. Poly (ionic liquid)-wrapped single-walled carbon nanotubes for sub-ppb detection of CO_2. Chem. Comm. 48: 8222–8224.

Liang, Y. X., Chen, Y. J., and Wang, T. H. 2004. Low-resistance gas sensors fabricated from multiwalled carbon nanotubes coated with a thin tin oxide layer. Appl. Phys. Lett. 85: 666–668.

Liao, Y., Zhang, C., Zhang, Y., Strong, V., Tang, J., Li, X. G. et al. 2011. Carbon nanotube/polyaniline composite nanofibers: facile synthesis and chemosensors. Nano Lett. 11: 954–959.

Lim, J. H., Park, J., Ahn, J. H., Jin, H. J., Hong, S., and Park, T. H. 2013. A peptide receptor-based bioelectronic nose for the real-time determination of seafood quality. Biosens. Bioelectron. 39: 244–249.

Liu, C., Chen, Y., Wu, C. Z., Xu, S. T., and Cheng, H. M. 2010. Hydrogen storage in carbon nanotubes revisited. Carbon. 48: 452–455.

Liu, G., and Lin, Y. 2005. Electrochemical sensor for organophosphate pesticides and nerve agents using zirconia nanoparticles as selective sorbents. Anal. Chem. 77: 5894–5901.

Liu, H., Zhang, W., Yu, H., Gao, L., Song, Z., Xu, S. et al. 2016. Solution-processed gas sensors employing SnO_2 quantum dot/MWCNT nanocomposites. ACS Appl. Mater. Interfaces. 8: 840–846.

Liu, Q., Zhang, F., Zhang, D., Hu, N., Wang, H., Hsia, K. J., and Wang, P. 2013. Bioelectronic tongue of taste buds on microelectrode array for salt sensing. Biosens. Bioelectron. 40: 115–120.

Liu, S. F., Petty, A. R., Sazama, G. T., and Swager, T. M. 2015. Single-walled carbon nanotube/metalloporphyrin composites for the chemiresistive detection of amines and meat spoilage. Angew. Chem. Int. Ed. 54: 6554–6557.

Liu, X., Luo, Z., Han, S., Tang, T., Zhang, D., and Zhou, C. 2005. Band engineering of carbon nanotube field-effect transistors via selected area chemical gating. Appl. Phys. Lett. 86: 243501.

Liu, Y., Liu, F., Pan, X., and Li, J. 2012. Protecting the environment and public health from pesticides. pp. 5658–5659. ACS Publications.

Lobez, J. M., and Swager, T. M. 2010. Radiation detection: Resistivity responses in functional poly (olefin sulfone)/carbon nanotube composites. Angew. Chem. Int. Ed. 49: 95–98.

Lu, G., Ocola, L. E., and Chen, J. 2009. Room-temperature gas sensing based on electron transfer between discrete tin oxide nanocrystals and multiwalled carbon nanotubes. Adv. Mater. 21: 2487–2491.

Lu, J. P. 1997. Elastic properties of carbon nanotubes and nanoropes. Phys. Rev. Lett. 79: 1297.

Lu, Y., Li, J., Han, J., Ng, H. T., Binder, C., Partridge, C., and Meyyappan, M. 2004. Room temperature methane detection using palladium loaded single-walled carbon nanotube sensors. Chem. Phys. Lett. 391: 344–348.

Mahajan, R., Templeton, A., Harman, A., Reed, R. A., and Chern, R. T. 2005. The effect of inert atmospheric packaging on oxidative degradation in formulated granules. Pharm. Res. 22: 128–140.

Mallakpour, S., and Soltanian, S. 2016. Surface functionalization of carbon nanotubes: fabrication and applications. RSC Adv. 6: 109916–109935.

Mann, J. A., Rodríguez-López, J., Abruña, H. D., and Dichtel, W. R. 2011. Multivalent binding motifs for the noncovalent functionalization of graphene. J. Am. Chem. Soc. 133: 17614–17617.

Martin, N., and Nierengarten, J. F. 2012. Supramolecular chemistry of fullerenes and carbon nanotubes. John Wiley & Sons.

Matsuo, Y., Tahara, K., and Nakamura, E. 2003. Theoretical studies on structures and aromaticity of finite-length armchair carbon nanotubes. Org. Lett. 5: 3181–3184.

Mehta, R. S., and Bassette, R. 1978. Organoleptic, chemical and microbiological changes in ultra-high-temperature sterilized milk stored at room temperature. J. Food Prot. 41: 806–810.

Michaels, R. A. 1999. Emergency planning and the acute toxic potency of inhaled ammonia. Environ. Health Perspect. 107: 617–627.

Mirica, K. A., Weis, J. G., Schnorr, J. M., Esser, B., and Swager, T. M. 2012. Mechanical drawing of gas sensors on paper. Angew. Chem. 124: 10898–10903.

Mortari, A., and Lorenzelli, L. 2014. Recent sensing technologies for pathogen detection in milk: a review. Biosens. Bioelectron. 60: 8–21.

Morton, J., Havens, N., Mugweru, A., and Wanekaya, A. K. 2009. Detection of trace heavy metal ions using carbon nanotube-modified electrodes. Electroanalysis: An International Journal Devoted to Fundamental and Practical Aspects of Electroanalysis. 21: 1597–1603.

Musameh, M. M., Hickey, M., and Kyratzis, I. L. 2011. Carbon nanotube-based extraction and electrochemical detection of heavy metals. Res. Chem. Intermed. 37: 675–689.

Mzoughi, N., Abdellah, A., Gong, Q., Grothe, H., Lugli, P., Wolf., B. et al. 2012. Characterization of novel impedimetric pH-sensors based on solution-processable biocompatible thin-film semiconducting organic coatings. Sens. Actuators B: Chem. 171: 537–543.

Noort, D., Benschop, H. P., and Black, R. M. 2002. Biomonitoring of exposure to chemical warfare agents: a review. Toxicol. Appl. Pharmacol. 184: 116–126.

Novak, J. P., Snow, E. S., Houser, E. J., Park, D., Stepnowski, J. L., and McGill, R. A. 2003. Nerve agent detection using networks of single-walled carbon nanotubes. Appl. Phys. Lett. 83: 4026–4028.

Olney, D., Fuller, L., and Santhanam, K. S. V. 2014. A greenhouse gas silicon microchip sensor using a conducting composite with single walled carbon nanotubes. Sens. Actuators B: Chem. 191: 545–552.

Ormsby, J. L., and King, B. T. 2004. Clar valence bond representation of π-bonding in carbon nanotubes. J. Org. Chem. 69: 4287–4291.

Paoletti, C., He, M., Salvo, P., Melai, B., Calisi, N., Mannini, M. et al. 2018. Room temperature amine sensors enabled by sidewall functionalization of single-walled carbon nanotubes. RSC Adv. 8: 5578–5585.

Park, J., Lim, J. H., Jin, H. J., Namgung, S., Lee, S. H., Park, T. H., and Hong, S. 2012. A bioelectronic sensor based on canine olfactory nanovesicle–carbon nanotube hybrid structures for the fast assessment of food quality. Analyst. 137: 3249–3254.

Peng, G., Trock, E., and Haick, H. 2008. Detecting simulated patterns of lung cancer biomarkers by random network of single-walled carbon nanotubes coated with nonpolymeric organic materials. Nano Lett. 8: 3631–3635.

Peng, S., and Cho, K. 2000. Chemical control of nanotube electronics. Nanotechnology. 11: 57.

Penza, M., Cassanom G., Rossi, R., Rizzo, A., Signore, M. A., Alvisi, M. et al. 2007. Effect of growth catalysts on gas sensitivity in carbon nanotube film based chemiresistive sensors. Appl. Phys. Lett. 90: 103101.

Pilan, L., and Raicopol, M. 2014. Highly selective and stable glucose biosensors based on polyaniline/carbon nanotubes composites. UPB Sci. Bull., Ser. B. 76: 155–166.

Popa, A., Li, J., and Samia, A. C. S. 2013. Hybrid platinum nanobox/carbon nanotube composites for ultrasensitive gas sensing. Small. 9: 3928–3933.

Qin, X., Peng, F., Yang, F., He, X , Huang, H., Luo, D. et al. 2014. Growth of semiconducting single-walled carbon nanotubes by using ceria as catalyst supports. Nano Lett. 14: 512–517.

Qu, F., Yang, M., Jiang, J., Shen, G., and Yu, R. 2005. Amperometric biosensor for choline based on layer-by-layer assembled functionalized carbon nanotube and polyaniline multilayer film. Anal. Biochem. 344: 108–114.

Rajavel, K., Lalitha, M., Radhakrishnan, J. K., Senthilkumar, L., and Rajendra Kumar, R. T. 2015. Multiwalled carbon nanotube oxygen sensor: enhanced oxygen sensitivity at room temperature and mechanism of sensing. ACS Appl. Mater. Interfaces. 7: 23857–23865.

Randeniya, L. K., Martin, P. J., and Bendavid, A. 2012. Detection of hydrogen using multi-walled carbon-nanotube yarns coated with nanocrystalline Pd and Pd/Pt layered structures. Carbon. 50: 1786–1792.

Rao, A. M., Eklund, P. C., Bandow, S., Thess, A., and Smalley, R. E. 1997. Evidence for charge transfer in doped carbon nanotube bundles from Raman scattering. Nature. 388: 257–259.

Richardson, M., Moulton, K., Rabb, D., Kindopp, S., Pishe, T., Yan, C. et al. 2016. Capnography for monitoring end-tidal CO_2 in hospital and pre-hospital settings: A health technology assessment.

Rigoni, F., Tognolini, S., Borghetti, P., Drera, G., Pagliara, S., Goldoni, A. et al. 2013. Enhancing the sensitivity of chemiresistor gas sensors based on pristine carbon nanotubes to detect low-ppb ammonia concentrations in the environment. Analyst. 138: 7392–7399.

Rocha, A. R., Rossi, M., Fazzio, A., and da Silva, A. J. R. 2008. Designing real nanotube-based gas sensors. Phys. Rev. Lett. 100: 176803.

Ruan, W., Li, Y., Tan, Z., Liu, L., Jiang, K., and Wang, Z. 2013. *In situ* synthesized carbon nanotube networks on a microcantilever for sensitive detection of explosive vapors. Sens. Actuators B: Chem. 176: 141–148.

Ruiz-Capillas, C., and Jiménez-Colmenero, F. 2005. Biogenic amines in meat and meat products. Crit. Rev. Food Sci. Nutr. 44: 489–599.

Ruoff, R. S., and Lorents, D. C. 1995. Mechanical and thermal properties of carbon nanotubes. Carbon. 33: 925–930.

Rushi, A. D., Datta, K. P., Ghosh, P. S., Mulchandani, A., and Shirsat, M. D. 2014. Selective discrimination among benzene, toluene, and xylene: Probing metalloporphyrin-functionalized single-walled carbon nanotube-based field effect transistors. J. Phys. Chem. C. 118: 24034–24041.

Sainato, M., Humayun, M. T., Gundel, L., Solomon, P., Stan, L., Divan, R., and I. Paprotny. 2016. Parts per million CH_4 chemoresistor sensors based on multi wall carbon nanotubes/metal-oxide nanoparticles. 2016 IEEE SENS.: 1–3.

Saito, R., Fujita, M., Dresselhaus, G., and Dresselhaus, M. S. 1992. Electronic structure of chiral graphene tubules. Appl. Phys. Lett. 60: 2204–2206.

Sayago, I., Terrado, E., Aleixandre, M., Horrillo, M. C., Fernández, M. J., Lozano, J. et al. 2007. Novel selective sensors based on carbon nanotube films for hydrogen detection. Sens. Actuators B: Chem. 122: 75–80.

Sekitani, T., Noguchi, Y., Hata, K., Fukushima, T., Aida, T., and Someya, T. 2008. A rubberlike stretchable active matrix using elastic conductors. Science. 321: 1468–1472.

Seo, K., Park, K. A., Kim, C., Han, S., Kim, B., and Lee, Y. H. 2005. Chirality-and diameter-dependent reactivity of NO_2 on carbon nanotube walls. J. Am. Chem. Soc. 127: 15724–15729.

Sharma, A. K., Kumar, P., Saini, R., Bedi, R. K., and Mahajan, A. 2016. Kinetic response study in chemiresistive gas sensor based on carbon nanotube surface functionalized with substituted phthalocyanines. AIP Conf. Proc. 1728: 20493.

Shirsat, M. D., Sarkar, T., Kakoullis, J., Myung, N. V., Konnanath, B., Spanias, A. et al. 2012. Porphyrin-functionalized single-walled carbon nanotube chemiresistive sensor arrays for VOCs. J. Phys. Chem. C 116: 3845–3850.

Shul'ga, Y. M., Vasilets, V. N., Baskakov, S. A., Muradyan, V. E., Skryleva, E. A., and Parkhomenko, Y. N. 2012. Photoreduction of graphite oxide nanosheets with vacuum ultraviolet radiation. High Energy Chem. 46: 117–121.

Shvetsov, A., Shvetsova, S., Kozyrev, V. A., Spharov, V. A., and Sheremet, N. M. 2017. The "car-bomb" as a terrorist tool at metro stations, railway terminals and airports. J. Transp. Saf. Secur. 10: 31–43.

Singh, N. B., Bhattacharya, B., and Sarkar, U. 2013. Nickel decorated single-wall carbon nanotube as CO sensor. Soft Nanoscience Lett. 2013.

Sinha, N., J. Ma, and Yeow, J. T. W. 2006. Carbon nanotube-based sensors. J.Nanosci. Nanotechnol. 6: 573–590.

Sinha, R. N., Tuma, D., Abramson, D., and Muir, W. E. 1988. Fungal volatiles associated with moldy grain in ventilated and non-ventilated bin-stored wheat. Mycopathologia. 101: 53–60.

Smalley, R. E. 2003. Carbon nanotubes: synthesis, structure, properties, and applications.

Song, S., Qin, Y., He, Y., Huang, Q., Fan, C., and Chen, H. Y. 2010. Functional nanoprobes for ultrasensitive detection of biomolecules. Chem. Soc. Rev. 39: 4234–4243.

Staii, C., Johnson, A. T., Chen, M., and Gelperin, A. 2005. DNA-decorated carbon nanotubes for chemical sensing. Nano Lett. 5: 1774–1778.

Star, A., Han, T., Joshi, V., Gabriel, J., and Grüner, G. 2004. Nanoelectronic carbon dioxide sensors. Adv. Mater. 16: 2049–2052.

Star, A., Tu, E., Niemann, J., Gabriel, J. C. P., Joiner, C. S., and Valcke, C. 2006. Label-free detection of DNA hybridization using carbon nanotube network field-effect transistors. Proc. Natl. Acad. Sci. 103: 921–926.

Suehiro, J., Zhou, G., and Hara, M. 2003. Fabrication of a carbon nanotube-based gas sensor using dielectrophoresis and its application for ammonia detection by impedance spectroscopy. J. Phys. D: Appl. Phys. 36: L109.

Sun, Y., and Wang, H. H. 2007. High-performance, flexible hydrogen sensors that use carbon nanotubes decorated with palladium nanoparticles. Adv. Mater. 19: 2818–2823.

Swager, T. M. 1998. The molecular wire approach to sensory signal amplification. Acc. Chem. Res. 31: 201–207.

Szinicz, L. 2005. History of chemical and biological warfare agents. Toxicology. 214: 167–181.

Takeda, S., Nakamura, M., Ishii, A., Subagyo, A., Hosoi, H., Sueoka, K. et al. 2007. A pH sensor based on electric properties of nanotubes on a glass substrate. Nanoscale Res. Lett. 2: 207–212.

Tanaka, T., Urabe, Y., Nishide, D., and Kataura, H. 2009. Continuous separation of metallic and semiconducting carbon nanotubes using agarose gel. Appl. Phys. Express. 2: 125002.

Tang, H., Chen, J., Yao, S., Nie, L., Deng, G., and Kuang, Y. 2004. Amperometric glucose biosensor based on adsorption of glucose oxidase at platinum nanoparticle-modified carbon nanotube electrode. Anal. Biochem. 331: 89–97.

Thom, S. R. 2002. Hyperbaric-oxygen therapy for acute carbon monoxide poisoning. In New England Journal of Medicine (Vol. 347, Issue 14, pp. 1105–1106). Mass Medical Soc.

Timmer, B., Olthuis, W., and Van Den Berg, A. 2005. Ammonia sensors and their applications—a review. Sens. Actuators B: Chem. 107: 666–677.

Timmermans, A. J. M., Ambuko, J., Belik, W., and Huang, J. 2014. Food losses and waste in the context of sustainable food systems. CFS Committee on World Food Security HLPE.

Usher, M. J., and Keating, D. A. 1996. Sensors and transducers: characteristics, applications, instrumentation, interfacing. Macmillan International Higher Education.

Varghese, O. K., Kichambre, P. D., Gong, D., Ong, K. G., Dickey, E. C., and Grimes, C. A. 2001. Gas sensing characteristics of multi-wall carbon nanotubes. Sens. and Actuators B: Chem. 81: 32–41.

Velusamy, V., Arshak, K., Korostynska, O., Oliwa, K., and Adley, C. 2010. An overview of foodborne pathogen detection: In the perspective of biosensors. Biotechnol. Adv. 28: 232–254.

Voggu, R., Rout, C. S., Franklin, A. D., Fisher, T. S., and Rao, C. N. R. 2008. Extraordinary sensitivity of the electronic structure and properties of single-walled carbon nanotubes to molecular charge-transfer. J. Phys. Chem. C. 112: 13053–13056.

Wang, F., Gu, H., and Swager, T. M. 2008b. Carbon nanotube/polythiophene chemiresistive sensors for chemical warfare agents. J. Am. Chem. Soc. 130: 5392–5393.

Wang, F., Yang, Y., and Swager, T. M. 2008a. Molecular recognition for high selectivity in carbon nanotube/polythiophene chemiresistors. Angew. Chem. 120: 8522–8524.

Wang, F., and Swager, T. M. 2011. Diverse chemiresistors based upon covalently modified multiwalled carbon nanotubes. J. Am. Chem. Soc. 133: 11181–11193.

Wang, H., Koleilat, G. I., Liu, P., Jiménez-Osés, G., Lai, Y. C., Vosgueritchian, M. et al. 2014. High-yield sorting of small-diameter carbon nanotubes for solar cells and transistors. ACS Nano. 8: 2609–2617.

Wang, J. 2005. Carbon-nanotube based electrochemical biosensors: A review. Electroanal.: Int. J. Dev. Fundam. Pract. Asp. Electroanal. 17: 7–14.

Wang, J., and Musameh, M. 2004a. Carbon nanotube screen-printed electrochemical sensors. Analyst. 129: 1–2.

Wang, J., Liu, G., and Jan, M. R. 2004b. Ultrasensitive electrical biosensing of proteins and DNA: carbon-nanotube derived amplification of the recognition and transduction events. J. Am. Chem. Soc. 126: 3010–3011.

Wang, L., Wang, X., Shi, G., Peng, C., and Ding, Y. 2012. Thiacalixarene covalently functionalized multiwalled carbon nanotubes as chemically modified electrode material for detection of ultratrace $Pb2+$ ions. Anal. Chem. 84: 10560–10567.

Wang, X., Ugur, A., Goktas, H., Chen, N., Wang, M., Lachman, N. et al. 2016. Room temperature resistive volatile organic compound sensing materials based on a hybrid structure of vertically aligned carbon nanotubes and conformal oCVD/iCVD polymer coatings. Acs Sens. 1: 374–383.

Wang, Y. T., Yu, L., Zhu, Z. Q., Zhang, J., Zhu, J. Z., and Fan, C. 2009. Improved enzyme immobilization for enhanced bioelectrocatalytic activity of glucose sensor. Sens. Actuators B: Chem. 136: 332–337.

Weis, J. G., Ravnsbæk, J. B., Mirica, K. A., and Swager, T. M. 2016. Employing halogen bonding interactions in chemiresistive gas sensors. ACS Sens. 1: 115–119.

Weis, J. G., and Swager, T. M. 2015. Thiophene-fused tropones as chemical warfare agent-responsive building blocks. ACS Macro Lett. 4: 138–142.

Weizmann, Y., Chenoweth, D. M., and Swager, T. M. 2010. Addressable Terminally Linked DNA– CNT Nanowires. J. Am. Chem. Soc. 132: 14009–14011.

Wepasnick, K. A., Smith, B. A., Bitter, J. L., and Fairbrother, D. H. 2010. Chemical and structural characterization of carbon nanotube surfaces. Anal. Bioanal. Chem. 396: 1003–1014.

Wismer, T. 2009. Chemical warfare agents and risks to animal health. In Handbook of Toxicology of Chemical Warfare Agents (pp. 721–738). Elsevier.

Wong, Y. M., Kang, W. P., Davidson, J. L., Wisitsora-At, A., and Soh, K. L. 2003. A novel microelectronic gas sensor utilizing carbon nanotubes for hydrogen gas detection. Sens. Actuators B: Chem. 93: 327–332.

Woods, L. M., Bǎdescu, S. C., and Reinecke, T. L. 2007. Adsorption of simple benzene derivatives on carbon nanotubes. Phys. Rev. B, 75: 155415.

Wu, G., Yang, Z., Zhang, Z., Ji, B., Hou, C., Li, Y., Jia, W. et al. 2021. High performance stretchable fibrous supercapacitors and flexible strain sensors based on CNTs/MXene-TPU hybrid fibers. Electrochim. Acta. 395: 139141.

Yang, J. S., and Swager, T. M. 1998. Fluorescent porous polymer films as TNT chemosensors: electronic and structural effects. J. Am. Chem. Soc. 120: 11864–11873.

Yari, A., and Sepahvand, R. 2011. Highly sensitive carbon paste electrode with silver-filled carbon nanotubes as a sensing element for determination of free cyanide ion in aqueous solutions. Microchim. Acta. 174: 321–327.

Yoon, H., Lee, S. H., Kwon, O. S., Song, H. S., Oh, E. H., Park, T. H. et al. 2009. Polypyrrole nanotubes conjugated with human olfactory receptors: high-performance transducers for FET-type bioelectronic noses. Angew. Chem. 121: 2793–2796.

Yu, G., Wu, W., Zhao, Q., Wei, X., X., and Lu, Q. 2015. Efficient immobilization of acetylcholinesterase onto amino functionalized carbon nanotubes for the fabrication of high sensitive organophosphorus pesticides biosensors. Biosens. Bioelectron. 68: 288–294.

Zaporotskova, I. V, Boroznina, N. P., Parkhomenko, Y. N., and Kozhitov, L. V. 2016. Carbon nanotubes: Sensor properties. A review. Mod. Electron. Mater. 2: 95–105.

Zeininger, L., He, M., Hobson, S. T., and Swager T. M. 2018. Resistive and Capacitive γ-Ray Dosimeters Based On Triggered Depolymerization in Carbon Nanotube Composites. ACS Sens. 3: 976–983.

Zhang, X., Dai, Z., Chen, Q., and Tang, J. 2014. A DFT study of SO_2 and H_2S gas adsorption on Au-doped single-walled carbon nanotubes. Phys. Scr. 89: 65803.

Zhang, X., Dai, Z., Wei, L., Liang, N., and Wu, X. 2013. Theoretical calculation of the gas-sensing properties of Pt-decorated carbon nanotubes. Sensors. 13: 15159–15171.

Zhao, G., Zhan, X., and Dou, W. 2011. A disposable immunosensor for Shigella flexneri based on multiwalled carbon nanotube/sodium alginate composite electrode. Anal. Biochem. 408: 53–58.

Zheng, M., A. Jagota, E. D. Semke, B. A. Diner, R. S. McLean, S. R. Lustig, et al. 2003. DNA-assisted dispersion and separation of carbon nanotubes. Nat. Mater. 2: 338–342.

Zhu, R., Azzarelli, J. M., and Swager, T. M. 2016. Wireless hazard badges to detect nerve-agent simulants. Angew. Chem. Int. Ed. 55: 9662–9666.

Zhu, R., Desroches, M., Yoon, B., and Swager, T. M. 2017. Wireless oxygen sensors enabled by Fe (II)-polymer wrapped carbon nanotubes. ACS Sens. 2: 1044–1050.

Zhu, Z. 2017. An overview of carbon nanotubes and graphene for biosensing applications. Nano-Micro Lett. 9: 1–24.

Zozulya, S., Echeverri, F., and Nguyen, T. 2001. The human olfactory receptor repertoire. Genome Biol. 2: 1–12.

CHAPTER 3

Graphene Based Chemical Sensors

Bhagwan G. Toksha,[1] Prashant Gupta[2] and Sagar E. Shirsath[3],*

Introduction

The area of chemical sensing has recently seen a lot of advancements, which makes it look like an endless tunnel (Steinberg et al. 2015, Schierenbeck and Smith 2017, Wang and Wolfbeis 2019). A lot of upcoming methods and procedures made possible due to inventions in physics, chemistry, material science, biochemistry, hardware, etc., have seen disruptive changes in currently established methods of performing work and doing activities. This is in combination with modifications leading to the fine-tuning of current methods in order to improve the yield, lower the cost, energy requirements, etc., of the process involved. The focus has always been on shifting towards smaller systems that are inexpensive compared to their counterparts, which are inoperable without the experts in the trade. Some of the modern day technological examples of smaller and smarter devices are pulse oximeter (Manta et al. 2020), home pregnancy tests (Gnoth and Johnson 2014), glucose meter (Gross et al. 2000), etc., which has had a telling impact over society. A lot many other such applicable devices await regulatory approvals and are in the pipeline (Enshaeifar et al. 2018, Talal et al. 2019). The media coverage over the latest electrical and electronic devices to extract complex environmental chemical data is well noticed. However, the overlooked fact remains the most important part of the devices, i.e., the sensor which is the accountable factor for the information quality of the data being generated (Swager and Mirica 2019).

[1] Electronics and Telecommunication Department, Maharashtra Institute of Technology, Maharashtra, India.
[2] Plastic and Polymer Engineering, Maharashtra Institute of Technology, Maharashtra, India.
[3] School of Materials Science and Engineering, University of New South Wales, Sydney, NSW 2052, Australia.
* Corresponding author: s.shirsath@unsw.edu.au

A chemical sensor can be considered as an analyzer which responds to a particular analyte in a reversible and selective manner and is responsible for transforming an input chemical parameter which is quantitative, such as concentration, composition, etc., into an analytic electrical signal. The information involved may be generated due to a chemical reaction/s by a biomaterial, a chemical compound, or an amalgamation of both clinging onto the surface of a transducer (physical) towards the specific analyte. The use of chemical sensors has been in so many applications due to two factors: first, due to its use in products that can be wore on the body or planted inside the body and second being the concerns surrounding the hazardous nature of these chemicals. Some of the noticeable areas where chemical sensors find their use are environmental and industrial process monitoring, analyzing gas composition, medicine, defence, waste management, biotechnology, etc. (Wen 2016).

Graphene is a wonder material with high thermal and electrical conductivities intrinsically (Geim 2009). Also, along with the same, its high aspect ratio to attain the percolation threshold at a low volume fraction of filler loading in the matrix materials increases suitability for use as sensing materials (Justino et al. 2017, Piras et al. 2021). The sheet atoms of graphene are surface-based and responsible for high sensitivity towards adsorbed molecules (Artiles et al. 2011, Chen et al. 2011). Graphene is a 2-D material in the nanoscale regime that offers good reproducibility for devices on large-scale use. Also, the volume of its monolayer increases the sensing capacity as it increases environmental exposure. The chemical/biological surface adsorption leads to a change in sensory device conductance while graphene functions as electron donor or acceptor (Wu et al. 2010). For instance, the analyte's electrochemical reactions are greatly enhanced on a graphene film, thus leading to an enhanced voltammetric response, due to its specific electronic mobility characteristics and high electrocatalytic activity (Li et al. 2020). The uses of additional functional materials, e.g., nano-catalysts, can enhance its electrochemical properties, thus making it versatile in terms of electrochemical sensory performance. Also, they exhibit high conductivity, ease of production, biocompatible nature, and abundance of the raw material precursors from which they can be synthesized (Artiles et al. 2011, Fan et al. 2011). Hence, it is assertive to contain the right characteristics that increase its suitability for advanced electrode materials which can further be integrated with required functional materials in order to fabricate sensing interfaces for electro-analysis (Wang et al. 2009, Li et al. 2011).

This chapter will give an account of graphene family materials, i.e., graphene oxide, graphite oxide, and reduced graphene oxide, which are being continually explored and researched upon for their synthesis and use in various sensor applications which are chemical-based. A detailed analogy of the characterization of such materials for their effectiveness as sensors will be given. Also, graphene in combination with other sensory materials such as CNT, conducting polymers, metal oxides, metal-organic hybrid, etc., will be discussed. Their effects on sensitivity, selectivity, the limit of detection, response time, etc., will be elaborated. The book chapter will also comprise case studies incorporating graphene-based sensory devices for sensing various chemicals.

Synthesis of Graphene Allotropes

The synthesis of graphene material to the commercialization scale faces many hurdles such as smaller quantity yield, and structural defects. The three step approach of exfoliation, intercalation, and expansion of raw graphite leading to graphene is one widely practiced approach of synthesis (Lin et al. 2019). The sonication and shearing energies are utilized to counter the interactive forces existing between the layers of the graphite (Atta et al. 2015, Madurani et al. 2020). The inherent problems due to the insolubility of the graphite material lead to wide thickness distribution and low exfoliation efficiency. The use of intercalation process, avoiding oxidation to expand graphite with increasing interlayer distance for a better-yield exfoliation could be one of the optimal improvements in this approach (Cai et al. 2012, Ilias et al. 2021). The graphene material obtained from liquid-phase is yet another variation of exfoliation. The unique synthesis methods involving conversion of cylindrical CNT shape into monolayer graphene under harsh acid treatment and appropriate thermodynamic conditions is reported in literature (Atta et al. 2015). The fabrication of graphene sheets (GSs) with well-defined dimensionalities has been reported by Dai and co-worker. In their approach, the structure of CNTs was considered to be seamless tubes and the argon plasma etching was utilized for controlled unzipping of CNTs. The use of acidic-basic treatment with sulphuric acid and potassium permanganate to produce GSs through the oxidization and the longitudinal unzipping of MWCNTs is reported in literature (Singh et al. 2011). There are several variations incorporated in the tube-sheet transition synthesis of graphene from the CNT (Dhakate et al. 2011, Singh et al. 2011, Tian et al. 2020).

Arc discharge is a method of synthesis of all carbon-based nanomaterials in general which banks on comparatively low-cost investments. The arc discharge method is known to produce the graphene sheets with good dispersibility in organic solvent, lesser defects, and higher thermal stability (Borand et al. 2021). The high grade-high purity graphite rods are subjected to voltage arcing. The extremely high voltage between the electrodes that are separated by very small distances develops very high electric fields which generate sparks. These sparks, when they cool down, in turn result in residue containing the product material (Wang et al. 2010). The reaction carried out in water-cooled stainless steel chamber filled with gases such as hydrogen, helium and CO_2 after evacuating the chamber by a mechanical pump contributes to quality of the final product (Wang et al. 2010, Kumar et al. 2013).

The chemical vapor deposition is a synthesis route to be explored for large-area production of high-quality graphene material for practical applications (Huang et al. 2012, Kobayashi et al. 2013, Son and Ham 2017). The basic approach in CVD techniques with many further modifications involves the decomposition of fluid (gas and liquid sprays) at high temperature. The final product may be in the form of either thin films or powders. The CVD grown graphene material is known to have a large detection area, making sensor device fabrication easier and hence suitable for sensing applications (Rigoni et al. 2017). The hydrogen gas sensor using epitaxial graphene covered with platinum by growing multi-layered graphene by CVD on a Si-polar 4 H-SiC substrate was reported by Chu et al. (Chu et al. 2011). In this study, the drain current change was in response to 1% hydrogen depending on time at a

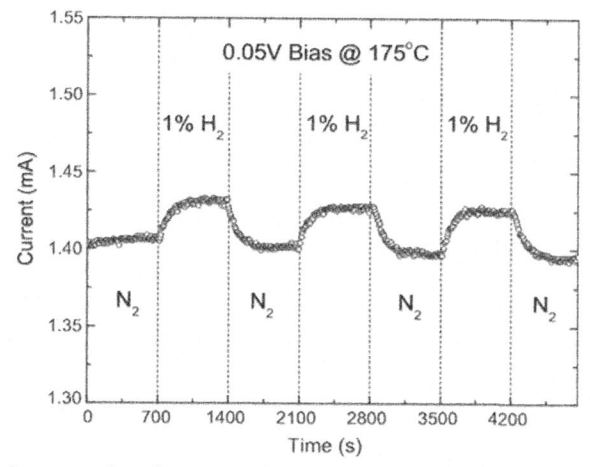

Figure 1. The hydrogen gas detecting sensor with durable rerun cycles of current response as a function of time. (Reprinted with permission from Chu et al. 2011).

constant bias of 0.05 V and temperature set at 175°C. In the first step, the nitrogen gas is filled in the testing chamber till 700s, then 1% hydrogen gas is released into the testing chamber. There was a spontaneous response from sensor in terms of continuous increase in current until equilibrium. In the next step, at 1400s, the supply of hydrogen was cut off, and pure nitrogen was flowing through the testing chamber again. This again resulted in the decrease in current following the end of hydrogen gas flow until it finally reached back to the initial value. It was reported in this study that the peak heights and response time to be same with few more iterations as shown in Figure 1. Beside these and many more approaches being explored for graphene manufacturing and processing for being used as sensors, the large-scale commercial utilization of graphene sensors is yet to be practiced to its fullest potential due to concerns such as high yield and economical production.

Mechanisms of Graphene Allotrope Sensors

The sensor based automatic controls improve the systems' response contributing to counter the greater damage that may occur in the absence of such systems and also contributes towards developments of energy efficiency projects. The growth in the sensor market worldwide is estimated at a compound annual growth rate (CAGR) of 11.4%, which nearly amounts to $193.9 billion which is 1.5 times of the year 2020 value which is going to grow to $332.8 billion by 2025 (Sensors Market Size, Share, Trend and Industry Analysis Report 2021). Adding to this, the commercial standpoint of the material of present interest is graphene which is going to touch $200 million by the end of 2022 (Global Graphene Market: Industry Size, Share, Demand, Drivers, Analysis and Forecast 2021). The time stamp of the beginning of the 21st century is related to many novel developments in the field of research and technology. The synthesis and immediate incorporation in the form of various devices of the graphene material for various applications is one such prominent development. In addition to many other fascinating properties, graphene has demonstrated properties for usage as a promising sensing element (Zhang et al. 2018). The detection of single molecule adsorption event is useful for chemical sensing (Hu et al. 2019). These features are useful for applications in various areas due to its large surface area and high signal-to-noise ratio arising from low intrinsic noise (Karaphun et al. 2021). The graphene material is an exceptionally low-noise material which leads to materials exhibiting step change behavior in resistance with adsorbed molecule whenever there is an event of modification in local carrier concentration. Moreover, it was demonstrated that the sensor shows concentration-dependent changes in electrical resistivity by adsorption of gases. The fundamental mechanism that could be understood is that the increase in gas adsorption increases the number of one type of charges, depending on whether the gas is an acceptor or donor. This change in the carrier concentration is the basic mechanism that governs the operation of all electrical conductivity-based graphene gas sensor devices. The gas being electron acceptor or electron donor affected the resistivity, magnitudes and the sign of the change. There exists a direct proportional relation between the conductivity and the product of the number of charge carriers and mobility; therefore, the change in conductivity accounts for the changes in the number density or mobility of carriers, or both. The extra charge carriers generated with the gas adsorption on sensor were also well supported through Hall-effect measurements.

The sensors with a technical name "chemiresistor" are the gas sensors based on the above-mentioned resistance change mechanism. The important features of such sensors are ease in their fabrication and measurements. The gaseous entity under consideration was detected by measuring the resistance changes of sensing layers induced by adsorbing the gas molecules while the sensor performances strongly depend on temperature. The structure of these sensors could be designed in various ways. One of the structures involves a four-point interlocked resistive mechanism and a micro-sized hotplate is introduced into the device to control the sensing temperature. The other structure involves two electrodes on the base having a graphene layer for sensing nitrogen gas. The diagram taken from the work of Ko et al. (Ko 2010) shows the schematic of the graphene-based gas sensors with the microscopic pictures shown in the insets at top and bottom. The resistor and FET based approaches are depicted in Figure 2 (Singhal et al. 2017). The figure reveals "chemiresistive" sensor mechanism of change in electrical resistance of the device due to exchange of electrons between the analyte gas and the sensitive material in which the selectivity towards specific

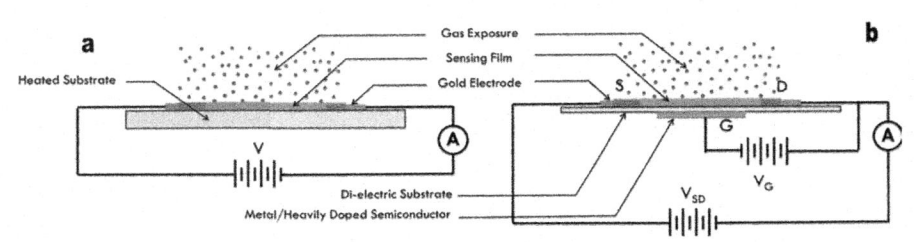

Figure 2. The symbolic representation of functionalized graphene-based gas sensor enabled with change in electric resistance (a) and change in electric field (b) modes. (Reprinted with permission from Singhal et al. 2017).

gas is depicted by 'a' and 'b'. The concentration dependent change in either charge carrier concentration or carrier mobility is transduced into an electrical signal. This enables sensors to detect the gas. The applications of such sensors are useful and not limited to detect nitrogen, ammonia, and toluene-based liquids.

The enzyme sensors built upon graphene for the detection of bio-molecules such as hydrogen peroxide, NADH, dopamine, glucose, ethanol, DNA sensors and heavy metal ion sensors enabled to detect lead and cadmium is a broad spectrum of applications in electrochemistry and environmental sector. The abilities of graphene regarding electron transfer, promoting towards enzymes and excellent catalytic behavior of bio-molecules make this material extremely attractive for sensor designing. The most critical parameters in this regard are electrochemical potential window, electron transfer rate, and redox potentials. The graphene based sensor exhibited glucose detection with the optimum performance at 0.2 V oxidation potential (Liu et al. 2020). The next parameter of the charge-transfer resistance on graphene is in the appropriate range useful for sensor applications (Zhou et al. 2009). The electron transfer behavior studies of graphene exhibit specific redox identifiers (Peng et al. 2018, Matysiak-Brynda et al. 2019). The reduction-oxidation on graphene electrodes is predominantly diffusion-controlled as revealed from the electrode peak currents' linear variation with the square root of the scan rate (Cardoso et al. 2009). The increased density of the electronic states for a well spread energy window turns out to be a unique electronic structure for graphene material enabling fast electron transfer (Velický et al. 2014). These certitudes indicate that the electronic structure and the physical-chemical surface parameters of graphene are beneficial for electron transfer (Fischer et al. 2004, McCreery 2008, Tang et al. 2009). The biometric electronic sensors which could be easily carried as the tattoo on the human body are a prospective biomedical application. The commercially available systems have yet not gone beyond rigid electrode-based devices to be worn on the wrist, chest, etc. A graphene sensor in the form of electronic tattoo mounted on skin, forehead, and chest with the performance matching commercial gel electrodes is reported by Kabiri Ameri and co-workers (Kabiri Ameri et al. 2017). The graphene sensor mounted on the skin and sustaining stretching and compression forces are depicted in Figure 3 (a to f) and the sensor worn on the forehand, forehead and chest by the subject is shown in Figure 3 (g to i).

Graphene Family Materials-Based Hybrid Sensors

Synthesis and processing

There may be two approaches to synthesizing graphene-based hybrids for use in sensory applications. The assembly method synthesizes graphene and carbon nanotubes, semiconducting nanomaterials, metallic nanomaterials, etc., distinctly first and later assembles them together. On the other hand, the *in situ* method grows these sensory materials directly over the structure of graphene. As given in Figure 4, we will discuss each synthesis route in detail for more understanding with the help of recent literature using these methods (Badhulika et al. 2015).

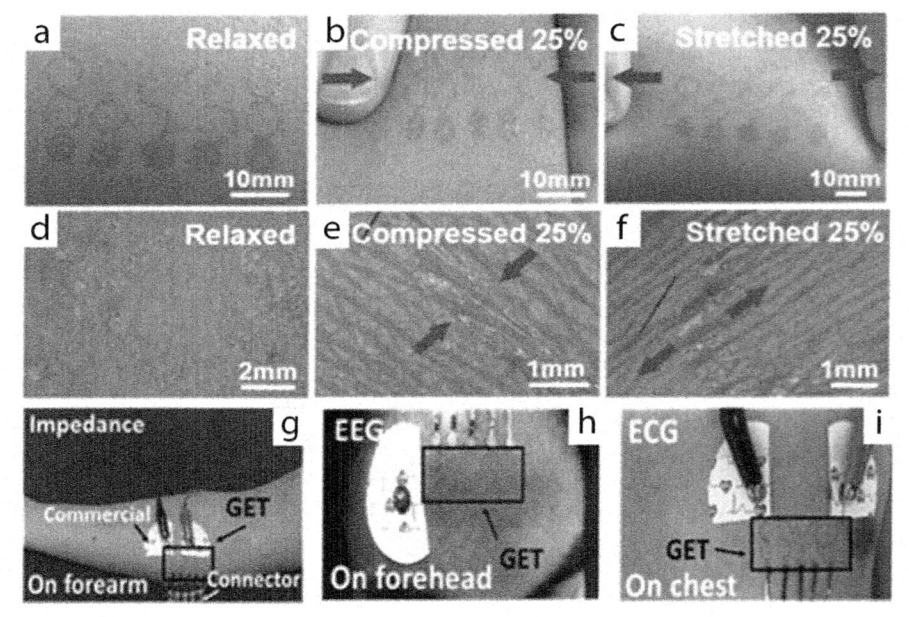

Figure 3. (a) Graphene sensor mounted on skin (b) Graphene sensor mounted on compressed by 25% (c) Graphene sensor mounted on compressed by 25% (d to f) Magnified photographs of graphene sensor on relaxed, compressed, and stretched skin (g) On forehand, sensor-skin contact performance matching commercial gel electrodes (h) On forehead, EEG brain wave graphene sensor and gel electrodes (i) On chest, ECG detection by graphene sensor and gel electrodes. (Reprinted with permission from Kabiri Ameri et al. 2017).

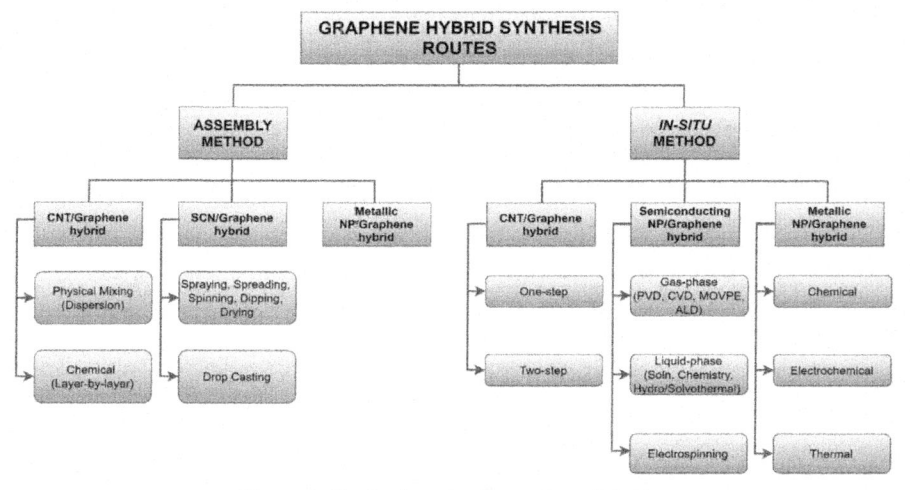

Figure 4. Synthesis routes for graphene-hybrids.

Assembly method

The assembly method is a popular method to generate graphene for the preparation of hybrids related to the self (spontaneous organization of components in liquid media via. dispersion) or chemical (chemical/electrostatic interaction-based layer by layer deposition) assembly.

A hybrid structure of graphene-CNT with a high surface area can be formed via the interactions due to Van der Waals forces of π-π interactions through their aromatic sp^2 structure. In the physical mixing/dispersion method, graphene oxide can be dissolved in water due to a host of oxygen-containing functional groups and can aid the dissolution of CNT in the liquid medium without the use of surfactant due to their π-π interactions. In turn, CNTs aid to avoid restacking of GO nanosheets (Sweetman et al. 2017, Musielak et al. 2019). A uniformly distributed network of CNTs along with a single conductive layer of chemically altered graphene can be achieved by reduction of GO after or during GO-CNT hybrid assembly. The preparation of several such hybrids is discussed (Lee et al. 2011, Salavagione et al. 2014). A carbon nanotube-graphene hybrid for direct electron transfer of glucose oxidase and glucose biosensor has been synthesized via the same approach (Chen et al. 2012). On the other hand, in the chemical method, the layer-by-layer assembly is done on the basis of their functionalities or by solvent wetting principle. Hong et al. reported see-through and flexible conducting hybrid multilayer thin films of multiwalled carbon nanotubes with graphene nanosheets in which MWCNT layers were incorporated in between r-GO nanosheets on the basis of their charges, i.e., positive and negative, respectively, via electrostatic interactions (Hong et al. 2010a). Another method, solid-phase stacking approach, was reported to grow graphene on Cu using CVD. The stacking on graphene/Cu foil was carried out after similar growth of CNT film on Ni mesh using the same technique. The copper foil was etched away with a drop of ethanol ensuring the adhesion between CNT and graphene which resulted in mechanically tough and transparent opto-chemical electrodes in sensing applications (Li et al. 2010).

Similarly, semiconductor nanomaterials (SCN) assembly with graphene in the form of a coating containing its dispersion/suspension includes a host of processes such as spreading via doctor blade, spraying, dipping, spinning, etc., and consecutive solvent drying steps. Also, drop casting can be employed to develop graphene-based sensors used in the electrochemical analysis by dropping r-GO over the surface of glassy carbon electrodes (Pham et al. 2016, Gao et al. 2018, Loan et al. 2018). A TiO_2/r-GO hybrid is reportedly fabricated by dispersion of TiO2 nanoparticles commercial paste into GO solution with a binding agent under stirring and ultrasonication for 30 minutes. The paste was then applied over tin oxide with fluorine doping using a doctor's blade. GO was reduced after washing the dried electrode via treating it with hydrazine vapors (Yang et al. 2010). Sun et al. reported the use of both physical and chemical assembly methods for the hybridization of single layer titania exfoliated from titanates and GO. The chemical methods in particular, responsible for the synthesis of these lamellar materials, are important for layouts enclosing newer 2-D materials such as BN, MoS_2, Bi_2Te_3, MnO_2, etc. (Sun et al. 2013).

Metallic nanomaterials-based graphene hybrids are largely synthesized by chemical reduction of metal salts. This, in turn, makes the surface electronegative, which upon further modification can be used as it is for self-assembly (Chang et al. 2011). There are reports of usage of pyrene butanoic acid (Hong et al. 2010b), or poly polyelectrolyte poly (diallyldimethyl ammonium chloride) (Fang et al. 2010), or polycationic protamine (Fu et al. 2012), which serve as a chemical linker. These hybrids improve the efficacy of metallic nanoparticles in diagnostic and biosensing applications (Madni et al. 2018).

In situ method

Even though the assembly methodology had a high surface area on offer along with a 3D conductive network of graphene hybrids such as graphene-CNT, there were some issues such as multiple steps involved in processing, absence of close interactions in between CNT and graphene, agglomeration causing difficulty in density/layers of graphene sheet in GO, etc. These issues cause problems with available surface area and overall conductivity for sensing applications. A solution to this is growing CNTs *in situ* on graphene. Though it is a complex process, the quality of materials produced is very high with lesser agglomeration, good contact, no residues, and uniform films. This results in better fabrication control regarding morphology, density, and orientation of synthesized structures.

CVD can be employed to grow graphene and/or CNT to ensure ohmic contact between the resulting hybrid in a one or two-step process. In the one-step process, the growth of CNT and graphene is carried out simultaneously in one step. The catalysts (Fe and Ni) are deposited over the copper foil, MgO, and nickel foam by evaporation of thin films or metal salt solution immersion (Li et al. 2019). The substrate coated with the catalyst is then exposed to CVD for the growth of graphene and CNTs simultaneously with the assistance of a carbon source. Zhu et al. reported the use of Fe/MgO and MgO for the growth of such hybrids wherein they were responsible for the growth of CNTs and graphene, respectively, resulting into a G-CNT hybrid (Zhu et al. 2012a). The two-step process is essential depending upon the metallic catalyst and substrate selection. The use of Fe catalyst on Cu foil may result in the formation of Fe/Cu alloy which hinders the active catalyst surface resulting in difficulties in CNT growth (Nguyen et al. 2012). The first step was growing a film with minimal thickness on the substrate which is capable of converting into metallic nanoparticles when exposed to higher temperatures. Upon passage of carbon-containing source in gaseous form, cracking occurs due to catalyst action of metallic nanoparticles to get carbon and hydrogen. The saturation of nanoparticles with carbon leads to the formation of CNT via precipitation. Graphene functions as a blocker to the formation of Fe/Cu alloy by facilitating the formation of active metallic nanoparticles. Zhu et al. reported a two-step synthesis wherein the first step, Fe catalyst (nanofilm) and alumina, with a particle size of 1 and 3 nm, respectively, were deposited on a graphene/Cu foil with CVD growth on it. The second step of CVD was then repeated for metal coated foil for the growth of CNT over the same. The floating buffer layer, i.e., alumina helped in covalent bond formation between graphene and CNT along with diameter control of CNT particles (Zhu et al. 2012b).

For graphene-SCN hybrid, the interface will be influenced by the SCN growth mechanism over the graphene substrate for *in situ* fabrication methodology. These processes may be classified on the basis of growth medium/phase as liquid and gaseous. The gas-phase process includes physical and chemical vapor deposition (PVD and CVD), atomic layer deposition (ALD), and metal-organic vapor phase epitaxy (MOVPE) for growing graphene-SCN hybrids. By virtue of a low melting and boiling point for zinc, ZnO is grown on graphene with the help of CVD. A variety of morphologies such as particles, ribbons, tetrapods, particles, etc., can be obtained with the process by altering the metallic catalyst treatments done prior to the actual CVD process. For instance, Au catalyst pre-treatment yields aligned nanorods whereas nanoribbons with random orientations are obtained without the pre-treatment as a large amount of Zn migrates over graphene surface with less coefficient of adhesion (Biroju et al. 2014). MOVPE has been employed for MoS2/graphene (Hoang et al. 2020), AiGaN/graphene (Munshi et al. 2018), ZnO/graphene (Kim et al. 2009), etc. ALD has been used for TiO_2 growth over graphene, via a gas-solid synthesis route in the layer-by-layer mode of growth. The precursors here, unlike CVD, make a direct reaction in the substrate and form a coating at temperatures around 150–160°C (Sun et al. 2012, Zhang et al. 2015). On the other hand, the liquid-phase process includes traditional solution chemistry, hydrothermal, solvothermal, electrochemical processes, etc. Also, advanced techniques such as electrospinning, successive ionic layer adsorption (SILAR) can be employed. Zhang et al. prepared Cu_2S-rGO hybrid via *in situ* method to be used as anodic materials in augmentation of electrochemical properties of lithium-ion battery. The stability is improved by the synthesis of a 3D matrix of rGO scaffold and Cu_2S nanoflowers by hydrothermal and freeze-drying process. The 3-D architecture could relieve the structural differences of nanoparticles during charge/discharge along with rGO which improves the specific capacity of Cu_2S nanoflowers by increasing the accessible area available to electrons and lithium ions by forming a bridge (Zhang et al. 2020). Hydro and solvothermal techniques are used over gas-phase methods in the case of polymer-based substrates for making flexible sensors as they have to be processed at lower operating temperatures. The hydrothermal/solvothermal process uses water/solvent as the liquid medium (Van Tuan et al. 2020, Kaveh and Alijani 2021). A mixture of benzyl alcohol and ethanol has been reported to be used in the solvothermal route for titanium isopropoxide, an organic precursor for TiO_2, leading to grafting over graphene by chemisorption (He et al. 2013). SILAR is employed largely for the synthesis of quantum dots based graphene hybrids such as a stacked triple composite of titania nanosheets, rGO, and CdS quantum dots (Yang et al. 2012, Wang et al. 2013). Electrospinning is less complex and economically feasible for preparing 1D nanostructures and has been demonstrated to prepare TiO2-graphene nanofiber mats by mixing the carrier solution with graphene powder and giving it to an electrospinning setup at pre-decided heat and voltage levels (Madhavan et al. 2012).

For *in situ* synthesis of graphene-metallic nanomaterials hybrid, chemical, electrochemical, or thermal reduction methods are employed for a variety of metals and precursor salts suspended with graphene (Basu and Hazra 2017, Xia et al. 2018, Song et al. 2019). Ma et al. performed and reported a comparative study on glucose sensors modified by β-cyclodextrin functionalized rGO-Au nanoparticle hybrids. β-cyclodextrin was added as capping agent in order to form 3D layered nanostructures in a one-step novel electrodeposition process *in situ* (Ma et al. 2017). Similar reports for Ag NP-graphene hybrid (Nancy et al. 2019) and Pd NP-graphene hybrid (Mangadlao et al. 2017), etc., have been reported. Other methods based on microwave aided one pot synthesis for Pt/Ru NP over graphene (Wang et al. 2011) and radio frequency catalytic CVD for Au/Ag-graphene hybrids (Pruneanu et al. 2013)respectively have been reported.

Characterization

The limiting features governing the performance of a chemical sensor includes selectivity, sensitivity, detection limit, response time and packaging. A direct reading chemical sensor will function by rapid detection and response to the analyte presence/concentration at the sensing interface (Feng et al. 2019). However, a lot of factors such as quality, cost, performance, etc., have to be traded off in terms of the characteristics listed at the start of this section and, hence, characterizing a graphene-hybrid sensor for its abilities after its synthesis and processing becomes imminent to be carried out. Table 1 summarizes some of the important characteristics for recent literature with their quantitative values for chemical sensing applications making use of a graphene hybrid.

Case Studies on Graphene Sensor-Based Devices

On the basis of research conducted over the applications of various materials that can be used for chemical sensor applications, there is an abundance of literature available. However, when the materials are weighed with respect to the value for money concept, i.e., performance v/s cost of the products that are/can be developed with their incorporation as sensory materials, not all may fit into the scheme. Therefore, it is very important to discuss graphene-based sensors that are not just studied as chemical sensing materials, but are prepared, fabricated, processed, and characterized as sensory devices.

Park et al. demonstrated the fabrication of a bioelectronic nose (B-nose), an ultrasensitive and flexible field-effect transistor (FET) system, made from plasma-treated bi-layer graphene conjugated with an olfactory receptor (Park et al. 2012). The B-nose was observed to work as a flexible and transparent testing device, which was understandably characterized using particular odorant amyl butyrate (AB) with single carbon atom resolution. The device was having high sensitivity and selectivity toward AB with a reported minimum detection limit (MDL) as low as 0.04 fM (10^{-15}; signal-to-noise ratio of 4.2). The p- and n-type behavior was exhibited by modified bilayer graphene (MBLG) after controlled treatments using oxygen and ammonia plasma. The human olfactory receptors 2AG1 (hOR2AG1: OR) were integrated

Table 1. Qualitative parameters during graphene hybrid chemical sensing use.

Sr. No	Hybrid	Sensing application	Quantitative Sensor parameters	References
1	Ag-S/r-GO	NO_2 gas sensor	Response: 74.6/50/0.2 (%/ppm/min); recovery: 0.33 min; range: 0.5–50 ppm; and bending/stretching sustainability: 1000 cycles (r = 1 cm)	(Huang et al. 2014)
2	Pt NF-graphene-CNT	Non-enzymatic glucose sensor	Linear response to glucose over the range of 1–7 mM ($R^2 = 0.978$) with a high sensitivity of 11.06 μAmM^{-1} cm^{-2}	(Badhulika et al. 2014)
3	NiO-graphene	H_2 gas sensor	High sensitivity of 52.4% at 2000 ppm of H_2 at temperature of 200°C, sensor response for NiO-G was 1.4 times that of pristine NiO; high sensor response level of 84% even after 4 cycles	(Kamal 2017)
4	SnO_2-graphene	NO_2 gas sensor	18% sensitivity to 5 ppm NO_2; response time and the recovery time for 5 ppm NO_2 is about 180 sec and 1800 sec at 30°C, respectively	(Zhang et al. 2019)
5	3D TiO_2/G-CNT	Toluene gas sensor	More than 7 times increase in response, higher selectivity and sensitivity in 50–500 ppm concentrations without this combination; more specifically, 42% in 500 ppm	(Seekaew et al. 2019)
6	Graphene/ Ag NP	Biochemical sensor for norovirus like particles	A linear response from 100 pg/mL to 10 μg/mL, limit of detection (LOD) was 92.7 pg/mL, 112 times lower than that of a conventional enzyme-linked immunosorbent assay (ELISA); the test sensitivity was 41 times higher as compared to commercial product	(Ahmed et al. 2017)
7	rGO/WS_2 nanoflakes	NH_3 gas sensor	121% response at 10 ppm ammonia concentration with response/recovery time of ~ 60 sec/300 sec at 33°C/20% RH	(Wang et al. 2019)
8	Au NP/ Graphene	Vanillin sensor	Wide linear response range to vanillin from 0.2 to 40 μM with a low limit of detection of 10 nM	(Gao et al. 2018)
9	rGO-ZnO, rGO-ZnO-Ag, rGO-ZnO-ZrO2, rGO-ZnO-Ag-Pd	H_2 gas sensor	rGO-ZnO-Ag-Pd film is found to have the highest sensitivity (~ 60%) towards hydrogen along with highest performance factor (sensitivity:cycle time ratio) with response time of 10 sec and recovery time of 14 sec	(Pal et al. 2021)
10	Titania/ Electro-Reduced Graphene Oxide	Allura red	Anodic peak currents are linear in 0.5–5.0 μM which change to a semi-logarithmic relationship with heavily concentrated region (5.0–800 μM); LOD was 0.05 μM at a signal-to-noise ratio of 3. The relative accuracy was more than 95% in the coexistence of 100-fold of potential interfering substances such as K^+, Na^+, Ca^{2+}, Mg^{2+}	(Li et al. 2020)

into the device, leading to the formation of the liquid-ion-gated FET-type platform. The equilibrium constants of OR-oxygen plasma-treated graphene (OR-OG) and ammonia plasma-treated graphene (-NG) are ca. 3.44×10^{14} and 1.47×10^{14} M^{-1}, respectively.

Vashist et al. reviewed and reported the use of graphene-based nano-electronic devices for DNA (single- and double-stranded, nucleobases and nucleotides) sensing, gas sensing (H, CO, NH_3, Cl_2, NO_2, O_2, NH_4, HCN), detection of environmental contaminants (paraoxon, hydroquinone, catechol, hydrazine, heavy metal ions, viz., Ag^+, Cd^{2+}, Ca^{2+} and Hg^{2+}, methyl jasmonate and nitromethane, etc.), pharmaceutical compounds (paracetamol, rutin, 4-aminophenol and aloe-emodin), and bacterial (*E. coli*) sensing (Vashist 2012).

Salvo et al. reviewed the use of graphene family materials for the fabrication of pH-sensitive devices, i.e., solution-gated FETs, solid-gate FETs, electrochemical sensors, and pH-sensitive quantum dots. They observed that a variety of configurations based on graphene are reported in the literature. However, only a few of them are commercially marketed due to various issues such as material degradation over time, low sensitivity, drift, and problems in manufacturing at a large-scale. Graphene as a pH-sensitive material is employed in pH ion-selective field effect transistors as it is advantageous in comparison to a glass electrode. For biological samples such as cells, etc., graphene quantum dots are more reliable due to smaller particle size and optical (non-contact) measurement. The commercial prospectus of graphene-based pH sensors will be based on experimentation to achieve desirable results with field effect transistors and graphene quantum dots. Even though there can be a blend of other pH-sensitive materials, graphene, GO and rGO are expected to considerably impact pH sensing devices domain as a standalone pH-sensitive material. It needs a thorough insight of the chemical interactions with target gas ions (Salvo et al. 2018).

Conclusion

Graphene in its all forms—pristine, oxide, reduced oxide—is effective in sensor devices. The research regarding sensors enabled with responding to a specific chemical entity, more cycles with even response of detection, and cost involved in synthesis and commercial product is still in progress. The inherent high electrical conductivity of this material makes it sensitive to a very small step change in humidity, heat, residual charge build-up, or impurities which need to be modulated by functionalization. The challenges faced by conventional sensors such as inability in miniaturization, limitations about temperature range, limited shelf life, low selectivity, short response time, and lower sensitivity could be effectively overcome by graphene based sensors.

References

Ahmed, S. R., Takemeura, K., Li, T.-C., Kitamoto, N., Tanaka, T., Suzuki, T., and Park, E. Y. 2017. Size-controlled preparation of peroxidase-like graphene-gold nanoparticle hybrids for the visible detection of norovirus-like particles. Biosens. and Bioelectron. 87: 558–565.

Artiles, M. S., Rout, C. S., and Fisher, T. S. 2011. Graphene-based hybrid materials and devices for biosensing. Adv. Drug Deliv. Rev. 63: 1352–1360.

Atta, N. F., Galal, A., and El-Ads, E. H. 2015. Graphene—a platform for sensor and biosensor applications. *In*: Rinken, T. ed. Biosensors - Micro and Nanoscale Applications. InTech.

Badhulika, S., Paul, R. K., Rajesh, Terse, T., and Mulchandani, A. 2014. Nonenzymatic glucose sensor based on platinum nanoflowers decorated multiwalled carbon nanotubes-graphene hybrid electrode. Electroanalysis, 26(1): 103–108.

Badhulika, S., Terse-Thakoor, T., Chaves Villarreal, C. M., and Mulchandani, A. 2015. Graphene hybrids: synthesis strategies and applications in sensors and sensitized solar cells. Frontiers in Chemistry, 3.

Basu, S., and Hazra, S. K. 2017. Graphene–noble metal nano-composites and applications for hydrogen sensors. C—Journal of Carbon Research, 3(4): 29.

Biroju, R. K., Giri, P. K., Dhara, S., Imakita, K., and Fujii, M. 2014. Graphene-assisted controlled growth of highly aligned ZnO nanorods and nanoribbons: growth mechanism and photoluminescence properties. ACS Appl. Mater. Interfaces. 6: 377–387.

Borand, G., Akçamlı, N., and Uzunsoy, D. 2021. Structural characterization of graphene nanostructures produced via arc discharge method. Ceram. Int. 47: 8044–8052.

Cai, M., Thorpe, D., Adamson, D. H., and Schniepp, H. C. 2012. Methods of graphite exfoliation. J. Mater. Chem. 22: 24992–25002.

Cardoso, W. S., Dias, V. L. N., Costa, W. M., de Araujo Rodrigues, I., Marques, E. P., Sousa, A. G., Boaventura, J., Bezerra, C. W. B., Song, C., Liu, H., Zhang, J., and Marques, A. L. B. 2009. Nickel-dimethylglyoxime complex modified graphite and carbon paste electrodes: preparation and catalytic activity towards methanol/ethanol oxidation. J. Appl. Electrochem. 39: 55–64.

Chang, H., Wu, X., Wu, C., Chen, Y., Jiang, H., and Wang, X. 2011. Catalytic oxidation and determination of β-NADH using self-assembly hybrid of gold nanoparticles and graphene. Analyst. 136: 2735–2740.

Chen, J., Zheng, X., Miao, F., Zhang, J., Cui, X., and Zheng, W. 2012. Engineering graphene/carbon nanotube hybrid for direct electron transfer of glucose oxidase and glucose biosensor. J. Appl. Electrochem. 42: 875–881.

Chen, L., Tang, Y., Wang, K., Liu, C., and Luo, S. 2011. Direct electrodeposition of reduced graphene oxide on glassy carbon electrode and its electrochemical application. Electrochem. Commun. 13: 133–137.

Chu, B. H., Lo, C. F., Nicolosi, J., Chang, C. Y., Chen, V., Strupinski, W., Pearton, S. J., and Ren, F. 2011. Hydrogen detection using platinum coated graphene grown on SiC. Sens. Actuators B: Chem. 157: 500–503.

Dhakate, S. R., Chauhan, N., Sharma, S., and Mathur, R. B. 2011. The production of multi-layer graphene nanoribbons from thermally reduced unzipped multi-walled carbon nanotubes. Carbon. 49: 4170–4178.

Enshaeifar, S., Zoha, A., Markides, A., Skillman, S., Acton, S. T., Elsaleh, T., Hassanpour, M., Ahrabian, A., Kenny, M., and Klein, S. 2018. Health management and pattern analysis of daily living activities of people with dementia using in-home sensors and machine learning techniques. PloS one, 13: e0195605.

Fan, Y., Liu, J. H., Yang, C. P., Yu, M., and Liu, P. 2011. Graphene–polyaniline composite film modified electrode for voltammetric determination of 4-aminophenol. Sens. and Actuators B: Chem. 157: 669–674.

Fang, Y., Guo, S., Zhu, C., Zhai, Y., and Wang, E. 2010. Self-assembly of cationic polyelectrolyte-functionalized graphene nanosheets and gold nanoparticles: a two-dimensional heterostructure for hydrogen peroxide sensing. Langmuir. 26: 11277–11282.

Feng, S., Farha, F., Li, Q., Wan, Y., Xu, Y., Zhang, T., and Ning, H. 2019. Review on smart gas sensing technology. Sensors. 19: 3760.

Fischer, A. E., Show, Y., and Swain, G. M. 2004. Electrochemical performance of diamond thin-film electrodes from different commercial sources. Anal. Chem. 76: 2553–2560.

Fu, X., Chen, L., and Li, J. 2012. Ultrasensitive colorimetric detection of heparin based on self-assembly of gold nanoparticles on graphene oxide. Analyst. 137: 3653–3658.

Gao, J., Yuan, Q., Ye, C., Guo, P., Du, S., Lai, G., Yu, A., Jiang, N., Fu, L., and Lin, C. T. 2018. Label-free electrochemical detection of vanillin through low-defect graphene electrodes modified with Au nanoparticles. Materials. 11: 489.

Geim, A. K. 2009. Graphene: status and prospects. Science, 324: 1530–1534.

Global Graphene Market: Industry Size, Share, Demand, Drivers, Analysis and Forecast, 2021. Global graphene market: industry size, share, demand, drivers, analysis and forecast 2016–2022 [online]. https://www.zionmarketresearch.com/report/graphene-market. Available from: https://www. zionmarketresearch.com/report/graphene-market [Accessed 29 Apr 2021].

Gnoth, C., and Johnson, S. 2014. Strips of hope: accuracy of home pregnancy tests and new developments. Geburtshilfe Frauenheilkd. 74: 661.

Gross, T. M., Bode, B. W., Einhorn, D., Kayne, D. M., Reed, J. H., White, N. H., and Mastrototaro, J. J. 2000. Performance evaluation of the MiniMed® continuous glucose monitoring system during patient home use. Diabetes Technol. Ther. 2: 49–56.

He, Z., Phan, H., Liu, J., Nguyen, T. Q., and Tan, T. T. Y. 2013. Understanding TiO_2 size-dependent electron transport properties of a graphene-TiO_2 photoanode in dye-sensitized solar cells using conducting atomic force microscopy. Adv. Mater. 25: 6900–6904.

Hoang, A. T., Katiyar, A. K., Shin, H., Mishra, N., Forti, S., Coletti, C., and Ahn, J. H. 2020. Epitaxial growth of wafer-scale molybdenum disulfide/graphene heterostructures by metal–organic vapor-phase epitaxy and their application in photodetectors. ACS Appl. Mater. Interfaces. 12: 44335–44344.

Hong, T. K., Lee, D. W., Choi, H. J., Shin, H. S., and Kim, B. S. 2010a. Transparent, flexible conducting hybrid multilayer thin films of multiwalled carbon nanotubes with graphene nanosheets. Acs Nano. 4: 3861–3868.

Hong, W., Bai, H., Xu, Y., Yao, Z., Gu, Z., and Shi, G. 2010b. Preparation of gold nanoparticle/graphene composites with controlled weight contents and their application in biosensors. J. Phys. Chem. C. 114: 1822–1826.

Hu, H., Yang, X., Guo, X., Khaliji, K., Biswas, S. R., García de Abajo, F. J., Low, T., Sun, Z., and Dai, Q. 2019. Gas identification with graphene plasmons. Nat. Commun. 10: 1131.

Huang, L., Chang, Q. H., Guo, G. L., Liu, Y., Xie, Y. Q., Wang, T., Ling, B., and Yang, H. F. 2012. Synthesis of high-quality graphene films on nickel foils by rapid thermal chemical vapor deposition. Carbon. 50: 551–556.

Huang, L., Wang, Z., Zhang, J., Pu, J., Lin, Y., Xu, S., Shen, L., Chen, Q., and Shi, W. 2014. Fully printed, rapid-response sensors based on chemically modified graphene for detecting NO_2 at room temperature. ACS Appl. Mater. Interfaces. 6: 7426–7433.

Ilias, S. H., Murshidi, J. A., and Ying, K. K. 2021. Effect of electrolyte concentration on the synthesis of graphene by electrochemical exfoliation process. IOP Conf. Ser.: Mater. Sci. Eng. 1106: 012013.

Justino, C. I., Gomes, A. R., Freitas, A. C., Duarte, A. C., and Rocha-Santos, T. A. 2017. Graphene based sensors and biosensors. TrAC Trends Anal. Chem. 91: 53–66.

Kabiri Ameri, S., Ho, R., Jang, H., Tao, L., Wang, Y., Wang, L., Schnyer, D. M., Akinwande, D., and Lu, N. 2017. Graphene Electronic tattoo sensors. ACS Nano. 11: 7634–7641.

Kamal, T. 2017. High performance NiO decorated graphene as a potential H_2 gas sensor. J. Alloys Compd. 729: 1058–1063.

Karaphun, A., Phrompet, C., Tuichai, W., Chanlek, N., Sriwong, C., and Ruttanapun, C. 2021. The influence of annealing on a large specific surface area and enhancing electrochemical properties of reduced graphene oxide to improve the performance of the active electrode of supercapacitor devices. Mater. Sci. Eng. B. 264: 114941.

Kaveh, R., and Alijani, H. 2021. An overview: recent development of semiconductor/graphene nanocomposites for photodegradation of phenol and phenolic compounds in aqueous solution. J. Asian Ceram. Soc. 9: 1–23.

Kim, Y. J., Lee, J. H., and Yi, G. C. 2009. Vertically aligned ZnO nanostructures grown on graphene layers. Appl. Phys. Lett. 95: 213101.

Ko, G. 2010. Graphene-based nitrogen dioxide gas sensors. Curr. Appl. Phys. 10: 3.

Kobayashi, T., Bando, M., Kimura, N., Shimizu, K., Kadono, K., Umezu, N., Miyahara, K., Hayazaki, S., Nagai, S., Mizuguchi, Y., Murakami, Y., and Hobara, D. 2013. Production of a 100-m-long high-quality graphene transparent conductive film by roll-to-roll chemical vapor deposition and transfer process. Appl. Phys. Lett. 102: 023112.

Kumar, R., Singh, Dr. R., Dubey, P., Kumar (Gangwar), P., Tiwari, R. adhey, and Oh, I.-K. 2013. Pressure-dependent synthesis of high- quality few-layer graphene by plasma- enhanced arc discharge and their thermal stability. J. Nanoparticle Res. 15: 1–10.

Lee, S. H., Lee, D. H., Lee, W. J., and Kim, S. O. 2011. Tailored assembly of carbon nanotubes and graphene. Adv. Funct. Mater. 21: 1338–1354.

Li, C., Li, Z., Zhu, H., Wang, K., Wei, J., Li, X., Sun, P., Zhang, H., and Wu, D. 2010. Graphene nano- "patches" on a carbon nanotube network for highly transparent/conductive thin film applications. J. Phys. Chem. C. 114: 14008–14012.

Li, F., Li, J., Feng, Y., Yang, L., and Du, Z. 2011. Electrochemical behavior of graphene doped carbon paste electrode and its application for sensitive determination of ascorbic acid. Sen. actuators B: Chem. 157: 110–114.

Li, G., Wu, J., Jin, H., Xia, Y., Liu, J., He, Q., and Chen, D. 2020. Titania/electro-reduced graphene oxide nanohybrid as an efficient electrochemical sensor for the determination of allura red. Nanomaterials. 10: 307.

Li, Y., Li, Z., Lei, L., Lan, T., Li, Y., Li, P., Lin, X., Liu, R., Huang, Z., Fen, X., and Ma, Y. 2019. Chemical vapor deposition-grown carbon nanotubes/graphene hybrids for electrochemical energy storage and conversion. FlatChem. 15: 100091.

Lin, L., Peng, H., and Liu, Z. 2019. Synthesis challenges for graphene industry. Nat. Mater. 18: 520–524.

Liu, B., Wang, X., Liu, H., Zhai, Y., Li, L., and Wen, H. 2020. 2D MOF with electrochemical exfoliated graphene for nonenzymatic glucose sensing: Central metal sites and oxidation potentials. Anal. Chim. Acta. 1122: 9–19.

Loan, P. T. K., Wu, D., Ye, C., Li, X., Tra, V.T., Wei, Q., Fu, L., Yu, A., Li, L.-J., and Lin, C.-T. 2018. Hall effect biosensors with ultraclean graphene film for improved sensitivity of label-free DNA detection. Biosens. Bioelectron. 99: 85–91.

Ma, M., Zhe, T., Song, W., Guo, P., Wang, J., and Wang, J. 2017. A comparative study on the glucose sensors modified by two different β-cyclodextrin functionalized reduced graphene oxide based Au nanocomposites synthesized through developed post immobilization and in situ growth technologies. Sens. Actuators B: Chem. 253: 818–829.

Madhavan, A. A., Kalluri, S., Chacko, D. K., Arun, T. A., Nagarajan, S., Subramanian, K. R. V., Nair, A. S., Nair, S. V., and Balakrishnan, A. 2012. Electrical and optical properties of electrospun TiO_2-graphene composite nanofibers and its application as DSSC photo-anodes. RSC Adv. 2: 13032–13037.

Madni, A., Noreen, S., Maqbool, I., Rehman, F., Batool, A., Kashif, P. M., Rehman, M., Tahir, N., and Khan, M. I. 2018. Graphene-based nanocomposites: synthesis and their theranostic applications. J. Drug Target. 26: 858–883.

Madurani, K. A., Suprapto, S., Machrita, N. I., Bahar, S. L., Illiya, W., and Kurniawan, F. 2020. Progress in graphene synthesis and its application: history, challenge and the future outlook for research and industry. ECS J. Solid State Sci. Technol. 9: 093013.

Mangadlao, J. D., Cao, P., Choi, D., and Advincula, R. C. 2017. Photoreduction of graphene oxide and photochemical synthesis of graphene–metal nanoparticle hybrids by ketyl radicals. ACS Appl. Mater. Interfaces. 9: 24887–24898.

Manta, C., Jain, S. S., Coravos, A., Mendelsohn, D., and Izmailova, E. S. 2020. An evaluation of biometric monitoring technologies for vital signs in the era of COVID-19. Clin. Transl. Sci. 13: 1034–1044.

Matysiak-Brynda, E., Sęk, J. P., Kasprzak, A., Królikowska, A., Donten, M., Patrzalek, M., Poplawska, M., and Nowicka, A. M. 2019. Reduced graphene oxide doping with nanometer-sized ferrocene moieties—New active material for glucose redox sensors. Biosens. Bioelectron. 128: 23–31.

McCreery, R. L. 2008. Advanced carbon electrode materials for molecular electrochemistry. Chem. Rev. 108: 2646–2687.

Munshi, A. M., Kim, D. C., Heimdal, C. P., Heilmann, M., Christiansen, S. H., Vullum, P. E., van Helvoort, A. T., and Weman, H. 2018. Selective area growth of AlGaN nanopyramid arrays on graphene by metal-organic vapor phase epitaxy. Appl. Phys. Lett. 113: 263102.

Musielak, M., Gagor, A., Zawisza, B., Talik, E., and Sitko, R. 2019. Graphene oxide/carbon nanotube membranes for highly efficient removal of metal ions from water. ACS Appl. Mater. Interfaces. 11: 28582–28590.

Nancy, P., Nair, A. K., Antoine, R., Thomas, S., and Kalarikkal, N. 2019. *In situ* decoration of gold nanoparticles on graphene oxide via nanosecond laser ablation for remarkable chemical sensing and catalysis. Nanomaterials. 9: 1201.

Nguyen, D. D., Tai, N. H., Chen, S. Y., and Chueh, Y. L. 2012. Controlled growth of carbon nanotube–graphene hybrid materials for flexible and transparent conductors and electron field emitters. Nanoscale. 4: 632–638.

Pal, P., Yadav, A., Chauhan, P. S., Parida, P. K., and Gupta, A. 2021. Reduced graphene oxide based hybrid functionalized films for hydrogen detection: Theoretical and experimental studies. Sensors International. 2: 100072.

Park, S. J., Kwon, O. S., Lee, S. H., Song, H. S., Park, T. H., and Jang, J. 2012. Ultrasensitive Flexible Graphene Based Field-Effect Transistor (FET)-Type Bioelectronic Nose. Nano Lett. 12: 5082–5090.

Peng, Y., Tang, Z., Dong, Y., Che, G., and Xin, Z. 2018. Electrochemical detection of hydroquinone based on MoS2/reduced graphene oxide nanocomposites. J. Electroanal. Chem. 816: 38–44.

Pham, T. S. H., Fu, L., Mahon, P., Lai, G., and Yu, A. 2016. Fabrication of β-cyclodextrin-functionalized reduced graphene oxide and its application for electrocatalytic detection of carbendazim. Electrocatalysis. 7: 411–419.

Piras, A., Ehlert, C., and Gryn'ova, G. 2021. Sensing and sensitivity: Computational chemistry of graphene-based sensors. Wiley Interdiscip. Rev. Comput. Mol. Sci. 11: e1526.

Pruneanu, S., Pogacean, F., Biris, A. R., Coros, M., Watanabe, F., Dervishi, E., and Biris, A. S. 2013. Electro-catalytic properties of graphene composites containing gold or silver nanoparticles. Electrochim. Acta. 89: 246–252.

Rigoni, F., Maiti, R., Baratto, C., Donarelli, M., MacLeod, J., Gupta, B., Lyu, M., Ponzoni, A., Sberveglieri, G., Motta, N., and Faglia, G. 2017. Transfer of CVD-grown graphene for room temperature gas sensors. Nanotechnology. 28: 414001.

Salavagione, H. J., Díez-Pascual, A. M., Lázaro, E., Vera, S., and Gómez-Fatou, M. A. 2014. Chemical sensors based on polymer composites with carbon nanotubes and graphene: the role of the polymer. J. Mater. Chem. A. 2: 14289–14328.

Salvo, P., Melai, B., Calisi, N., Paoletti, C., Bellagambi, F., Kirchhain, A., Trivella, M. G., Fuoco, R., and Di Francesco, F. 2018. Graphene-based devices for measuring pH. Sens. and Actuators B: Chem. 256: 976–991.

Schierenbeck, T. M., and Smith, M. C. 2017. Path to impact for autonomous field deployable chemical sensors: a case study of *in situ* nitrite sensors. Environ. Sci. Technol. 51: 4755–4771.

Seekaew, Y., Wisitsoraat, A., Phokharatkul, D., and Wongchoosuk, C. 2019. Room temperature toluene gas sensor based on TiO$_2$ nanoparticles decorated 3D graphene-carbon nanotube nanostructures. Sens. and Actuators B: Chem. 279: 69–78.

Sensors Market Size, Share, Trend & Industry Analysis Report, 2021. Sensors Market Size, Share, Trend & Industry Analysis Report 2016–2022 [online]. https://www.bccresearch.com/market-research. Available from: https://www.bccresearch.com/market-research/instrumentation-and-sensors/sensors-technologies-markets-report.html# [Accessed 29 Apr 2021].

Singh, V., Joung, D., Zhai, L., Das, S., Khondaker, S. I., and Seal, S. 2011. Graphene based materials: Past, present and future. Prog. Mater. Sci. 56: 1178–1271.

Singhal, A. V., Charaya, H., and Lahiri, I. 2017. Noble metal decorated graphene-based gas sensors and their fabrication: A review. Crit. Rev. Solid State Mater. Sci. 42: 499–526.

Son, M., and Ham, M. H. 2017. Low-temperature synthesis of graphene by chemical vapor deposition and its applications. Flat. Chem. 5: 40–49.

Song, H., Zhang, X., Liu, Y., and Su, Z. 2019. Developing graphene-based nanohybrids for electrochemical sensing. Chem. Rec. 19: 534–549.

Steinberg, M. D., Kassal, P., Kereković, I., and Steinberg, I. M. 2015. A wireless potentiostat for mobile chemical sensing and biosensing. Talanta. 143: 178–183.

Sun, P., Ma, R., Wang, K., Zhong, M., Wei, J., Wu, D., Sasaki, T., and Zhu, H. 2013. Suppression of the coffee-ring effect by self-assembling graphene oxide and monolayer titania. Nanotechnology. 24: 075601.

Sun, X., Xie, M., Wang, G., Sun, H., Cavanagh, A. S., Travis, J. J., George, S. M., and Lian, J. 2012. Atomic layer deposition of TiO$_2$ on graphene for supercapacitors. J. Electrochem. Soc. 159: A364.

Swager, T. M., and Mirica, K. A. 2019. Introduction: Chemical Sensors. Chem. Rev. 119: 1–2.

Sweetman, M. J., May, S., Mebberson, N., Pendleton, P., Vasilev, K., Plush, S. E., and Hayball, J. D. 2017. Activated carbon, carbon nanotubes and graphene: materials and composites for advanced water purification. C—Journal of Carbon Research. 3: 18.

Talal, M., Zaidan, A. A., Zaidan, B. B., Albahri, A. S., Alamoodi, A. H., Albahri, O. S., Alsalem, M. A., Lim, C. K., Tan, K. L., and Shir, W. L. 2019. Smart home-based IoT for real-time and secure remote health monitoring of triage and priority system using body sensors: Multi-driven systematic review. J. Med. Syst. 43: 42.

Tang, L., Wang, Y., Li, Y., Feng, H., Lu, J., and Li, J. 2009. Preparation, Structure and Electrochemical Properties of Reduced Graphene Sheet Films. Adv. Funct. Mater. 19: 2782–2789.

Tian, J., Liu, L., Zhou, K., Hong, Z., Chen, Q., Jiang, F., Yuan, D., Sun, Q., and Hong, M. 2020. Metal–organic tube or layered assembly: reversible sheet-to-tube transformation and adaptive recognition. Chem. Sci. 11: 9818–9826.

Van Tuan, P., Phuong, T. T., Tan, V. T., Nguyen, S. X., and Khiem, T. N. 2020. *In-situ* hydrothermal fabrication and photocatalytic behavior of ZnO/reduced graphene oxide nanocomposites with varying graphene oxide concentrations. Mater. Sci. Semicond. Process. 115: 105114.

Vashist, S. K. 2012. Advances in Graphene-Based Sensors and Devices. J. Nanomed. Nanotechnol. 4: e127.

Velický, M., Bradley, D. F., Cooper, A. J., Hill, E. W., Kinloch, I. A., Mishchenko, A., Novoselov, K. S., Patten, H. V., Toth, P. S., Valota, A. T., Worrall, S. D., and Dryfe, R. A. W. 2014. Electron transfer kinetics on mono- and multilayer graphene. ACS Nano. 8: 10089–10100.

Wang, K., Wan, S., Liu, Q., Yang, N., and Zhai, J. 2013. CdS quantum dot-decorated titania/graphene nanosheets stacking structures for enhanced photoelectrochemical solar cells. RSC adv. 3: 23755–23761.

Wang, S., Jiang, S. P., and Wang, X. 2011. Microwave-assisted one-pot synthesis of metal/metal oxide nanoparticles on graphene and their electrochemical applications. Electrochim. Acta. 56: 3338–3344.

Wang, X., Gu, D., Li, X., Lin, S., Zhao, S., Rumyantseva, M. N., and Gaskov, A. M. 2019. Reduced graphene oxide hybridized with WS2 nanoflakes based heterojunctions for selective ammonia sensors at room temperature. Sens. and Actuators B: Chem. 282: 290–299.

Wang, X., and Wolfbeis, O. S. 2019. Fiber-optic chemical sensors and biosensors (2015–2019). Anal. Chem. 92: 397–430.

Wang, Y., Li, Y., Tang, L., Lu, J., and Li, J. 2009. Application of graphene-modified electrode for selective detection of dopamine. Electrochem. Commun. 11: 889–892.

Wang, Z., Li, N., Shi, Z., and Gu, Z. 2010. Low-cost and large-scale synthesis of graphene nanosheets by arc discharge in air. Nanotechnology. 21: 175602.

Wen, W. 2016. Introductory Chapter: What is Chemical Sensor? Progresses in Chemical Sensor. IntechOpen.

Wu, W., Liu, Z., Jauregui, L. A., Yu, Q., Pillai, R., Cao, H., Bao, J., Chen, Y. P., and Pei, S.-S. 2010. Wafer-scale synthesis of graphene by chemical vapor deposition and its application in hydrogen sensing. Sens. and Actuators B: Chem. 150: 296–300.

Xia, Y., Li, R., Chen, R., Wang, J., and Xiang, L. 2018. 3D architectured graphene/metal oxide hybrids for gas sensors: A review. Sensors. 18: 1456.

Yang, N., Zhai, J., Wang, D., Chen, Y., and Jiang, L. 2010. Two-dimensional graphene bridges enhanced photoinduced charge transport in dye-sensitized solar cells. ACS Nano. 4: 887–894.

Yang, N., Zhang, Y., Halpert, J. E., Zhai, J., Wang, D., and Jiang, L. 2012. Granum-like stacking structures with TiO2-graphene nanosheets for improving photo-electric conversion. Small 8: 1762–1770.

Zhang, H., Shuang, S., Wang, G., Guo, Y., Tong, X., Yang, P., Chen, A., Dong, C., and Qin, Y. 2015. TiO$_2$–graphene hybrid nanostructures by atomic layer deposition with enhanced electrochemical performance for Pb (ii) and Cd (ii) detection. RSC adv. 5: 4343–4349.

Zhang, L., Shi, J., Huang, Y., Xu, H., Xu, K., Chu, P. K., and Ma, F. 2019. Octahedral SnO$_2$/graphene composites with enhanced gas-sensing performance at room temperature. ACS Appl. Mater. Interfaces. 11: 12958–12967.

Zhang, X., Duan, L., Zhang, X., Li, X., and Lü, W. 2020. Preparation of Cu$_2$S@ rGO hybrid composites as anode materials for enhanced electrochemical properties of lithium ion battery. J. Alloys Compd. 816: 152539.

Zhang, Y., Zhao, J., Sun, H., Zhu, Z., Zhang, J., and Liu, Q. 2018. B, N, S, Cl doped graphene quantum dots and their effects on gas-sensing properties of Ag-LaFeO$_3$. Sens. Actuators B: Chem. 266: 364–374.

Zhou, M., Zhai, Y., and Dong, S. 2009. Electrochemical sensing and biosensing platform based on chemically reduced graphene oxide. Anal. Chem. 81: 5603–5613.

Zhu, X., Ning, G., Fan, Z., Gao, J., Xu, C., Qian, W., and Wei, F. 2012. One-step synthesis of a graphene-carbon nanotube hybrid decorated by magnetic nanoparticles. Carbon. 50: 2764–2771.

Zhu, Y., Li, L., Zhang, C., Casillas, G., Sun, Z., Yan, Z., Ruan, G., Peng, Z., Raji, A. R. O., and Kittrell, C. 2012. A seamless three-dimensional carbon nanotube graphene hybrid material. Nat. Commun. 3: 1–7.

CHAPTER 4
Hydrogel Based Chemical Sensors

Moumita Shee,[1] Piyali Basak,[2] Amit K. Das[3] and Narayan Ch. Das[4,1,]*

Introduction

Hydrogels are hydrophilic polymeric networks formed by physical or chemical crosslinking via gelation process in aqueous medium. Chemically cross-linked hydrogels exhibit greater extent of covalent bonding between the polymer chains possessing high mechanical strength, whereas, physically crosslinked hydrogels are derived through non-covalent interactions among polymers containing Van der Waals/electrostatic/dipole-dipole forces, hydrogen bonding, π-π stacking or entanglement with polymeric network. Significant volume change of hydrogel shows some unique properties, which are controlled by physical factors such as temperature, electrical voltage, magnetic field, etc., as well as chemical factors (pH value) too. Swelling kinetics is an important key factor to make hydrogel as suitable candidate for being used in sensors and actuators. Typically, hydrogels are placed into the signal transduction system to create responses by varying the interaction with analytes to observe the physicochemical/biochemical molecular recognition process. Nanomaterials integrated polymer hydrogel-based chemical sensors and bioreceptors embedded biosensors also have been proposed for electrochemical detection of different analytes (Sinha et al. 2019).

High amount of water absorption facilitates the hydrogel's strong elastic compatibility or softness. Biological and elastic compatibility are the required features for the application of wearable chemical(bio) sensors (Zhang et al. 2018). It also provides mechanical flexibility, which is comfortable to the body. Hydrogel's swelling and shrinking properties are used to measure the concentration of the analytes by mass sensing based on the principle of quartz crystal microbalance

[1] School of Nanoscience and Technology, Indian Institute of Technology Kharagpur.
[2] School of Bioscience and Engineering, Jadavpur University, Kolkata.
[3] Department of Biotechnology, Indian Institute of Technology Kharagpur.
[4] Rubber Technology Centre, Indian Institute of Technology Kharagpur, West Bengal, India.
* Corresponding author: ncdas@rtc.iitkgp.ac.in

(QCM). A real time shift is the result of the mass change in the smart hydrogel system controlling the QCM fundamental resonant frequency and mass with good precision (Gupta et al. 2015, 2011). These smart hydrogels were developed for the biosensors and other microfluidic systems by modifying the elements with variation of functionalities. Reversible swelling-shrinking or geometrical differences of the hydrogels occur because of the changes of the equilibrium electrostatic force present among the polymeric chains by varying the concentrations of their target in environment (Zhou et al. 2018).

Electroconductive hydrogels contain hydrated structure with unique property like electronic functionality, which is suitable for biomaterial field. Conductive polymers support electron transport across the interface while the porous hydrogel provides a larger surface area with greater diffusivity. In the biosensor platform, flexibility and process ability of the hydrogel facilitates the functionalities by the conducting electrons with chemical modifications. Doping/de-doping mechanism of the conductive hydrogel provides the alternation in the current or voltage, which can be regulated to gauge response to concentrations. Impedance value between the electrode and environment decreases due to the ionic conductivity of electroconductive hydrogels. Electroconductive hydrogels have been utilized for the detection of vitamins (Wang et al. 2015), glucose (Homma et al. 2014, Yang et al. 2017), human metabolites [cell viability] and function, lactate (Babeli et al. 2021), DNA, dopamine, peptide, tumors, and hydrogen peroxide (Tavakoli and Tang 2017). On the other hand, electrochemical enzyme-immobilized biosensors are redox-reactive, so electron transport occurs via the redox reactions across the electroconductive hydrogel. It induces current or alters potential to produce voltage.

In glucose sensor, the chemical reaction occurs in the presence of glucose oxidase via catalyzed oxygen. This type of sensors work on the principle of interaction between glucose and oxygen molecules and the subsequent change in the signal due to the oxidation of glucose. Oxidation of glucose converts it into gluconic acid and hydrogen peroxide causing the reduction of partial pressure of oxygen in presence of glucose oxidase as shown below (Wang et al. 2017).

$$\text{Glucose} + O_2 \xrightarrow{\text{Glucose Oxidase}} \text{Gluconic acid} + H_2O_2$$

Changes in chemical components provide the concentration measurement of glucose. Glucose oxidase is encapsulated by polyacrylamide gel in this kind of sensor. Immobilization of enzyme onto electrode surface helps in biorecogniton component to the transducer. It has been observed that the response time of the bioelectrode is slow and this can only able to sense when the biological elements are in the close proximity to the transducer. Although enzymatic glucose sensor has some disadvantages, some alternative approaches have been proposed to develop a glucose sensor. Fluoresence-based and reversible competitive affinity sensor can be used for the immobilization technique. Here, sensing element comprises of 3 mm hollow dialysis tube connected to a fluorimeter via a sensing optical fiber. It consists Concanavalin A, carbohydrate receptor immobilized on the surface and a fluorescein-labelled indicator, which acts as a competing agent (Mastrototaro et al. 1991).

In the following sections, we have detailed the various chemical sensing applications of the functionalized hydrogels.

Operational principle

The operational principle of the hydrogel-based chemical sensors depends on the external stimulus, which decreases or increases an energy barrier between a deswollen or stable state and a swollen state or metastable state of the gel. The signal response and the reproducibility of the sensor are influenced by the forward and backward transition causing a difference in that energy barrier height (Guenther et al. 2008).

Depending upon the swelling behavior of hydrogels, the following principle for measurement and detection of environmental parameters is used in sensors:

1. Shifts of the resonance frequency of a quartz crystal microbalance in microgravimetric sensors.
2. Changes in the holographic diffraction wavelength in optical Bragg grating sensors.
3. Swelling of hydrogel is connected to a swelling pressure that can be determined via deflection of a membrane or bending plate in capacitor or inductor micromachined resonator and in piezoresistive pressure sensors.
4. Hydrogel layers are deposited on bilayer cantilevers that can cause bending on the plate (e.g., micromechanical).

A conversion of chemical input value into an appropriate measuring output signal occurs in chemical sensors. The signals are categorized by the transducer and material recognizing element. Transformation of the non-electric measuring value into electric measuring value occurs by transducer. The swelling or shrinking process of the hydrogel is monitored by corresponding changes in the peizoresistance of a fabricated Wheatstone bridge, which is constructed within a silicon membrane. A change in resistivity of the resistors and a stress state change are caused by the deflection of silicon membrane. This is proportional to the sensor's output voltage (U_{out}) of the sensor. The operational mechanism of the hydrogel-based chemical sensor has been summarised in Figure 1 (Guenther and Gerlach 2009).

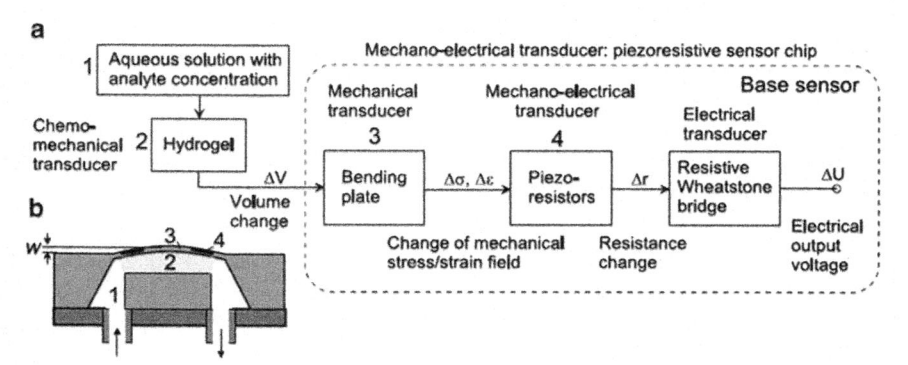

Figure 1. Operational principle **(a)** and cross-section **(b)** of hydrogel-based chemical sensor: 1 measuring solution; 2 hydrogel; 3 Si bending plate; 4 piezoresistors. (Reprinted with permission from Guenther and Gerlach 2009).

Sensor design

For the designing of chemical sensors, commercially available pressure sensor chips (Aktiv Sensor GmbH, Stahnsford, Germany) with a thin distortable silicon membrane are used for the transformations of electrical output signal from membrane deflection in transducers. At the backside of the silicon chip, the hydrogel itself is brought into a cavity and it is also enclosed with a cover. A silicon nitride mask is used as an etch resist at that cavity. The front side of the electronic component is always protected from the specific measuring species and only the backside of the chip comes into close contact with the substance. The stability of the hydrogel characteristics determines the long-term stability of the sensors. It shows excellent stable properties.

The sensor chip is attached with a socket connecting with inlet and outlet flow channels. The aqueous solution which is to be measured was pumped within the inlet tubes into the cavity of silicon chips. The cavity is coated with a 200 nm thick PECVD (Phase Electro Chemical Vapor Deposition) silicon nitride film to prevent chemical corrosion. For the detailed schematic representation of the set-up, readers may follow these references (Gerlach et al. 2004, 2005, Guenther et al. 2005, 2006, 2007a).

Sensor calibration

The silicon plate bending or deflection $w = f(p)$, $w = f(U_{out})$, $p = f(U_{out})$ is obtained by the controller of the pressure and it is actually the change (delta d) of the gel layer thickness by means of a two beam laser interferometer (Guenther et al. 2007b, 2008). The regression equation of defection w or delta d (micrometer) vs U_{out} (mv) was obtained by fitting with the experimental curve.

Generally, a 50 to 100 μm thick hydrogel is placed on a silicon platform to achieve a small gap between dry (absolutely unswollen) hydrogel and the Si bending plate.

An example of sensor calibration with a proper material was given. A layer of hydrogel was spin coated at the Si wafer and it was enclosed with two layers: one was 550 nm thick silicone oxide layer, another was 17 nm thick adhesion promoter layer. The hydrogel material was dried and cross-linked. Dimension of the dried and cross-linked (250 nm) hydrogel was measured (1 mm × 1 mm) and cut into pieces. A foil was inserted into the chip cavity and attached to a socket.

Based on the operational principle of sensor, a thin layer hydrogel was directly inserted onto the backside of the bending plate, which was covered with 220 nm thick PECVD nitride film within a 17 nm adhesion promoter layer. The final thickness of the dried, cross-linked hydrogel was measured and it was about 4 to 50 μm (Gerlach 2019).

Sensor response time

The response time of the sensor is one of the most important parameters to execute a successful sensor operation. It solely depends on the gel swelling/deswelling kinetics. If the bending plate of the sensor chip gives significant dynamic properties, a sensor response time of less than 1 second is obtained.

The ionic strength at acidic range has a great influence on the response time and it can be observed on both the pH sensors and microsensors within a range of 300 nm thickness. The square of the gel sample dimension is proportional to the time response related to the volume changed by the material. A 500 ms response time is the result when quartz crystal microbalance was experimented at high ionic strength. From the above experiments, it can be said that scaling to micro-dimension enriches the time response. In order to achieve an optimum sensitivity and a higher sensor signal, a decrease of the hydrogel thickness is restricted by the necessity. But, in case of polyelectrolyte gel, various methods of swelling kinetic occur with different additive concentrations (Guenther et al. 2013).

Sensor with force compensation

A very systematic path to cut off the response time and to enhance long term stability is the usage of the force compensation principle. Because of the second sensitivity, swelling gets compensated by applying counterforce. This converts the hydrogel to stay in its initial state with a constant volume, and the signal is measured from the microactuator by the counterforce.

Due to the constant volume, diffusion of analyte solution does not occur into and out of the hydrogel for avoiding the creep effects and shortening the response time. Miniaturization of force compensation sensors can be achieved by using a second hydrogel (e.g., temperature sensitive) as actuators and it controls the volume constancy through a corresponding temperature change (Gerlach et al. 2020).

Integration of the sensors and the actuator hydrogel in one single set up is one of the novel approaches, known as bi-sensitive hydrogels.

In this case, mechanical stability and sensitivity is appropriately shown with respect to the interpenetrating network.

Sensor transducers

Sensor transducers are the components which can transfer the non-electrical changes of the properties of stimuli responsive hydrogel into a measurable signal or electrical signal. In gel sensors, two basic principles are used:

- transducers based on mechanical performance by hydrogel swelling and shrinking, and
- transducers observing changes in properties (e.g., densities, mass, volume, stiffness) of free swelling gels.

Fundamental transducers principle

Transducers of free swelling gels

Transducers using free swelling gels can directly determine the changes in one or more hydrogel properties and these transducers act as a sensing material. Recently, optical, oscillating, and conductometric transducer principle was used.

Optical transducers

Optical transducers can measure the variation in optical properties of the hydrogel. Due to the swelling kinetics of the hydrogels, some different approach is applied based on surface coating or special fillings (Peppas and Van Blarcom 2016).

Optical transmission

Optical transmission varies due to different state of hydrogels. It has been observed that a low optical transmission results from opaque hydrogels in their shrunken state, whereas homogeneous or clear structure of hydrogel is obtained in its swollen state. Increase of water absorption enhances the transmission in colored microscope. Turbidity of the sensing elements occurs due to the shrinking of the microspheres. These variations of the optical properties can be determined as a transmission measurement using a miniature fiber optic spectrometer or other standard spectrophotometer.

Refractive index

Change in the refractive index of swelling or shrunken state of hydrogel depends on the change in the optical transmission of the hydrogel. Dual-channel chirped grating pad membrane sensor is used for the pH detection of the hydrogel membrane. Here, one channel is for pH detection of the membrane while another is used as an on-chip refractometer for the determination of the refractive index of the test solutions (Richter et al. 2008).

Conductometric transducers

In conductometric sensor, an inter-digitated electrode array is coated with hydrogel. This concept was introduced by Sheppard and his co-workers (Sheppard Jr. et al. 1995). The electrode impedance value is resistive at the frequency value of 100 Hz to 100 kHz. Swelling of hydrogel layer enhances the conductivity of the hydrogel by decreasing the resistance simultaneously.

Oscillating transducers

In oscillating transducers, change of the resonance frequency value is observed. A shift of this frequency occurs due to the change of load accompanying the change of signal amplitude.

Quartz crystal microbalance

Quartz crystal microbalance (QCM) is generally used as a transduction element to control the changes in the thin layer hydrogel properties. Density of the hydrogel decreases by swelling that increases the volume or mass of the hydrogel. As a result, surface load decreases during the swelling of the gels and mechanical changes of the hydrogel coated with quartz results in the change in signal amplitude. Therefore, decreasing resonance frequency, surface load of quartz crystal increases (Eichelbaum et al. 1999).

Common Polymers for Hydrogel Preparation

In case of sensing applications, polymers based hydrogels are reportedly derived from poly(acrylamide), poly(N-isopropylacrylamide), poly(vinyalcohol), poly(vinyl pyrolidine), poly(ehylene glycol), etc. (Bhattacharya and Samanta 2016).

Biopolymer based hydrogels include different functional groups and unique properties like high protein affinity, ease of surface modification which make them suitable for the application in biosensors (Tavakoli and Tang 2017). Some examples of biopolymers of polysaccharides and polypeptide origins are chitin, alginate, pectin, chitosan, agarose, cellulose, dextran, collagen, etc., which have also been used for the hydrogel based sensors (Tavakoli and Tang 2017).

Different crosslinked hydrogel material preparation and their characterization

Poly (vinyl alcohol)/Poly(acrylic acid) (PVA/PAAc) blends

The swelling characteristics of these polyanionic gels show a sharp change with the pH value of measuring species. In acidic medium, the value is maximum and minimum in basic medium (Günther et al. 2007).

Poly(N-Isopropyl acrylamides) (NIPAam)

Poly(N-isopropyl acrylamide) (PNIPAam) is one of the best studied thermoresponsive materials. It consists of hydrophilic amino and carbonyl groups as well as hydrophobic isopropyl groups and exhibits large and sharp changes in its swelling property in water, increasing the temperature above its volume phase transition temperature (T_{cr} = 33°C) (Richter et al. 2008).

The thermally induced reversible collapse of PNIPAam can be limited by varying the use of mixed solvents (Guenther et al. 2008), Addition of metal ions (Guenther et al. 2008), and surfactants (Guenther and Gerlach 2009), changes the T_{cr}-shift. By the co-polymerization of PNIPAam with anionic (Rahman et al. 2008), cationic (Günther et al. 2007), neutral monomers, chemical modification is possible for practical applications. Volume phase transition temperature generally varies from 25°C to 58°C by the composition of the polymer and it also allows to prepare different thermoresponsive gels with variation in the response temperature (Gerlach et al. 2005). In the microfabricated thermoresponsive hydrogels, scaling to micro-dimension is very much effective in decreasing the response time (Günther et al. 2007). The crosslinked PNIPAAm-hydrogel were prepared by free radical polymerization of NIPAAm with N,N-methylene-bisacrylamide (BIS; BIS content 4%) as the cross-linking agent.

Enhancement of the response time of gel system is specifically related to application of smart hydrogel in microsystem.

Co-polymer of N,N-dimethyl aminoethyl methacrylate (DMAEMA)

DMAEMA polycationic pH and temperature sensitive gel consists of particular co-ordination binding site for the transition metal ions.

Photo cross-linkable PNIPAAm-terpolymer

Photo cross-linkable PNIPAAm-DMAAm-DMIAAm terpolymers were used to prepare thin PNIPAAm layers. These polymers were synthesized by co-polymerization of N-isopropylacrylamide (NIPAAm), dimethyl-acrylamide (DMAAm) and 2-(dimethyl maleimido)-N-ethyl-acrylamide (DMIAAm) as chromophore (Guenther and Gerlach 2009) (Vo et al. 2002).

Hydrogel Material Conditioning

After the sensor preparation method, an initial gel conditioning process is generally performed. Typically, the formed thermoresponsive hydrogel has been swollen in de-ionized water at a low temperature (T < Tc) for 24 h and after that the hydrogel has to experience a cyclic swelling-deswelling process at an increasing and decreasing temperatures respectively. At the time of the first operation of the hydrogel, sensor gives a bad repeat accuracy and a drift to the sensor parameter. This happens because of the changes in the microscopic structure of the polymer network. By repeating the number of conditioning cycles, sensor signal precision may be achieved. Here, the conditioning process was continued after 5–7 swelling cycles. The output voltage of the sensor was measured during the swelling/deswelling of the layer of the hydrogel under different ambient conditions, variable temperature, and humidity uncertainties (Guenther et al. 2007a).

Swelling/deswelling kinetics in aqueous solutions

In chemical sensor, one of the most important factors is response time determination during the operation in aqueous solution and it can be obtained by the thickness of the polymer layers with additives as well as by the hydrogel swelling kinetics. Diffusion rates of the additives and the water present in the gel structure is very important with respect to determine the swelling kinetics. As the mass transfer processes occur at the same time simultaneously, the kinetics of swelling-deswelling is too complex to detect (Plum et al. 2006).

Swelling kinetics of the dry gel in water

Diffusion coefficient of water in a crosslinked gel is estimated during the swelling of the gel in dry state. The output voltage of the sensor consisting three parts considers the time dependent swelling process. In general, the slow gel softening with retardation time caused by the diffusion of water is shown in the first part and in the second part, water diffusion into the soft gel at a constant rate within time is controlled (Guenther et al. 2007a).

Co-operative diffusion and sensor response

A two-step mechanism is generally required for the response of a hydrogel changing its environmental properties. In the first step, the stimulus causing the swelling or shrinking must penetrate the gel. As the first step is faster, the rate of this step is regulated either by heat transfer (for temperature sensitive polymers) or mass transfer ions and solvents. Because of this reason, thermo-sensitive gels can respond faster than chemoresponsive ones.

The second step is categorized by the change of the degree of the swelling which is controlled by the movement of the network chains of the polymer, the so-called cooperative diffusion. The response time depends on the square of the dimensions like length, thickness, etc., and this is inversely proportional to the co-operative diffusion co-efficient (Schulz et al. 2012). This concept was first introduced by Tanaka, Hocker, and Benedek within the THB theory (Li and Tanaka 1990).

According to Li and Tanaka, a shear process at the time of the co-operative swelling, it decreases the anisotropic deformation (Li and Tanaka 1990).

For symmetric hydrogel, shear energy is very low basically for the spherical geometry, and for the other geometries like slab, it may be more important for the sensor applications. In contrast, shear energy has to be considered leading to the minimized speed of the co-operative diffusion process (Schulz et al. 2012).

For the asymmetric gel geometries, diffusion does not take place in all directions because of the non-zero shear modulus. This procedure leads to the shear relaxation process and reduces the speed of the co-operative diffusion process. So, the decreasing rate of the speed is most important for the round or disc-like hydrogel structure as only uniaxial diffusion occurs here. The rest of the directions are subjected to shear relaxation.

To prepare the hydrogel, both the co-operative diffusion and the diffusion of the analyte are to be measured into and out of the hydrogel as fast as possible. This can be executed in various ways.

The structure and size of the hydrogel should be thin and small. By means of MEMS technology, this requirement of miniature can be easily implemented. An optimum sensitivity between sensor signal amplitude and the sensor response time can be achieved by using composites as well as hybrid material. A significant depletion (72% compared to the homogeneous hydrogel) of the sensor response time was obtained in case of the hybrid hydrogel incorporating hygroscopic fibers that accelerated the diffusion of the solution in the gel and subsequently the gel swelling or deswelling (Guenther et al. 2007a). Hydrophilic porous fibers give a faster and increased solution uptake.

This method determines the analyte concentration in the solution by measuring the time. In the solution, concentration of the analyte should be shortened from the time to reach a full saturation of the solution uptake which helps to determine the initial rate determination.

Optimum porosity and mechanical stability are the main features of the hydrogel. Therefore, homogeneous pore distribution and narrow size distributions with the hydrogels should be targeted.

Polymer Hydrogel Based Chemical Sensors and Biosensors

By utilizing various suitable electrochemical transduction mechanisms, the responsive character of hydrogel interacting with analyte is used in terms of electrochemical signals. Different polymer-based hydrogels and their nano(bio)composites improved transducers have been developed to attain the electrochemical process.

Conductive polymers possessing large surface area proceeds the electron transport process within porous hydrogel matrix. However, some electroconductive polymers like polyaniline, polypyrrole, poly(ethylene dioxythiophene) have also been withdrawn significant attention in preparing of hydrogel sensors.

Hydrogel based plasmonic sensors

Chemical and biochemical measurements are broadly spread depending on the optical effects for the sensors. They execute the advantages of fast response at low costs and it works on very small volumes of analyte and very thin layers of biochemical recognition structure is required.

Sensors which depend on molecular dyes basically help to determine the concentration of analyte molecules. So, dye-based experiments face various difficulties (particularly inhomogeneous development of color on the sensor surface, limited number of combinations between fluorophores and analytes and leaching of the fluorophores), which has many limitations for the regeneration of active sensor area.

In this aspect, hydrogels are very promising materials because of changing different parameters with respect to the biochemical properties present in the surroundings. Photonic bandgaps is one of the ways by which structural coloring can be achieved into hydrogels through periodic refractive index variations on the nanoscale.

Surface plasmon is a collective oscillation of conductive electrons and proceeds to spectrally selective absorption followed by the scattering of the light. This can be excited in nanoscale dimension or nanostructured materials (Kreibig and Vollmer 2013). Noble metals (especially gold) are used to make the oscillation of electrons because of the high conductivity and chemical inertness.

Refractive index of the surrounding medium is responsible for changing the oscillation and also resonance wavelength. It leads to the variation in spectral distribution of scattered and absorbed light indicating the change of the dielectric refractive index. This is the process how plasmonic materials in a hydrogel can behave as reporters for the special change induced by external stimulus. By incorporating noble metal particles into the hydrogel or placing hydrogel in nanostructured metal surface, the above process can be done (Tokarev et al. 2008). The latter approach is preferred than that of the former, because there is no risk of nanoparticles when it is being washed out from the hydrogel sensor layer. It has high importance in the application of medical and food production.

In commercial products, this sensor concept has reached at a good stage. In this aspect, some challenges like reproducibility and cost-effective fabrication of the plasmonic transducer structure, reliability, adhesion of hydrogel layer on metal surface, and reduction of sensor response time need to be addressed by plasmonic transducer (Tavakoli and Tang 2017).

Optical transducer can be used in any kind of hydrogel and can be covered onto the surface of the structure practically. One of the works based on an ethanol-sensitive (polyacrylamide, PAM) and a pH sensitive (HPMA/DMEMA/TEGDMA) gel was reported (Daikuzono et al. 2017). In this layer set up, gold structure with

N,N'-bis(acrylol)cystamine (BAC) was pretreated before spin coating process of the pregel solution to achieve the covalent binding of the hydrogel on the transducer. UV lamp was used to enhance the crosslinking via the photoinitiator containing the respective gel. The refractive index sensitivity of the nanostructure is a very concerned factor with respect to the very vicinity of the surface structure (few tens of the nanometer perpendicular to the surface structure) (Härtling et al. 2008). Therefore, very thin layered hydrogel structure (1μm thickness and below) can be used for very fast response time of the setup.

Piezoresistive hydrogel based sensors

The operational principle of a hydrogel based piezoresistive chemical sensors depends on the measuring solution, hydrogel, Si plate bending, and piezoresistors (Scarpa et al. 2020). In this sensor, the solution which is to be measured is pumped through the inlet cavity of the silicon chip. Therefore, the respective ions cause the swelling or shrinking of the hydrogel. As a result, a swelling pressure initiates the deflection of the bending plate. This results in a mechanical stress in the plate and hence an electrical output voltage is generated via the piezoresistive effect. This principle has a number of advantages.

Piezoresistive Si sensor element shows excellent long term behavior with respect to the chemical sensors and Si sensors are also cost effective (Tadigadapa and Mateti 2009). Here, signal processing is very easy and simple. By selecting the appropriate type of hydrogel, sensitivity, selectivity and long term stability as well as the sensor properties must be adaptive for particular measurement and the above set up of the figure can be used as universal sensor platform (Gerlach et al. 2020). In Table 1, we have summarized list of some typical hydrogels, used for the sensing of chemical analytes.

Biomorphic structures

The use of bimorph structure (particularly bimorphic strips and plates) has an important role for the chemical and biochemical sensors for the measurement of the swelling of the hydrogels. The change in volume or length of a layer in a layered composite affects the entire structure changing the temperature also. Different co-efficient of metal strips like thermal expansions and temperature change is evaluated by interconnecting the metal strips (Gerlach 2019).

Bimorph based structures have different characteristics than those based on volume swelling. The diffusion paths are short until the water absorption is completed by the thin hydrogel layers. On the other hand, small out of plane deflections approach the lower sensitivity destructing the bimorph effect with respect to the volume expansion. Here, a trade-off must be created between the fastest response behavior and a high sensitivity. Generally, sensors applications need a space or separation between the quantity to be measured in a fluid or gas and the electronic components. This can be better accomplished when bimorph plates are clamped on all sides in the sensor structure. Therefore, an appropriate change in the sensitivity of a MEMS based plates is expected in this kind of sensor (Gulnizkij and Gerlach 2020).

Table 1. Hydrogels for the measurement of chemical analyte (Gerlach et al. 2020).

Measure and analyte	Hydrogel	References
pH value	poly(vinyl alcohol)/poly(acrylic acid) (PVA/PAAc) HPMA/DMAEMA/TEGDMA	(Gerlach and Arndt 2009, Günther 2009, Gerlach et al. 2005, Guenther et al. 2014)
Temperature	poly(N-isopropylacrylamide) (PNIPAAM	(Günther 2009, Gerlach and Arndt 2009)
Ethanol	PNIPAAm(MBAAm 4) poly(acrylamide-bisacrylamide) (Aam/Bis)	(Gerlach and Arndt 2009, Günther 2009, Guenther et al. 2008, Erfkamp et al. 2018, 2019a)
Salt Concentration NaCl,NaI	P2Vblock-P(NIPAAm-co-DMAAm) PNIPAAm(MBAAm 4),PNIPAAm-DMAAm-DMIAAm)	(Günther et al. 2007)
Transition metal ions	P2VP-block-P(NIPAAm-co-DMIAAm), PDMAEMA-DMIMA	(Günther et al. 2007)
Ammonia	poly(acrylic acid/2-dimethylamino) ethyl methacrylate) (Aac/DMAEMA)	(Erfkamp et al. 2019b)

Due to the swelling of the hydrogel in gas with relative humidity, bimorph elements (both plates and beams) act as a spring element with a torsional compliance representing torsional moment. The largest deflection is obtained when these two compliances are same. In MEMS devices, compliances are determined by the plate material as Young Modulus of the hydrogel is very small. From these occurrences, it can be said that maximum deflection can be obtained when the ratio between hydrogel-covered and hydrogel uncovered part of the plate is almost equal (Muralikrishna et al. 2017).

When the hydrogel performs mechanical work or swelling, deflection of the bending plates also occurs mechanically. It means that no electrical energy is required and the energy supplied by environment helps to operate the sensor autonomously. The devices can be used for switching on and off, respectively, since the swelling and deswelling processes are reversible.

Hydrogel based biosensors

Hydrogel biosensors are useful for understanding the molecular recognition process and also analyte response towards physicochemical or biochemical aspects. Immobilization on hydrogel surface is very important for designing the bioreceptors with its binding type (covalent or non-covalent), biocompatibility, stability and sensing performance. There are different strategies proposed for the purpose of conjugation and selection of encapsulation. Bioreceptors should be stable under various conditions (such as temperature, pH, solvents, etc.) during polymerization. These receptors are encapsulated with hydrogel matrix and can be acrylated also. Some bioreceptors are used for protein molecules specifically. Those delicate structures can be denatured because of the harsh reaction condition or polymerisation. So, it can be performed under normal reaction condition for suitable loading of bimolecular in gel forms (Yan et al. 2013). But, biomolecules which are prepared in rigid condition can be incorporated by equilibrium partitioning (Knipe et al. 2015).

Both hydrogels and bimolecules should possess specific functionalities for conjugation or attachment. This conjugation is followed by crosslinking, affinity-ligand binding, physical adsorption, encapsulation, etc. (Tavakoli and Tang 2017). The stability between hydrogel and bioreceptor is caused by covalent bonding preferably. When the average intermolecular distance of hydrogel is less than that of the bioreceptor, encapsulation is applied for the conjugation. Moreover, crosslinking is also an effective method. Affinity-ligand binding and physical adsorption includes Van der Waals interaction, hydrogen bonding or salt-linkage for conjugation. Other techniques are chelation and establishment of disulfide linkage.

Bioreceptor embedded hydrogels are well-established molecular recognition factor. But, it is too expensive to perform. It has the lowest possible detection capacity as well as biosensor sensitivity faces challenging issues. There are lots of interesting literature are available on bioreceptor embedded hydrogels such as polyacrylate based nucleic acid (DNA) embedded hydrogels (Liu et al. 2018, Cai et al. 2017), PANI (Zhai et al. 2013), and biopolymer (Zhang and Ji 2010) based enzymes embedded hydrogels, which are useful for electrochemical biosensing, and different proteins and polymer-based hybrid hydrogels such as collagen protein embedded PPY (Ravichandran et al. 2018) hydrogel and bovine serum albumine based PNIPAAm hydrogel biosensors (Wei et al. 2018), etc.

Nucleic acid embedded hydrogel biosensors

Nucleic acids have the capabilities of specific recognition sites to the target molecules. Structural changes may occur when targeted molecules are trapped by the hydrogel interacting with analytes. DNA aptamers are encapsulated on the basis of competitive binding of targets with respect to the structural design of the entrapped hydrogel matrix.

Various nucleic acids modified hydrogel sensor was also reported. Liu et al. developed a hybrid DNA hydrogel based sensor, which was immobilized on the indium tin oxide/polyethylene terephthalate (ITO/PET) electrode and the hybrid hydrogel was used for the detection of cancer biomarker microRNA-21(miR-21) (Liu et al. 2018). Here, ferrocene tagged DNA recognition probes were used for crosslinking. DNA grafted polyacrylamide polymer was used to form the hybrid DNA hydrogel and immobilization on 3-(mithoxysilyl) propyl methacrylate was performed for the fabrication of biosensor. In this sensing process, the loss of ferrocene tags and subsequent reduction in the current was detected by Cyclic Voltammetry (CV) and Differential Pulse Voltammetry (DPV). DNA impedance biosensors was also developed for the detection of heavy metal ion such as Hg^{2+} (Cai et al. 2017).

In oligonucleotide, thymine-thymine (T-T) mismatched base pair was active and highly attracted towards the Hg^{2+} ions reportedly. So, T-Hg^{2+}, -T co-ordination chemistry was applied for this heavy metal ion detection. Two hairpin DNAH3 and H4 modified polyacrylamide polymer chain was linked to enhance the activity of Hg^{2+} and activate Mg^{2+} specific DNAzyme strategies for biosensor modification. Self-assembly of DNA hydrogel was completed by this fabrication at GCE surface and it was used for Hg^{2+} sensing. Detection limit of Hg^{2+} is up to 0.042 pM and the impedance shows significant sensitivity of the sensor (Sinha et al. 2019).

Enzyme embedded hydrogel biosensor

Enzymes are incorporated into the hydrogel matrices by the covalent-binding or encapsulation (Tavakoli and Tang 2017). For example, multifunctional hydrogel biosensor was prepared constituting polyacrylic acid, -rGO vinyl substituted PANI and leutetium phthalocyanine ($LuPc_2$) (Al-Sagur et al. 2017) incorporating glucose oxidase(GO_x). Methylene bis-acrylamide was used as a crosslinker and ammonium persulfate as an initiator in the free radical polymerization reaction of PAA with –rGO and vinyl substituted PANI. In a typical synthesis method, on the glassy carbon electrode (GCE) surface, a multifunctional hydrogel consisting of PAA/-rGO and PANI/LUPc2 was fabricated and GOx was immobilized for the amperometric biosensing of Glucose. Reduced form of enzyme activates the electron transfer between the immobilized GO_x and the electrode surface through the oxidation process. Sensitivity of this hydrogel biosensor (PAA/rGO/PANI/LuPc$_2$) was observed at the value of 15.31 uAmM^{-1} cm^{-2} for the glucose detection range at 0.3 V over the concentration range of 2–12 Mm with the limitation upto 25 μM with rapid response time of 1 s. Figure 2 explains the general examples of the polymer composite hydrogels and examples of some common gelators.

Another hydrogel biosensor (GO_x immobilized PANI/PtNPs nanoconjugate) was reported for glucose biosensors. Platinum coated electrode was used and mixture of phytic acid, aniline monomer, and APS was added to form the hydrogel (Zhai et al. 2013).

Hydrogel-based chemical sensors

Different hydrogel altered transducers have been selected for analyte detection. Several metal nanoparticles and carbon nanomaterials are incorporated into the hydrogel matrices. Thus, the formation of hydrogel nanocomposites is done by suitable host-guest interactions. These composite materials possess good functionalities, electronic and surface properties than that of the individual or native hydrogel material. Here, the polymer gelators act as host and the nanomaterials perform as guest interacting with each other through Van der Waals force via some aliphatic chain (π-stacking among the aromatic sites, hydrogen bonding, dipolar and electrostatic interactions through polar moieties) (Balla et al. 2019).

Molecular design of the polymer gelators is the key factor to control the host-guest interactions. But, aliphatic gelator cannot interact with the aromatic skeleton (e.g., fullerene, GR) due to the insufficient pi-pi interaction. So, nanomaterials having aromatic residues are the best suited to aromatic gelators for the suitable π-π interaction (Bhattacharya and Samanta 2016).

For sugar (glucose and fructose) detection, polyacrylamide (PAAm) based hydrogel was prepared as reported (Daikuzono et al. 2017). The modification of a thin layer containing acrylamide co-polymerisation with 3-(acrylamido) phenyl boronic acid (PBA) is included in inter-digitated carbon screen printed paper in the sensor. The sensor platform, which is consisting of the copolymerized PBA-based

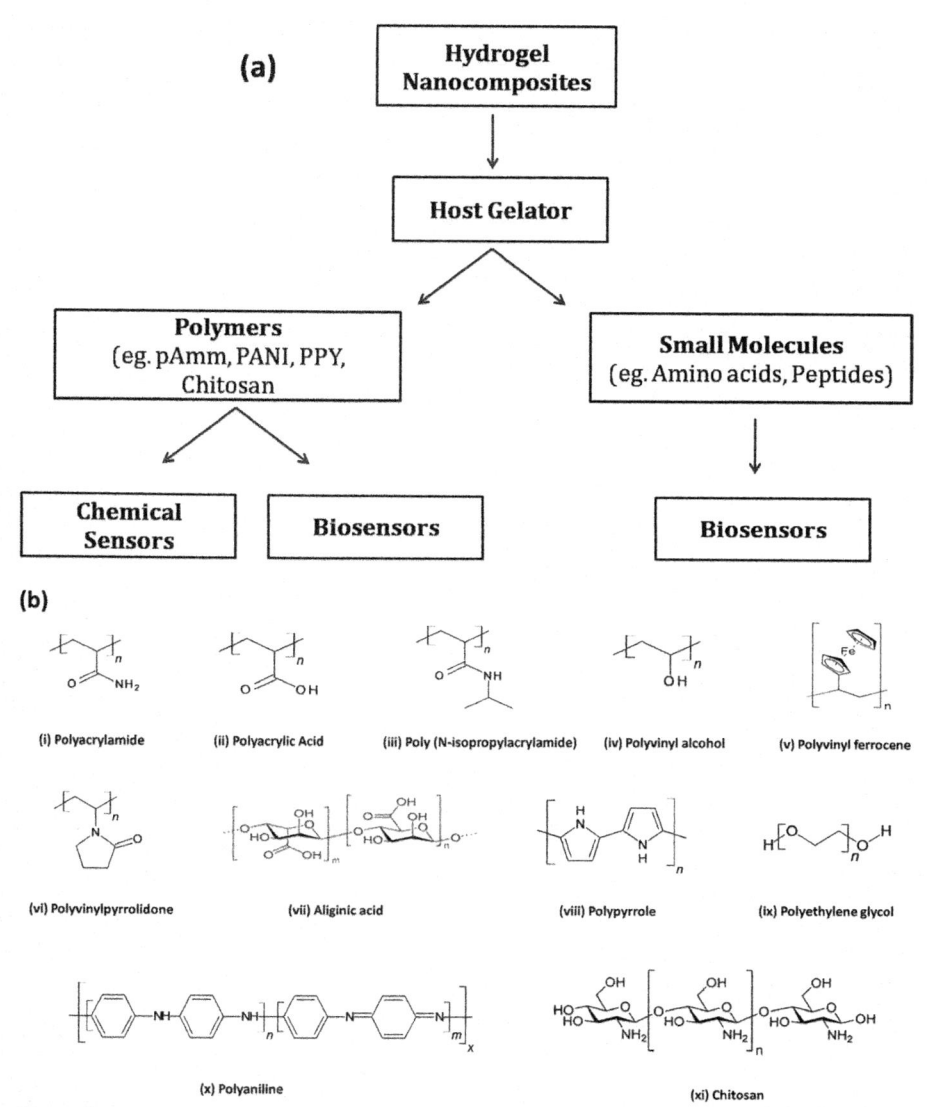

Figure 2. (a) General presentation of hydrogel nanocomposites, (b) chemical structure of some common polymer gelators. (Reprinted with permission from Sinha et al. 2019).

hydrogel is a very useful sensor for sugar detection. Fabrication of sensor is followed by the direct immobilization of hydrogel on its surface due to the porous structure of the paper.

Increased concentration of boronate ions resulting in osmotic swelling leads to further swelling of the hydrogel in presence of different concentration of sugar molecules. Some electrical properties such as impedance and capacitance were measured through the hydrogel matrix preceded by the increased mobility or ion diffusion. Variation of impedance showed a reduced value with increasing the sugar concentration (0.5 Mm). Relative capacitance changes were previously reported for fructose and glucose. This sensor was found for the sugar detection in sweat samples. For quantification of different types of sugar molecules present in one sample, this type of sensor is not suitable. A poly(N-isopropyl amide) (PNIPAAm) matrix containing covalent bonding PBA and $Ru(bpy)^{3+/2+}$ redox centers was utilized as a fructose sensing platform in this same approach (Li et al. 2018).

Redox activities of electrochemical sensors and biosensor have been developed to improve the sensitivities. Polymers having biological origin, abundant oxygen and other functionalities modify the performance for better activity of the sensors.

In a study by Fu et al., a hydrogel electrochemical sensor was prepared using chitosan (one of the best-known biological origin polymers) and silver ions were taken as the cross-linking agent for antioxidants (ascorbic acid, uric acid and luteolin) screening. Moreover, the activity of the silver ion was quite inhibited due to the complexation of the moieties and crosslinking also. Hydrogen peroxide (H_2O_2) was added to ease the activity and depolymerization was induced by producing hydroxyl (●OH) radicals. Hydroxyl radical attracted the glycosidic bonds of chitosan polysaccharide that is caused by depolymerization and recovered the activity of the silver ions.

In an electrolytic system, H_2O_2 was incorporated in a mixture solution of antioxidant to continue the screening process. H_2O_2 acts as a function of antioxidant concentration by increasing the concentration of it, resulting in gradual decrease of silver ion peaks. Based on the similar technique, a chitosan hydrogel and zinc ion composite sensing activity was also organized for hydroxyl radical ●OH and H_2O_2 detection (Fu et al. 2018).

A pectin (PT) biopolymer based hydrogel sensor has been reported for the detection of dopamine (DA) and paracetamol (Kokulnathan et al. 2018).

At GCE, this hydrogel sensor was prepared by normal drop casting of Pt/rGO (reduced graphene) nanocomposite suspension. The catalytic activity and synergy were improved and modified for the detection of DA and PAC by the functional groups (-COOH, $-C_2H_3O$, -OH) of pectin forming ionic bonds with the functional group of reduced graphene oxide. Detection value was obtained as 1.5 nM and 1.8 nM, respectively, for DA and PAC at PT(rGO/GCE) hydrogel sensor by using LSV (Linear sweep voltammetry). In pharmaceutical samples and human serum, this kind of sensor provides excellent result for the real time screening of DA and PAC.

For heavy metal (Pb^{2+}) detection, sensor was prepared by using GO fabrication and conducting polymers such as polypyrrole (PPY) and polyaniline (PANI) (Suvina et al. 2018).

In presence of GO, *in situ* polymerization of pyrrole and aniline was carried out for the preparation of PPY/GO (Muralikrishna et al. 2017) and PANI/GO hydrogel sensors (Suvina et al. 2018).

pH Sensor

In case of wearable platforms, colorimetric and electrochemical (bio)sensors are employed for monitoring sweat samples, although they have low stability of the sensitive element. On the other hand, mass (bio)sensors are also used for the detection of analyte because of their rigidity. To get over these limitations, a flexible mass (bio)sensor was reported for pH measurement. This type of device facilitates piezoelectric membrane fabricated on a polymer substrate integrated to the sensitive properties of pH responsive hydrogel based on PEG-DA/CEA molecules. At different pH, due to the swelling or deswelling kinetics of the hydrogel, resonance frequency shift is differentiated. In the advancement of micro and nano electrochemical systems [(MEMSs) and (NEMSs)], commercialization of wearable and portable biosensors is available and suitable for checking the health status (Steinhubl et al. 2015). Sweat is one of the best components for wearable (bio) sensing among various bio-fluids. It is secreted from local areas of the skin and has less risk of contamination or degradation. Sweat constitutes a different range of metabolites (urea, amino acid, glucose, lactate, etc.), electrolytes (sodium, chloride, potassium, etc.), antigens, antibodies, xenobiotics, ethanol and drugs. Composition of different bio-fluids changes due to pathological or diseased condition (Jadoon et al. 2015). pH of sweat sample is an important factor to detect any abnormalities of individual's health because of the variation of its value in physiological or pathological condition. Cystic fibrosis detection is one of the example and it is determined by checking the chloride level in sweat (Beauchamp and Lands 2005). Due to the lack of reabsorption of bicarbonate, sweat value of pH 9 indicates the cystic fibrosis. For healthy body, the pH of the physiological sweat is used to remain in between 4.0 to 6.8 (Patterson et al. 2000). However, these methods are generally utilized to fabricate selective and sensitive (bio) sensors, and some disadvantages are linked to reusability of sensors. Biological elements are affected by some environmental factors like temperature, pH, etc., in response to the stability. More stable and sensitive molecules are required to enhance the stability of biological entities which are incorporated into the polymeric chain network called hydrogels. So, some of the smart hydrogels have shown the ability of selective response for target analytes and it may provide more effective and stable alternative to the standard biological sensing elements (Sinha et al. 2019).

Application Potential of Hydrogel Based Sensors

The features of hydrogels such as increase in the volume and other material limiting factors depending on the external variables can be easily used for chemical or biochemical sensors. This can be executed by coupling the hydrogel to a corresponding transducer which transfers the variation in hydrogel characteristics into an electrical signal. Hydrogel can simply be monitored so that they are particularly measurable. The transducers can then be used as a platform technology serving as a family of sensors for different type of biochemical species, such as the measurements of the swelling pressure by means of pressure of optical sensors. The smart characteristics of the hydrogel-based chemical sensors are as follows:

(1) Any other counter and reference electrodes are not needed, which are the source of long-term instabilities except the electrochemical sensor.

(2) The separation between electronic components and the chemical part can easily be obtained while measuring the swelling pressure.

MEMS (Micro electromechanical system)-based sensors are cost efficient and scaled-down. In addition to good sensitivity chemical sensors, it should also manifest long-term stability, short response time and high selectivity (Guenther et al. 2009).

Summary

The operational principle of hydrogel-based chemical sensors has been explained. Polyelectrolyte gels as well as neutral gels are used for chemo mechanical transducers in various experiments reportedly. In aqueous solution, analyte dependent swelling of hydrogel is generally determined by the micro-fabrication of pressure sensor chip combining to the gel. The sensitivity of the hydrogel with respect to the concentration of H^+ ions (pH sensor), transition metal ions, solvents, and organic compounds is also shown in different observations. So, it can be said that sensitivity depends on the polymer composition as well as on the polymer crosslinking degree. Higher sensitivity can be achieved with increasing the concentration of ionizable groups for polyelectrolyte gels. Signal reproducibility and long term stability are the important factors for an effective sensor implementation. This long-term stability is determined by the hydrogel characteristics and the strength of the sensor chip also. Hydrogel material preparation and its conditioning are necessary for signal production and stability measurement. Gel swelling/deswelling kinetics can be regulated to obtain an optimum value of sensor signal amplitude and response time.

Hydrogel based different biosensors are also useful for molecular recognition process and analyte response towards the physicochemical/biochemical behavior. Suitable encapsulation, immobilization, and cross-linking methods are applied for hydrogel based chemical sensors and biosensors through the process of incorporation of various nanomaterials into polymer matrix. Small biomolecular based hydrogels like peptide, di-peptide and amino acid derived gels have been modified for biosensing application. On the other hand, nanocarbon based hydrogel sensors have been developed for the electrochemical sensing applications. Some metal-based nanoparticles such as titanium oxide, silver bromide, bismuth phosphate, nickel-manganese spinel-oxide, etc., have been used to develop pH responsive hydrogels, which have pH sensitivity ranges up to 10^{-5} pH units. It has the capability to measure the real time. Although it cannot provide results in a large range of pH, it has an advantage of sensors determining the tailored sensor solution for many applications.

References

Al-Sagur, H., Komathi, S., Khan, M. A., Gurek, A. G., and Hassan, A. 2017. A novel glucose sensor using lutetium phthalocyanine as redox mediator in reduced graphene oxide conducting polymer multifunctional hydrogel. Biosens. Bioelectron. 92: 638–645.

Babeli, I., Puiggalí-Jou, A., Roa, J. J., Ginebra, M.-P., García-Torres, J., and Alemán, C. 2021. Hybrid conducting alginate-based hydrogel for hydrogen peroxide detection from enzymatic oxidation of lactate. Int. J. Biol. Macromol. 193: 1237–1248.

Balla, P., Sinha, A., Wu, L., Lu, X., Tan, D., and Chen, J. 2019. Co_3O_4 nanoparticles supported mesoporous carbon framework interface for glucose biosensing. Talanta. 203: 112–121.

Beauchamp, M., and Lands, L. C. 2005. Sweat-testing: a review of current technical requirements. Pediatr. Pulmonol. 39: 507–511.

Bhattacharya, S., and Samanta, S. K. 2016. Soft-nanocomposites of nanoparticles and nanocarbons with supramolecular and polymer gels and their applications. Chem. Rev. 116: 11967–12028.

Cai, W., Xie, S., Zhang, J., Tang, D., and Tang, Y. 2017. An electrochemical impedance biosensor for Hg2+ detection based on DNA hydrogel by coupling with DNAzyme-assisted target recycling and hybridization chain reaction. Biosens. Bioelectron. 98: 466–472.

Daikuzono, C. M., Delaney, C., Tesfay, H., Florea, L., Oliveira, O. N., Morrin, A., and Diamond, D. 2017. Impedance spectroscopy for monosaccharides detection using responsive hydrogel modified paper-based electrodes. Analyst. 142: 1133–1139.

Eichelbaum, F., Borngräber, R., Schröder, J., Lucklum, R., and Hauptmann, P. 1999. Interface circuits for quartz-crystal-microbalance sensors. Rev. Sci. Instrum. 70: 2537–2545.

Erfkamp, J., Guenther, M., and Gerlach, G. 2018. Hydrogel-based piezoresistive sensor for the detection of ethanol. J. Sens. Sens. Syst. 7: 219–226.

Erfkamp, J., Guenther, M., and Gerlach, G. 2019a. Hydrogel-based sensors for ethanol detection in alcoholic beverages. Sensors. 19: 1199.

Erfkamp, J., Guenther, M., and Gerlach, G. 2019b. Piezoresistive Hydrogel-Based Sensors for the Detection of Ammonia. Sensors. 19: 971.

Fu, L., Wang, A., Lyu, F., Lai, G., Yu, J., Lin, C.-T., Liu, Z., Yu, A., and Su, W. 2018. A solid-state electrochemical sensing platform based on a supramolecular hydrogel. Sens. Actuators B Chem. 262: 326–333.

Gerlach, G., Guenther, M., Suchaneck, G., Sorber, J., Arndt, K., and Richter, A. 2004. Application of sensitive hydrogels in chemical and pH sensors. Macromol. Symp. 210: 403–410.

Gerlach, G., Guenther, M., Sorber, J., Suchaneck, G., Arndt, K.-F., and Richter, A. 2005. Chemical and pH sensors based on the swelling behavior of hydrogels. Sens. Actuators B Chem. 111: 555–561.

Gerlach, G., and Arndt, K.-F. 2009. Hydrogel sensors and actuators: engineering and technology (Vol. 6). Springer Science and Business Media.

Gerlach, G. 2019. Hydrogel-based chemical and biochemical sensors. Zeszyty Naukowe Wydziału Elektrotechniki i Automatyki Politechniki Gdańskiej.

Gerlach, G., Guenther, M., and Härtling, T. 2020. Hydrogel-based Chemical and Biochemical Sensors-A Review and Tutorial Paper. IEEE Sens. J. 21: 12798–12807.

Guenther, M., Gerlach, G., Sorber, J., Suchaneck, G., Arndt, K. F., and Richter, A. 2005. pH sensors based on polyelectrolytic hydrogels. In Smart Structures and Materials. 2005. Electroactive Polymer Actuators and Devices (EAPAD) (Vol. 5759, pp. 540–548). SPIE.

Guenther, M., Gerlach, G. U., Kuckling, D., Kretschmer, K., Corten, C., Weber, J., Sorber, J., Suchaneck, G., and Arndt, K.-F. 2006. Chemical sensors based on temperature-responsive hydrogels. In Smart structures and materials: smart sensor monitoring systems and applications (Vol. 6167, p. 61670T). International Society for Optics and Photonics.

Guenther, M., Gerlach, G., Corten, C., Kuckling, D., Muller, M., Shi, Z., Sorber, J., and Arndt, K. 2007a. Application of polyelectrolytic temperature-responsive hydrogels in chemical sensors. Macromol. Symp. 254: 314–321.

Guenther, M., Gerlach, G., and Wallmersperger, T. 2007b. Modeling of the nonlinear effects in pH sensors based on polyelectrolytic hydrogels. In Electroactive Polymer Actuators and Devices (EAPAD) 2007 (Vol. 6524, p. 652417). International Society for Optics and Photonics.

Guenther, M., Gerlach, G., Corten, C., Kuckling, D., Sorber, J., and Arndt, K.-F. 2008. Hydrogel-based sensor for a rheochemical characterization of solutions. Sens. Actuators B Chem. 132: 471–476.

Guenther, M., and Gerlach, G. 2009. Hydrogels for chemical sensors. In Hydrogel sensors and actuators (pp. 165–195). Springer.

Guenther, M., Gerlach, G., and Wallmersperger, T. 2009. Non-linear effects in hydrogel-based chemical sensors: experiment and modeling. J. Intell. Mater. Syst. Struct. 20: 949–961.

Guenther, M., Gerlach, G., Wallmersperger, T., Avula, M. N., Cho, S. H., Xie, X., Devener, B. V., Solzbacher, F., Tathireddy, P., and Magda, J. J. 2013. Smart hydrogel-based biochemical microsensor array for medical diagnostics. Adv. Sci. Technol. 85: 47–52.

Guenther, M., Wallmersperger, T., and Gerlach, G. 2014. Piezoresistive chemical sensors based on functionalized hydrogels. Chemosensors. 2: 145–170.

Gulnizkij, N., and Gerlach, G. 2020. Modelling and model verification of an autonomous threshold sensor for humidity measurements. J. Sens. Sens. Syst. 9: 1–6.

Günther, M., Kuckling, D., Corten, C., Gerlach, G., Sorber, J., Suchaneck, G., and Arndt, K.-F. 2007. Chemical sensors based on multiresponsive block copolymer hydrogels. Sens. Actuators B Chem. 126: 97–106.

Günther, M. 2009. Anwendung polymerer Funktionsschichten in piezoresistiven chemischen und Feuchtesensoren. TUDpress, Verlag der Wiss.

Gupta, V. K., Ganjali, M. R., Norouzi, P., Khani, H., Nayak, A., and Agarwal, S. 2011. Electrochemical analysis of some toxic metals by ion-selective electrodes. Crit. Rev. Anal. Chem. 41: 282–313.

Gupta, V. K., Karimi-Maleh, H., and Sadegh, R. 2015. Simultaneous determination of hydroxylamine, phenol and sulfite in water and waste water samples using a voltammetric nanosensor. Int. J. Electrochem. Sci. 10: 303–316.

Härtling, T., Alaverdyan, Y., Hille, A., Wenzel, M. T., Käll, M., and Eng, L. M. 2008. Optically controlled interparticle distance tuning and welding of single gold nanoparticle pairs by photochemical metal deposition. Opt. Express. 16: 12362–12371.

Homma, T., Sumita, D., Kondo, M., Kuwahara, T., and Shimomura, M. 2014. Amperometric glucose sensing with polyaniline/poly (acrylic acid) composite film bearing covalently-immobilized glucose oxidase: A novel method combining enzymatic glucose oxidation and cathodic O2 reduction. J. Electroanal. Chem. 712: 119–123.

Jadoon, S., Karim, S., Akram, M. R., Kalsoom Khan, A., Zia, M. A., Siddiqi, A. R., and Murtaza, G. 2015. Recent developments in sweat analysis and its applications. Int. J. Anal. Chem. 2015.

Knipe, J. M., Chen, F., and Peppas, N. A. 2015. Enzymatic biodegradation of hydrogels for protein delivery targeted to the small intestine. Biomacromolecules. 16: 962–972.

Kokulnathan, T., Ramaraj, S., Chen, S.-M., and Han-Yu, Y. 2018. Eco-friendly synthesis of biocompatible pectin stabilized graphene nanosheets hydrogel and their application for the simultaneous electrochemical determination of dopamine and paracetamol in real samples. J. Electrochem. Soc. 165: B240.

Kreibig, U., and Vollmer, M. 2013. Optical properties of metal clusters (Vol. 25). Springer Science and Business Media.

Li, H., Voci, S., Ravaine, V., and Sojic, N. 2018. Tuning electrochemiluminescence in multistimuli responsive hydrogel films. J. Phys. Chem. Lett. 9: 340–345.

Li, Y., and Tanaka, T. 1990. Kinetics of swelling and shrinking of gels. J. Chem. Phys. 92: 1365–1371.

Liu, S., Su, W., Li, Y., Zhang, L., and Ding, X. 2018. Manufacturing of an electrochemical biosensing platform based on hybrid DNA hydrogel: Taking lung cancer-specific miR-21 as an example. Biosensors and Bioelectronics. 103: 1–5.

Mastrototaro, J. J., Johnson, K. W., Morff, R. J., Lipson, D., Andrew, C. C., and Allen, D. J. 1991. An electroenzymatic glucose sensor fabricated on a flexible substrate. Sens. Actuators B Chem. 5: 139–144.

Muralikrishna, S., Nagaraju, D. H., Balakrishna, R. G., Surareungchai, W., Ramakrishnappa, T., and Shivanandareddy, A. B. 2017. Hydrogels of polyaniline with graphene oxide for highly sensitive electrochemical determination of lead ions. Anal. Chim. Acta. 990: 67–77.

Patterson, M. J., Galloway, S. D. R., and Nimmo, M. A. 2000. Variations in regional sweat composition in normal human males. Exp. Physiol. 85: 869–875.

Peppas, N. A., and Van Blarcom, D. S. 2016. Hydrogel-based biosensors and sensing devices for drug delivery. Journal of Controlled Release. 240: 142–150.

Plum, T. J., Saxena, V., and Jessing, R. J. 2006. Design of a MEMS capacitive chemical sensor based on polymer swelling. 2006 IEEE Workshop on Microelectronics and Electron Devices. WMED'06. pp-2.

Rahman, M. D., Kumar, P., Park, D.-S., and Shim, Y.-B. 2008. Electrochemical sensors based on organic conjugated polymers. Sensors. 8: 118–141.

Ravichandran, R., Martinez, J. G., Jager, E. W. H., Phopase, J., and Turner, A. P. F. 2018. Type I collagen-derived injectable conductive hydrogel scaffolds as glucose sensors. ACS Appl. Mater. Interfaces. 10: 16244–16249.

Richter, A., Paschew, G., Klatt, S., Lienig, J., Arndt, K.-F., and Adler, H.-J. P. 2008. Review on hydrogel-based pH sensors and microsensors. Sensors. 8: 561–581.

Scarpa, E., Mastronardi, V. M., Guido, F., Algieri, L., Qualtieri, A., Fiammengo, R., Rizzi, F., and De Vittorio, M. 2020. Wearable piezoelectric mass sensor based on pH sensitive hydrogels for sweat pH monitoring. Sci. Rep. 10: 1–10.

Schulz, V., Ebert, H., and Gerlach, G. 2012. A closed-loop hydrogel-based chemical sensor. IEEE Sens. J. 13: 994–1002.

Sheppard, Jr., N. F., Lesho, M. J., McNally, P., and Francomacaro, A. S. 1995. Microfabricated conductimetric pH sensor. Sens. Actuators B Chem. 28: 95–102.

Sinha, A., Kalambate, P. K., Mugo, S. M., Kamau, P., Chen, J., and Jain, R. 2019. Polymer hydrogel interfaces in electrochemical sensing strategies: A review. TrAC - Trends Anal. Chem. 118: 488–501.

Steinhubl, S. R., Muse, E. D., and Topol, E. J. 2015. The emerging field of mobile health. Sci. Transl. Med. 7: 283rv3-283rv3.

Suvina, V., Krishna, S. M., Nagaraju, D. H., Melo, J. S., and Balakrishna, R. G. 2018. Polypyrrole-reduced graphene oxide nanocomposite hydrogels: A promising electrode material for the simultaneous detection of multiple heavy metal ions. Mater. Lett. 232: 209–212.

Tadigadapa, S., and Mateti, K. 2009. Piezoelectric MEMS sensors: state-of-the-art and perspectives. Meas. Sci. Technol. 20: 92001.

Tavakoli, J., and Tang, Y. 2017. Hydrogel based sensors for biomedical applications: An updated review. Polymers. 9: 364.

Tokarev, I., Tokareva, I., and Minko, S. 2008. Gold-nanoparticle-enhanced plasmonic effects in a responsive polymer gel. Adv. Mater. 20: 2730–2734.

Vo, C. D., Kuckling, D., Adler, H.-J., and Schönhoff, M. 2002. Preparation of thermosensitive nanogels by photo-cross-linking. Colloid Polym. Sci. 280: 400–409.

Wang, L., Zhang, Y., Xie, Y., Yu, J., Yang, H., Miao, L., and Song, Y. 2017. Three-dimensional macroporous carbon/hierarchical Co3O4 nanoclusters for nonenzymatic electrochemical glucose sensor. Appl. Surf. Sci. 402: 47–52.

Wang, Z., Wang, Z., Zhang, H., Duan, X., Xu, J., and Wen, Y. 2015. Electrochemical sensing application of poly (acrylic acid modified EDOT-co-EDOT): PSS and its inorganic nanocomposite with high soaking stability, adhesion ability and flexibility. RSC Adv. 5: 12237–12247.

Wei, Y., Zeng, Q., Hu, Q., Wang, M., Tao, J., and Wang, L. 2018. Self-cleaned electrochemical protein imprinting biosensor basing on a thermo-responsive memory hydrogel. Biosens. Bioelectron. 99: 136–141.

Yan, L., Zhu, Z., Zou, Y., Huang, Y., Liu, D., Jia, S., Xu, D., Wu, M., Zhou, Y., and Zhou, S. 2013. Target-responsive "sweet" hydrogel with glucometer readout for portable and quantitative detection of non-glucose targets. J. Am. Chem. Soc. 135: 3748–3751.

Yang, M., Jeong, J.-M., Lee, K. G., Lee, S. J., and Choi, B. G. 2017. Hierarchical porous microspheres of the Co3O4@ graphene with enhanced electrocatalytic performance for electrochemical biosensors. Biosens. Bioelectron. 89: 612–619.

Zhai, D., Liu, B., Shi, Y., Pan, L., Wang, Y., Li, W., Zhang, R., and Yu, G. 2013. Highly sensitive glucose sensor based on Pt nanoparticle/polyaniline hydrogel heterostructures. ACS Nano. 7: 3540–3546.

Zhang, Y.-Z., Lee, K. H., Anjum, D. H., Sougrat, R., Jiang, Q., Kim, H., and Alshareef, H. N. 2018. MXenes stretch hydrogel sensor performance to new limits. Sci. Adv. 4: eaat0098.

Zhang, Y., and Ji, C. 2010. Electro-induced covalent cross-linking of chitosan and formation of chitosan hydrogel films: Its application as an enzyme immobilization matrix for use in a phenol sensor. Anal. Chem. 82: 5275–5281.

Zhou, M., Guo, J., and Yang, C. 2018. Ratiometric fluorescence sensor for Fe^{3+} ions detection based on quantum dot-doped hydrogel optical fiber. Sens. Actuators B Chem. 264: 52–58.

CHAPTER 5

Luminescent Metal-Organic Frameworks as Chemical Sensors

Yogeshwar D. More, Sahel Fajal, Subhajit Dutta and
*Sujit K. Ghosh**

Introduction

Sensing

Environmental pollution due to rapid industrialization and urbanization all over the world has become the prime concern globally in recent years (Mukherjee et al. 2021, Wang et al. 2018, Resolution of United Nations General Assembly 2010). Adding to that, approximately 54% of the global population (~ 3.9 billion) lives in urban areas currently (World Urbanization Prospects, UN Department of Economic and Social Affairs 2014). Such quick urban expansion will result in additional strain on environment, which is already under high stress. Particularly the environmental pollution, along with its associated health risks in the developing world, is currently contributing to more than 20% of the overall diseases (Zhou et al. 2014). Such adverse effects and the level of exposure may rise even more with the growing cities in the developing world. Hence, development of effective detection methods and efficient sensory materials to monitor potentially harmful environmental toxins and pollutants is of prime demand (Lustig et al. 2017, Karmakar et al. 2019, Rasheed and Nabeel 2019). In addition, with the increase in amplitude of global terrorism, the ability to detect the existing explosive materials and its precursors instantaneously has become crucial as ever (Bennett 2003, Bobbitt et al. 2017). Moreover, precise sensory control over biologically relevant species along with monitoring the pH and temperature by the advanced biological devices hold much importance for basic understanding of several biological activities and in imaging applications (Dalgliesh

Department of Chemistry, Indian Institute of Science Education and Research (IISER) Pune, Dr. Homi Bhabha Road, Pashan, Pune-411008, Maharashtra, India.
* Corresponding author: sghosh@iiserpune.ac.in

1951, Le Floc'h et al. 2011, Mutihac et al. 2011, Wu et al. 2019). A sensor is typically composed of a sensing unit along with a transduction unit, which in combination translate the photonic information into a different type of signalling pathway, e.g., an optical or electrical signal. For a sensory material, the transduction mechanism is the most crucial aspect, which typically correlates with the changes in electrical, optical, mechanical or photophysical properties of the probe upon interacting with the analytes (Falcaro et al. 2016, Wang et al. 2015). An advanced sensory material should comprise several important characteristics such as selectivity, sensitivity, response time, long-term stability, reusability, and cost-effectiveness (Moldovan et al. 2015, Samanta et al. 2020). Hence, precise design and selection of the sensing material used for the sensor platform holds importance in terms of the sensory efficiency. Thus far, several micro- and nanomaterials with various intrinsic properties have been utilized as sensory platforms to monitor relevant environmental issues, e.g., metals and metal oxides, nanocarbon materials (graphene and carbon nanotube), quantum dots, semiconducting materials and polymers (Fang and Wang 2013, Mao et al. 2015, Bo et al. 2018, Zhou et al. 2008, Xu et al. 2010, Zhang et al. 2016). However, lack of effectiveness in terms of selectivity and sensitivity necessitates the development of new sensory materials with enhanced efficiency to address the pressing environmental issues. Since MOFs encompass a very broad spectrum, LMOF are selectively discussed in the present chapter with reference to their peculiar role as a chemical sensor.

Metal-organic Frameworks (MOFs)

Porous coordination polymers (PCPs) or metal-organic frameworks (MOFs) form an important subset in the field of coordination chemistry, which are generally considered as the coordination networks with intrinsic voids. Fabricated from organic ligands, which further extend into an infinite periodic architecture with the assistance of metal nodes, MOFs are considered as an advanced class of porous material family (Li et al. 2009, Stavila et al. 2014, Easun et al. 2017). The judicious choice of building units along with the help of combinatorial chemistry provides the liberty of synthesis of on demand architectures to achieve targeted structure-property correlation towards a particular application. Such exciting aspects along with benefits of high surface areas and high porosity have made MOFs the frontrunners over the other contemporary porous solid materials, such as activated carbon and zeolites Guo et al. 2019, Huang et al. 2017, Dutta et al. 2020, Sharma et al. 2021, Ding et al. 2019). In addition, the ability to control pore size and properties, modulate the nanospace architecture, utilization of reticular chemistry, access mesoporosity and incorporate active functional sites in porous voids as per requirement have triggered the MOFs based research interest across several disciplines (Li Burtch et al. 2020, Wang and Cohen 2009, Rojas and Horcajada 2020, Burtch et al. 2014).

The spectrum of applicability of MOFs has broadened with the continuous development of this field, which conspicuously include sensing and photonics-based applications, gas separation and storage, industrially relevant hydrocarbon separation, ion-conduction, drug delivery, heterogeneous catalysis, etc. (Kreno et al. 2012, Wang et al. 2019, Cai et al. 2019). Apart from the conventional applications, recently MOFs are typically utilized to address several pressing environmental issues and found affirmative potential in recognition, remediation and detoxification of hazardous environmental toxins, which exist in various media including soil, water and air (Let et al. 2020, Wu et al. 2021, Kumar et al. 2020, Sen et al. 2018).

Luminescent Metal-organic Frameworks (L-MOFs)

Luminescent metal-organic frameworks (LMOFs) are considered as one of the most important sub-class of MOFs in which upon irradiation, emission of photon occurs followed by absorption of the radiative energy. Generally, this happens via fluorescence mechanisms which involve an emission originated due to the transition between singlet excited state (S_1) and ground state (S_0), resulting in a typically short excited state lifetime (1–100 ns). On the other hand, phosphorescence mechanism is another very common phenomenon, in which intersystem crossing from the singlet state (S_1) to the triplet excited state (T_1) happens and subsequently a forbidden photon-emitting transition occurs to the S0, resulting in the excited state lifetimes as ≥ 1 ms (Allendorf et al. 2009, Rocha et al. 2011, Cui et al. 2012). In addition, L-MOF based chemical sensing has become a promising application owing to its superior advantages such as large surface area, permanent porosity, and numerous active sites of MOFs, which induce the surface host–guest interactions. Moreover, the high sensing selectivity and sensitivity along with its excellent reversibility and recyclability make them the most promising candidates towards real-time applicability (Razavi et al. 2020, Lustig et al. 2018, Rasheed et al. 2019). The photoluminescence properties in the L-MOFs can originate from a variety of mechanisms owing to their structural diversity and complex molecular compositions including inorganic ions and clusters, different types of organic ligand molecules and guest or ion molecules. In case of ligand-centered emission, the same ligand moiety is responsible for both the photon absorption and photon emission processes. In addition, these photon transition processes can also involve two or more separate regions inside the framework with non-radiative energy, which is generally outlined as Förster-Dexter theory. However, the most common mechanisms include ligand to metal charge transfer (LMCT), ligand to ligand charge transfer (LLCT), metal-to-metal charge transfer (MMCT) and metal-to-ligand charge transfer (MLCT). Additionally, the guest molecules inside the pores of the L-MOFs also contribute in these processes, termed as guest-sensitization and guest centered emission (Allendorf et al. 2009, Shustova et al. 2011, Dai et al. 2002). Detailed discussions of all these mechanisms are in the following sections.

Design Principles of LMOF as Chemosensors

Synthetic strategies used in designing LMOFs

Metal-organic frameworks have achieved considerable attention on grounds of target oriented synthetic emancipation and pertinence in the diversified application domain covering various disciplines in the material chemistry, including selective sensing and recognition-based applications. The freedom of modulating the coordination geometry or nanospaces considering unlimited choice available for linkers and metal nodes with range of oxidation states grant a superior advantage in order to depict the desired architectural attributes to the framework (Bitzer and Kleist 2019).

Virtues of LMOFs can be utilized in order to explore them as sensory material by deploying rational design strategies (Hu et al. 2014). Modular building block approach is the most suited approach for the development of sensor materials and in photonics as well (Yi et al. 2016, Medishetty et al. 2017). Strategies based on incorporation of multiple functionalities into one framework has gained considerable attention in recent few years. The incorporation of a variety of metals or multiple-linker molecules, or both inside one framework offer a potential tool for the modification of metal-organic framework materials and thus tune their properties as per desired application (Burrows et al. 2011, Qin et al. 2017, Fei et al. 2013). Therefore, the resultant MOFs are even more specialized for particular applications and can be further used as multifunctional materials to perform a variety of functions simultaneously. Meticulous designing of pore dimension on the molecular level encompasses precision for the adsorption of the target analyte (Chen et al. 2010, Li et al. 2013b). Control over pore size serves as a selection rule for the molecules, which are smaller or slimmer than the pore. Thus, porosity of the MOFs can be used as a versatile podium, which can be subjected to several chemical manipulations. Therefore, pore size control appears to be an effective step in designing LMOF based sensors (Hu et al. 2014).

In order to attain and realize sensing behavior of LMOFs, realization of their electronic properties can serve a crucial role. As transfer of either electrons or energy, or both at the same time, is observed between analyte and LMOF to be the prime causes for the fluorescent response to occur and therefore LMOFs designing aspect should rationally account at promoting these features. Aromatic or conjugated analyte is expected to interact or entice towards conjugated linkers of the framework through π-π interactions; hence, such conjugated linkers should be used for rational design and synthesis of LMOFs (Chaudhari et al. 2013). It is essential to note that by incorporation of the electron, withdrawing or donating groups in the linkers can be used as an effective tool for tailoring of relative orbital energies of the LUMO or conduction band (CB) (Karmakar et al. 2016). It has been realized that the notable decrease in the fluorescence intensity is observed as an outcome of the overlap of the absorption spectrum of specific analyte (analyte under study) with the LMOFs emission spectrum (Nagarkar et al. 2010).

Several physicochemical properties in the nanospace of the MOFs can be fine-tuned, such as polarity, hydrophobicity, polarizability, acidity, affinity towards protons and other ions. Selective capture and detection of the desired target analyte can be achieved by controlling the chemical environment within the nanospace. Several studies suggest that open metal sites (OMS) can be effectively utilized for small molecule sensing (Chen et al. 2007). Selective detection of small molecules like NH_3 have been achieved by preferred binding of the guest molecule with the OMS (Shustova et al. 2013). For example, functionalization of the pore surface with Lewis basic sites (LBS) facilitates the attraction of electron deficient moieties such as 2,4,6-Trinitrophenol (TNP) and metal cations within the pore (Nagarkar et al. 2010, Chen et al. 2009). LMOFs could act as effective sensors for various anions by permitting hydrogen bonding between terminal solvent molecules and anions

(Chen et al. 2008). Noteworthy sensitivity and distinguished selectivity are the real time requirements of the LMOFs in the practical world. Since extensive research is carried in the field of sensing-based applications by using LMOFs, it perpetuated to the incremental discovery of LMOF based sensors with passing time and enabled researchers to have a largely expanding library of sensory materials. Utilizing the same as a virtue, it foreordains the possibility for explicit identification of a targeted analyte by following the method of cross-referencing out of a sequential set of LMOFs selected specifically from a large and growing library of sensory materials.

Origin and mechanisms of luminescence in MOFs

The term luminescence or photoluminescence is generally ascribed to the process of spontaneous emission of light induced by the absorption of photons or energy. Fluorescence and phosphorescence are the two types of luminescence processes observed predominantly, which are commonly depicted by "Jablonski Diagram" (Lustig et al. 2017, Allendrof et al. 2009). Substantially, fluorescence is a short-lived process which in turn is very fast and transpires with very short timescale. On the other hand, phosphorescence is a relatively slow, long-lived one and thus the overall process accounts for a comparatively prolonged timescale, which can last over few microseconds to minutes and sometimes over hours for a few molecular systems (Lustig et al. 2017). On mechanistic grounds, luminescence in MOFs occur when electrons from the excited singlet state returns back to ground state by means of radiative emission, i.e., photon emission.

Process and the extent of luminescence can either be enhanced, referred to as "turn-on" or can be quenched and called as "turn-off" mechanism depending on the extent of analyte absorption. Utilization of luminescence quenching behavior of LMOFs as a probe in the presence of electron withdrawing analytes such as nitro-aromatics is perceived through distinct mechanisms such as electron exchange, photo-induced electron transfer (PET), Förster resonance energy transfer (FRET) and intramolecular charge transfer (ICT) as well which have been speculated in the literature and depicted in Figure 1 (Lustig et al. 2017, Allendrof et al. 2009). Origin of luminescence in the MOFs bearing transition-metal ions as metal nodes can be attributed to both, either ligand or metal ion based. Luminescence from transition metal ions-based MOFs is particularly centered on the linker instead of the metal ions; however, 'antenna effect' causes framework comprised of transition metal

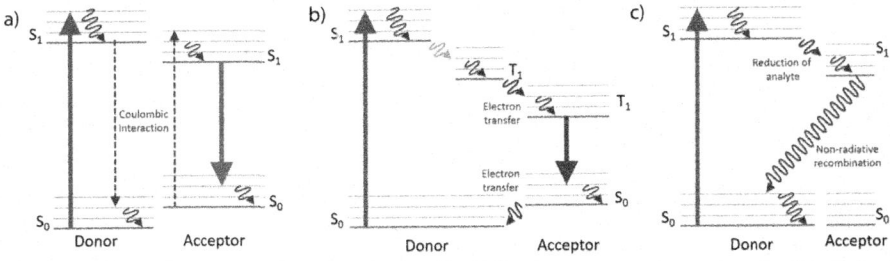

Figure 1. Schematic representation of mechanism for (a) Förster resonance energy transfer, (b) electron exchange, and (c) photoinduced electron transfer (Reprinted with permission from Lustig et al. 2017).

ions to exhibit metal-centric luminescence by linking it to π-conjugated systems (Allendorf 2009). Reports related to MOFs exhibiting metal centered fluorescence and comprising of transition metals as metal nodes are considerably rare and quite a few examples are reported which shows metal centered luminescence. For instance, consider examples of chromium complexes in which observed luminescence is generally rendered to "d-d" transition, which is often less intense or weak. A framework bearing paramagnetic transition metals as metal nodes generally does not show intense or strong emissions. It is due to 'ligand field d-d' type of transition involving electron/energy transfer from the partially filled d-orbital of metal resulting in the quenching or strong re-absorption resulting in the decrease in the fluorescence originating from the organic ligand/linker. Frameworks comprising of diamagnetic transition metals, especially those having 'd^{10}-configuration (Cd^{+2}, Zn^{+2}, In^{+3}, Zr^{+4}...)', exhibit strong emission (Pamei and Puzari 2019).

Ligand-centered emission, Ligand to metal charge transfer (LMCT), Ligand to ligand charge transfer (LLCT), and Metal to ligand charge transfer (MLCT) are the processes that are generally ascribed to the operating mechanisms for MOF based luminescence. Similarly, luminescence induced by guest and luminescence which is stemmed by sensitizer, shown in Figure 2, are discussed in detail under respective topics followed by present discussion (Lustig et al. 2017, Ghosh 2019).

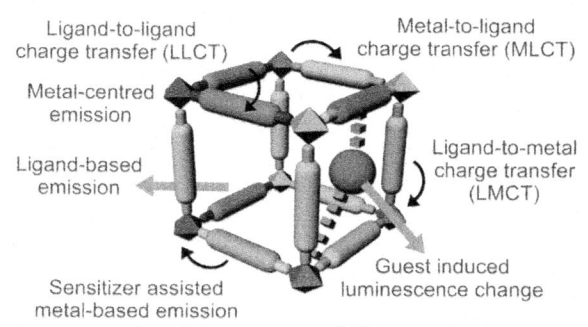

Figure 2. Schematic representation of the various possibilities contributing to the emission of MOFs. (Reprinted with permission from Lustig et al. 2017).

Ligand to ligand charge transfer (LLCT)

The distinctly covetable redox active MOFs can offer unparalleled insights into the fundamental level by means of charge-transfer (CT) in the coordination space. In such systems, redox active ligands do act as desired and exceed building blocks when compared to that of redox active metal-center, as redox activity at metal center might result into destruction of framework, thus paving the path for ligands here to decidedly foster redox activity through stable radical species which originate from consequent oxidation-reduction processes (Jiang et al. 2020). LLCT is predominantly realized in the LMOFs containing mixed ligand systems and controlled by several factors (Huo et al. 2016). Thereof, research on LMOFs inclusive of mixed ligand systems is still growing and has vast scope to cover. LLCT in the LMOFs stemmed from selective choice of linkers, i.e., carefully selecting the electron rich donor (D) and electron deficient acceptor (A) linker. In such a system, metal coordination (M) with that of electron deficient acceptor and electron rich donor should be earlier to D-A combinations and M-D or M-A separation. Emission behavior of organic lumophores varies in the solid state when compared with that of in the solution due to lumophore-interactions in the latter state (Cornil et al. 1998). Organic molecules with considerable extent of conjugation are ofttimes strongly emissive and absorbing as well, and electronic transitions which contain extended π systems are generally accountable for the origin of aforementioned properties. In solid state, these molecules tend to exhibit π-stacking owing to extended conjugation. Owing to their rigidity, organic compounds with fused rings are often used as a choice of linkers in the MOFs. In the highly ordered solid state (crystalline form), lumophores can be brought close together by means of molecular interactions, wherein electronic interactions are enabled between the lumophores and allow LLCT. Because of this emission broadening, spectral shifts, increased lifetimes for emission and loss of vibronic structure, etc., can be observed. Thus, selective tuning of this ligand–ligand (lumophore) interactions becomes a valuable tool for the applications that demand tenability in emission colors and charge transport. Owing to above mentioned facts, as a result, MOFs have an inherent advantage in terms of the degree of structural certainty that is established and the organised crystalline framework that pivots the path for maximising structure-property relationships.

Ligand to metal charge transfer (LMCT)

Origin of Ligand to metal charge transfer (LMCT) in the MOF is an outcome of the electron transfer from molecular orbital (MO) with ligand like character to those with metal like character, i.e., MO of the metal center. Such a type of electron or charge transfer is typically predominant in the MOFs having metal nodes comprised of low-lying empty orbitals and ligands bearing comparably high-energy lone pairs such as in O, S or Se. Fluorescence can originate through LMCT after absorption of light by the ligands or linkers present in the LMOFs. Since LMCT bestow the luminescent property of the MOFs, the governing factors, which influence the photoluminescence behavior of LMOFs, are found to be the electronic configuration and electron densities of metal ions and that of ligands, respectively, spatial arrangement of ligands within MOFs, inter-ligand distances and coordination environment, etc. (Allendorf et al. 2009).

Range of MOFs exhibits LMCT including majority of Zn(II) and Cd(II) MOFs consisting of linkers from benzene derivatives exhibiting distinctive green colored emission which is commonly observed in such systems. For example, an atypical supramolecular ribbon-candy-like architecture of Cd(II) having 1,4-BDC linkers and two-fold interpenetration have been found to produce intense emission at 474 nm upon excitation at 292 nm (Dai et al. 2002). LMCT can get influenced by the variations in the MOF structure; generally, MOFs comprised of benzene dicarboxylic acid (BDC) and benzene tricarboxylic acid (BTC) are the examples which illustrate this point. Minute changes in the linker can be utilized as a tool for understanding whether the emission is due to LMCT or linker-based. Thus, miniscule modification in linker can demonstrate a consequent structural effect on the transition between linker-based emission and LMCT. Consider the example that when 1,3,5-BTC linker is altered to 1,2,4-BTC, it produces a MOF $Cd_3(1,2,4-BTC)_2-(H_2O)_6\cdot 3H_2O$ having

3-D pillared structure which, upon excitation at 328 nm, produces notable intense blue luminescence (λ_{max} = 436 nm), which can be assigned to LMCT rather than that of solely linker based (Fan et al. 2003). Remarkably, in some cases where MOFs don't have conjugated ligands in their framework, they can still exhibit an emission profile that is similar to that of a similar MOF with conjugated ligands.

Metal to ligand charge transfer (MLCT)

Both luminescent metal-organic frameworks (LMOFs) and transition metal complexes (TMCs) bear significant relevance in diverse sensing and optoelectronic applications, prime reason being their assorted photochemical and photophysical properties (Fumanal et al. 2020). Along with TMCs, LMOFs are also center of interest for optoelectronic applications as LMOFs can conveniently combine light absorption in the visible spectrum with consequent processes such as charge-transfer (CT) or efficient electron-transfer (ET), like TMC (De Cola and Belser 1998). These features encompass LMOFs to be desired photosensitizers and can be deployed to perform in notably diverse environments such as solution, surfaces, crystalline material or proteins. Exceedingly, MOFs withhold all the potential capabilities of TMCs that are embedded in an orderly-crystalline, occasionally dynamic/flexible and usually porous structure.

Metal to ligand charge transfer (MLCT) is generally ascribed to a process of electronic transition taking place from a metal-localized/centered orbital to a ligand-localized (linker) orbital. MLCT is often observed in the MOFs comprised of metal nodes having d^{10} systems, i.e., having filled valence d orbital (Pamei and Puzari 2019). MOFs bearing metal nodes like Ag(I) and Cu(I), which are easily oxidized transition metal ions and reducible ligands as linkers, usually exhibit MLCT (Pamei and Puzari 2019, Song et al. 2017). Several examples are reported in the literature for MLCT mechanism which indicates presence of Cu(I) and Ag(I) at metal nodes of the MOFs; however, when highly reducing ligands are deployed in the LMOFs, then choice of metal node is seen to be Zn for MLCT process. The reason is that the properties of the organic linkers in the MLCT state are closely coherent with the properties of reduced linkers. Thus, in d^{10} systems like Cu(I) and Ag(I) based LMOFs, excitation due to charge transfer to ligand from the metal centers is usually observed. In the LMOFs bearing aforementioned metal nodes such as Ag(I) and Cu(I), MLCT is enabled by valence d-electrons of the metal center whereas in metal centers such as Cd(II) and Zn(II), these orbitals are usually core-alike. On the other hand, ligand-based luminescence can be quenched by incorporation of paramagnetic (d^9) metal ions like Cu(II), commonly observed comparisons among numerous Zn(II) and/or Cd(II) MOFs which is also similar to copper(II) complexes (Bitzer and Kleist 2019). Metal-based emissions are also known to be quenched by Cu(II) incorporated MOFs. As an illustrative example, compare the Cu(II)-doped analogue $UO_2(3,5$-pyrazole dicarboxylate)(H_2O) with its original uranyl complex $UO_2(C_5H_2N_2O_4)\cdot H_2O$ (Frischa and Cahill 2005). The latter displays characteristic emission spectra of UO_2^{2+}. When uranyl units or linkers are subjected to excitation, then strikingly the first compound shows lack of emission upon copper addition and this observation is irrespective of excitation wavelength. Likewise, the homo-metallic MOF $Nd(C_8H_6NO_2)_3H_2O$ and the hetero-metallic MOF, $Cu_3(trans$-3-(3-pyridyl)acrylate$)_6Nd_2(NO_3)_6$ exhibits identical absorption spectra; however, existence of Cu(II) in the latter causes quenching in the luminescence (Bitzer and Kleist 2019, Gunninga and Cahill 2005). Few reports comprising of emissions from the luminescent silver MOFs have demonstrated the existence of charge-transfer (CT) character. Intense green emission having peak maximum at 530 nm was observed for a 2-D layered architecture Ag(4-(2-pyrimidylthiomethyl)benzoate) after excitation at $\lambda = 370$ nm. The spectral origin was altered by metal-centered (ds/dp) states possessing Ag-Ag interactions, which were originally attributed to LMCT and/or MLCT (Han et al. 2006). Another notable example of MLCT includes $[Ag(4,4'-bipy)]_n[Ag(1,2,4-HBTC)]_n$ complex which exhibits strong fluorescent emission feature having emission maxima at 502 nm when excited at $\lambda = 410$ nm as a result of MLCT (Zhang et al. 2007).

Ligand-based photoluminescence in LMOFs

It is a well-known fact that π-electron rich conjugated organic linkers/ligands absorb light and as a consequence of this, radiative transition from excited or higher energy state to the low energy state imparts the existence of the photoluminescence behavior in the compound. For the aforementioned photoluminescence behavior to be seen in the π-electron rich conjugated systems (organic linkers) having minuscule spin-orbit coupling, selection rules are directed by the symmetry of the singlet ground state and excited. Owing to this fact, luminescence behavior of the LMOFs, which is comprised of organic linker, is dominantly driven via fluorescence process and transition resembles the lone or discrete organic molecules in the solution which comprise of either π→π* or n→π* transitions (Bitzer and Kleist 2019). Although similarities exist in the photoluminescence behavior of the free organic lumophores (ligands) and that of LMOFs, other properties like quantum yields (Φ) and wavelength corresponding to emission maxima (λ_{max}) of the free lumophores can vary in LMOFs. This distinct behavior is assigned to an outcome of random interaction taking place between freely orienting organic lumophores in a liquid state; it is contrary to those lumophores, which are orderly oriented in the MOFs blended by an intrinsic structural feature of rigid and well-organized manner of the framework. Such orderly and consistent orientation of linkers within crystalline solids of MOFs leads to the efficient luminescent behaviors by reducing the nonradiative decay and stems the notable increment in the quantum yield, etc., of LMOFs (Bitzer and Kleist 2019).

For linker based LMOF, further choice of metal ions to design MOFs with desired features becomes crucial. Therefore, metal ions are chosen accordingly which generally do not exhibit d-d transition, either due to lack of electrons in the outer d-orbital, i.e., metal ions with vacant outer d-orbitals orbitals (for example Ce(IV) and Zr(IV)), or metal ions in which outer-d orbitals are completely inhabited by electrons (such as Zn(II), Cd(II) and Ag(I)). Further, upon light absorption by LMOFs linkers, various pathways can lead to the origin of fluorescence such as interligand charge transfer (ILCT), ligand-to-metal charge transfer (LMCT) and metal to ligand charge transfer (MLCT) (Sun et al. 2013). Since ILCT, LMCT, MLCT, etc., enriches the luminescence features of the LMOFs, respective factors such as metal ions electronic configuration, ligands electron densities, coordination environment as well as ligand arrangement within the MOFs and interligands distances, etc., can straightway influence the photoluminescence outcome of LMOFs. In a report by

Ghosh and coworkers, MOF (UiO-66@NH$_2$) bearing Zr(IV) at metal node were deployed for nitric oxide (NO) sensing, where LMCT in the given MOF has been ascribed to the fluorescence origin (Desai et al. 2015). Reaction with nitric oxide resulted in deamination of NH$_2$BDC (BDC: 1,4-benzene dicarboxylic acid) and ligand changed into BDC. It resulted into reduced electron density on the ligand as a consequence of the deamination of ligand, which further caused decrement to the extent of the LMCT process and thus ultimately quenched the fluorescence.

Metal node or metal cluster-based luminescence in LMOFs

Majority of metal node-based luminescent MOFs are comprised of lanthanide (Ln) metal ions and are fabricated into distinctly luminescent materials by means of various synthetic strategies (Eliseevaa and Bunzli 2010, Cable et al. 2013). Lanthanides have low quantum yield due to inefficient excitation as f-f transition is forbidden by Laporte selection rule. This pertinent problem needs to be tamed by means of complexation of lanthanide metal ions with desired linkers enabled for selective absorption of light (wavelength) and can concord energy transfer by means of the process termed as "antenna effect" or "sensitization", thus resulting in appreciable enhancement to the luminescence property (Allendorf et al. 2009). As a result of vibronic coupling between linker and lanthanide metal nodes, charge transfer is induced from the accessible excited state associated with linker to the attainable energy state associated with that of the lanthanide metal ion. Considering the applications aimed at selective sensing of diversely ranged hazardous and toxic metal ions to the domain of tuneable optoelectronic properties, lanthanide MOFs comprising of organic ligands-derived coordination polymeric networks have seen upsurge development in the recent decade.

Ln-MOFs are one of the prime materials exhibiting tuneable luminescent properties, and have gained significant heed for the strategic design and depiction of sensory materials aimed at sensing of diverse toxic and hazardous species/analytes and temperature as well (Heine and Muller-Buschbaum 2013, Yan 2017). In the same regard, in an illustrative example reported by Dong et al., they prepared Tb(III)-MOF having formula unit $\{[Tb_4(OH)_4(DSOA)_2(H_2O)_8]\cdot(H_2O)_8\}_n$ (where, Na$_2$H$_2$DSOA: disodium-2,2'-disulfonate-4,4'-oxydibenzoic acid), for selective sensing of Fe(III)-ion (Dong et al. 2015). Tb(III)-MOF, after excitation at 350 nm, produced emissions at 487, 542, 580, and 619 nm, which are related to the transitions $^5D_4{\rightarrow}^7F_6$, $^5D_4{\rightarrow}^7F_5$, $^5D_4{\rightarrow}^7F_4$, and $^5D_4{\rightarrow}^7F_3$, respectively. Thus, potential of Tb-MOF towards selective sensing of metal ion/Fe(III) over other competent ions is successfully demonstrated by aforementioned study. Besides metal-based luminescence in the Ln-MOFs, other LMOFs exhibiting luminescence origin ascribed to metal nodes/clusters are also reported. Such LMOFs exhibiting metal-cluster led luminescent property encompasses enormous potential for sensing applications (Dong et al. 2015).

Guest induced emission in LMOFs

Besides organic linkers of the framework, it is worthwhile to address the role played by guest molecules/ions within the framework. Guests, be they organic or inorganic, have found a distinctly advantageous role in sensing related applications. MOFs are crystalline solids and intrinsically well-organized rigid/flexible host structures with inherent synthetic potential to avail controlled pore sizes, thus acting as preferential choice for the encapsulation of the luminescent guest species. Inclusion of guest molecules inside the voids or nanochannels of the host MOFs has achieved significant recognition in the view of fundamental aspects and the development of the range of sensing-based applications. Guest molecules can be selected diligently from the array, both organic or inorganic in nature, which can either bear fluorescent features or can modulate the host MOFs' luminescence properties (Cui et al. 2012, Mukherjee et al. 2016). For an illustration of inorganic guest induced emission, in a report by An et al. series is prepared by doping MOFs with lanthanide ions by means of cation exchange on pristine bio-MOF-1 in order to obtain Ln^{3+}@Bio-MOF-1 (Ln^{3+} = Eu^{3+}, Sm^{3+}, Tb^{3+}, or Yb^{+3}) (An et al. 2011). The doped MOFs emitted visually distinctive colors (Eu^{3+}, red; Sm^{3+}, orange-pink; Tb^{3+}, green) upon excitation at 365 nm, which can be seen with naked eyes (Figure 3). Emissions at 970, 640, 614, and 545 nm were exhibited by Yb^{3+}@bio-MOF-1, Sm^{3+}@bio-MOF-1, Eu^{3+}@bio-MOF-1 and Tb^{3+}@bio-MOF-1, respectively. For all of these reported MOFs doped with lanthanide ions, generally an emission at 340 nm is observed which indeed denotes that the energy migration enroutes through the identical electronic levels that are located in the chromophoric structure of the MOF. Remarkably, it is noted that bio-MOF-1 scaffold amply protects and adequately sensitizes lanthanide ions. As a result, characteristic luminescence was observed and lanthanide ions were detected in aqueous environments. Strikingly, the luminescence feature remains unaffected by strong quenching effects of water molecules, therefore retaining high quantum yields for lanthanide emission. Thus, overall observation indicates that the energy transfer from MOF to lanthanide ions occurred efficiently and adequate protection of lanthanide ions within the confined pores of the bio-MOF-1. In another report, Luo et al. synthesized the Tb^{3+}, Eu^{3+} doped MOFs aimed at sensing of metal ions, and which can exhibit tuneable luminescence properties (Luo and Batten 2010).

In an illustrative example of an organic guest induced emission, Taylor et al. reported an observation with rhodamine B dye as a fluorescent guest, wherein fluorescent proteins are decorated on the MOF surface (Taylor et al. 2008). This report represents target-specific imaging by means of surface functionalization

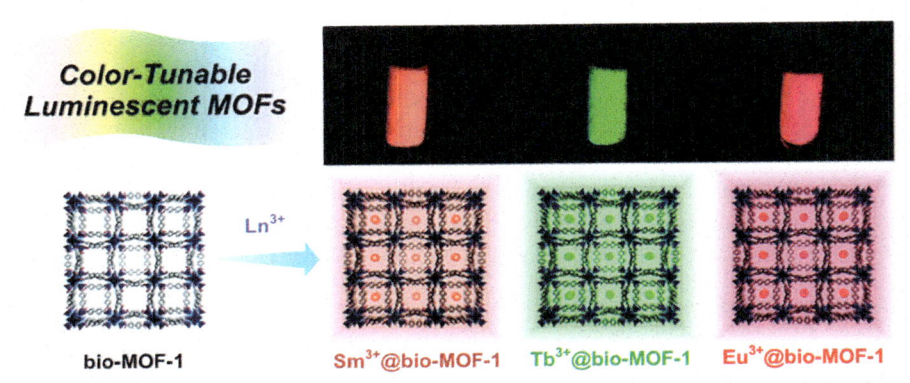

Figure 3. Luminescence of lanthanide ions doped in bio-MOF-1. (Reprinted with permission from An et al. 2011).

approach, whereas being guest, Rhodamine B furnishes a characteristic fluorescence, which is utilized for optical imaging. Two Mn-based NMOFs were prepared, namely $Mn_3(BTC)_2-(H_2O)_6$ and $Mn(1,4\text{-}BDC)(H_2O)_2$. MOFs were further coated with a thin shell comprised of silica and then functionalized with cyclically oriented peptide c(RGDfK) and a guest moiety (Rhodamine B) which indeed is a fluorescence dye. Such approach based on guest-induced luminescence enables LMOFs to be capable of acting as efficient environmental probes and utilizable for molecular detection applications (Mukherjee et al. 2016).

Other/Sensitizer assisted luminescence in LMOFs

In general, variation in the luminescence intensity profile or shift in the intensity pattern can act as an effective probe to indicate presence and nature of the interacting species or a guest. For LMOFs, nature and electronic features of the guest involve tunes or moderate luminescence outcome by means of the peculiar mechanism followed by either or both energy transfer and electron transfer, resulting in quenching or enhancement of luminescence profile and peak shifts (Pramanik et al. 2011).

Direct involvement of the sensitizer in the interaction processes with the guest can result in luminescence intensity changes. In case of lanthanide containing MOFs, it is commonly observed that the HOMO of the lanthanide cations can receive energy from organic ligands. As any of the incoming guests can interact with organic ligands, such interaction results in the contingent changes to the lanthanide cations/nodes observed emission intensities further causing either enhancement or quenching in the luminescence intensities (Wu et al. 2011, Heine and Muller-Buschbaum 2013, Xu et al. 2011, Nagarkar et al. 2013). However, when compared between changes observed in the intensity to that of over the emergence of entirely new peak or vanishing of the peak chaperoned by an entirely new emission peak, the latter is found to be more reliable and efficient pathway as the former can appear due to variation in the local surrounding or setting of the fluorescing species resulting in false or inaccurate observation. In LMOFs, it is observed that incoming guest molecules can affect preexisting luminescence intensity, which originated due to energy transfer from organic ligand to HOMO of lanthanide cations. In such cases, an intensity change (quenching or enhancement) in the preexisting emission peak can be seen. Also, a completely new emission can appear, or the existing peak can vanish and then be replaced by a entirely new emission peak. Such 'turn-on' response generally denotes the selectively judicious binding to the luminescent moiety and acts as a sensitive method of detection (Douvali et al. 2015, Shustova et al. 2013, Meyer et al. 2015). Alternate approach to observe 'turn-on' signal is commonly referred to as 'chemodosimeter' wherein secondary functional sites in appended ligand systems which are already bearing weakly-emitted or originally dark signal are subjected to strategic modulation. Chemodosimeter is a stauncher route, both in terms of highly sensitive and selective response. In order to attain an impeccably idyllic material for the desired sensing application, a number of factors are generally taken into account for the MOF and analyte both, such as pH, ionic character, hydrophobicity, pore aperture and polarizability. Although in most of the cases, these properties are affected by choice of metal, ligand and guest, in certain cases these properties can be seen because of the polymeric network formation. Linkers can play a crucial role in the radiative decay process; deliberate choice of linker embedded with π-conjugated organic moiety can shorten the HOMO-LUMO effectively and thus promote radiative decay. Modulation of pore surface has become a mode of choice in the design strategy, which allows MOF to segregate between contending guest molecules by preferential tendencies for the binding (Hu et al. 2014, Lustig 2017). Generally, sensors having emission in the visible region are favored. Therefore, organic ligands with π-conjugated systems are selected so that emission in the red and green region is attained.

Sensing Pathways for Diverse Sensing Applications

Gas sensors

Over the past few years, there is a serious change in the natural air composition that has been observed due to the rapid increment of various sources of air pollution. Different sources of toxic gases include uncontrolled release of green-house gases, geological hazards, several human activities and especially the excessive emission of toxic industrial relevant gases such as ammonia (NH_3), carbon monoxide (CO), nitric oxide (NO), hydrogen sulfide (H_2S), nitrogen dioxide (NO_2), sulfur dioxide (SO_2), etc. The direct contact of these toxic gases with the natural air significantly disturbs the indoor air quality (IAQ) which causes serious problems to human health as well as to the entire environment. Now, in order to maintain the air quality as well as to monitor the global environmental condition, selective detection of such gases is an urgent research topic which motivates the research community to design MOF based sensing materials as a smart probe for the detection of such aforementioned toxic gases. In this section, we will discuss a few representative examples of LMOFs based on various gas sensing applications.

MOF based sensors are the effective and ideal platform for the sensing of different gases including, formaldehyde, hydrochloric acid (HCl), H_2S, NH_3, O_2, NO, NO_2, SO_2, CO, water vapor, etc., as reported in the literature (Li et al. 2020a). Ammonia is an example of toxic gas, causing dangerous destruction to human health, which can be detected by LMOFs as demonstrated by a few groups (Campbell et al. 2015, Yao et al. 2020, 2017, Rubio-Gimenez et al. 2018). Dinca and co-workers reported the ammonia vapor sensing with a 2D MOF, for the first time in 2014 (Campbell et al. 2015). For this work, the group fabricated MOF with high electrical conductivity, namely, $Cu_3(HITP)_2$ (where the ligand HITP is 2,3,6,7,10,11-hexaiminotriphenylene) based device to detect ammonia. The high value of conductivity (0.2 S cm^{-1}) of the 2D MOF originated from its extended pi-conjugation network as well as delocalized charge of the 2D sheets of the MOF. They exposed the device to an ammonia-N_2 mixture, where the concentration of the NH_3 was 0.5 ppm, and found a rapid enhancement in the current profile. This experiment indicates the sensitive detection of ammonia by $Cu_3(HITP)_2$ Moreover, to evaluate the reversibility of the detection process, they replaced the NH_3-N_2 mixture with pure N_2 gas, which restored the current to its baseline position, indicating reusable ammonia sensing nature of the device. Recently, in 2019, Kitagawa et al. reported sensing of ammonia in chemiresistive way over a wide range of concentration (starting from 1 ppm up to

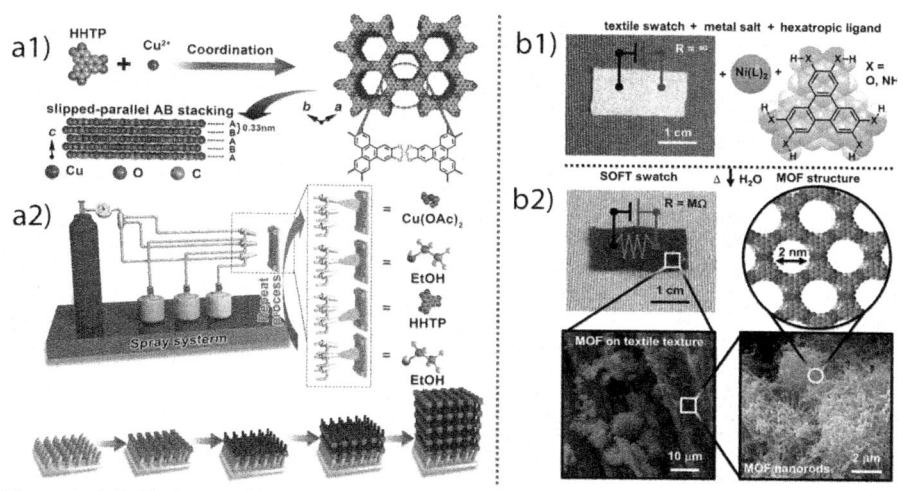

Figure 4. (a1) Single crystal structure of $Cu_3(HHTP)_2$, (a2) The construction of $Cu_3(HHTP)_2$ thin film gas sensors. (b1) Raw materials for SOFT-sensors [fabric swatch, metal (Ni-$(OAc)_2$ or $NiCl_2$) and organic ligand (triphenylenebased (HHTP or HATP))], (b2) Fabrication of SOFT devices (textiles coated with MOF). (a1 and a2: Reprinted with permission from Yao et al. (2017). b1 and b2: Reprinted with permission from Smith and Mirica (2017)).

saturated vapor) with a dual-ligand based MOF, $Cu_3(HHTP)$ (THQ) (where, HHTP is 2,3,6,7,10,11-hexahydroxytriphenylene and THQ is tetrahydroxy-1,4-quinone) (Yao et al. 2020). In another study, a Cu-MOF based high quality thin-film was fabricated in order to investigate the ammonia sensing, as reported by Xu and co-workers (Yao et al. 2017). The spray layer-by-layer liquid-phase epitaxial method was adopted to develop the high quality $Cu_3(HHTP)_2$, MOF based thin film (Figure 4a). Thus, fabricated MOF based film exhibits very high sensitivity as well as superior selectivity towards ammonia gas sensing even in presence of other interfering gases. The thin-film shows an ultrafast response with a superior low limit of detection (LOD = 0.5 ppm) along with good reproducibility, which demonstrated its potential towards real-time ammonia detection process.

Nitric oxide (NO) gas and nitrogen dioxide (NO_2) gas are another two extremely toxic gases. Mirica et al. developed a 2D conductive MOF based device for the detection of NO with a limit of detection approximately 1 ppb (Meng et al. 2019). The group finds out the probable mechanism behind such sensitive NO detection, which may arise due to the redox reaction occurs between the MOF (host) and NO (guest). On the other hand, Schaate et al. synthesized a novel calixarene containing MOF constructed with (Zr-cal, $[Zr_6O_4(OH)_4(FA)_6]_2(cal)_3$, where, FA is formate and cal is 1,3-*alt*-25,26,27,28-tetrakis[(carboxy)methoxy]calixarene) for the detection of NO_2, visually (Schulz et al. 2018). The 3D MOF was able to incorporate NO_2 molecules into the porous cavity, which was decorated with the calixarene functionality. The visual color change occurs due to the formation of charge-transfer complex between the host-MOF and guest-NO_2 gas (Figure 5a). There are few other literature reports which explore the promising application of MOF in the field of nitrous oxides sensing (Desai et al. 2015, Nickerl et al. 2015, Yuvaraja et al. 2020).

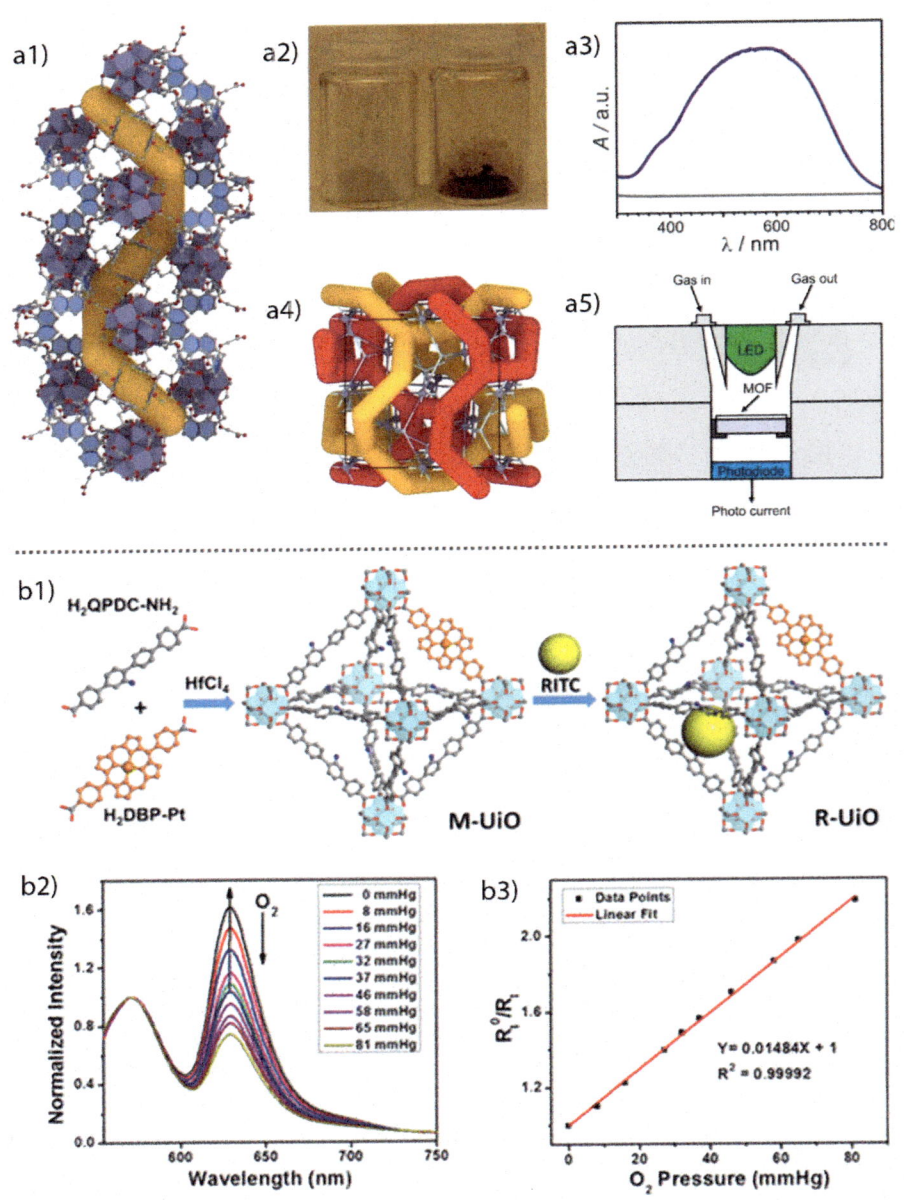

Figure 5. (a1) Helical channel structure of the system, (a2) Images of the Zr-cal MOF before (left) and after (right) NO_2 exposure, (a3) UV-vis spectra of the Zr-cal MOF before (below) and after (above) NO_2 exposure, (a4) Schematic topological representation of the Zr-cal MOF, (a5) Schematic diagram of the sensor setup, (b1) Synthesis of M-UiO NMOF and postsynthetic modification of M-UiO to constructed R-UiO NMOF, (b2) Emission spectra of R-UiO-1 under different oxygen partial pressures in HBSS buffer, (b3) Plot of R_I^0/R_I as a function of oxygen pressure. (a: Reprinted with permission from Schulz et al. 2018. b: Reprinted with permission from Xu et al. 2016).

Sulfur dioxide (SO_2) is also considered as a toxic gas pollutant. MOFs based sensors for the detection of SO_2 gas are well explored in the literature (Wang et al. 2018, DMello et al. 2019, Zhang et al. 2018a). Taking the advantage of interaction between amine groups and sulfur dioxide molecules, Cao et al. utilized an amino grafted luminescence MOF (MOF-5-NH_2) for the detection of SO_2 gas (Wang et al. 2018). Utilizing the similar strategy, Kalidindi and coworkers reported NH_2-UiO-66 (UiO: University of Oslo) MOF as an efficient sensory probe for SO_2 gas sensing (DMello et al. 2019). Another well-known poisonous gas is hydrogen sulfide (H_2S). Mirica and coworker developed Ni-metal based conductive MOFs' coated electronic textiles (e-textiles) as chemiresistive sensors for H_2S gas sensing (Smith and Mirica 2017). Thus, for the first time, fabricated self-organized frameworks on textile (SOFT) devices show multifunctional characteristics due to the presence of both mesoporous and microporous features (Figure 4b). In addition, the MOFs [Ni_3HITP_2 and Ni-CAT (synthesized with the ligand, 2,3,6,7,10,11-hexahydroxytriphenylene (HHTP) ligand)] coated e-textiles exhibit superior sensitivity (in ppm level) towards H_2S and NO gases with the theoretical LOD values of 1–80 ppm for H_2S gas. Further, H_2S gas detection with few other LMOFs have also been found in the literature (Cui et al. 2014, Smith et al. 2016).

LMOFs are also flourished as an efficient sensor for hydrochloric acid (HCl) gas sensing. As discussed earlier, taking the advantage of 'antenna effect' in the lanthanide-based MOFs, Lu et al. studied the fluorescence quenching phenomena of a newly developed Eu-MOF with the exposure of HCl vapor (Zhang et al. 2015). Almost 90% of the luminescence intensity of the MOF, [$EuH(L)_2(NO_3)_2$ (where, L = 2-(pyridin-2-yl)quinoline-4-carboxylate)], was found to be quenched upon two-minute exposure of HCl gas vapor. As evidence from the single crystal XRD, IR, and ^1H NMR data, protonation of the Lewis basic pyridyl sites in presence of HCl gas results in disturbance in the antenna effect which causes quenching in the emission of the MOF.

The most important gas is oxygen (O_2). Oxygen is an essential element not only for human beings but also for any living organism present in the environment. Apart from its biological importance, oxygen is also one of the significant species in many industries, such as the chemical industry, medicine industry and other. Therefore, the selective recognition of such an important gas (O_2) is of paramount interest and can be detected by the virtue of LMOFs. Recent past years are the witness for the hotspot research on oxygen sensing with several novel luminescent MOFs (Xu et al. 2016, Zhang et al. 2017b, Zhang et al. 2018b). Lin et al. reported the ratiometric detection of oxygen with strategically developed MOF, (R-UiO) (Xu et al. 2016). R-UiO was constructed with two distinct organic linkers, one is Pt-5,15-di-(p-Benzoato)porphyrin (DBP-Pt) which is sensitive towards O_2, another one is rhodamineB isothiocyanate (RITC)-conjugated quaterphenyldicarboxylate (QPDC) which is independent of O_2. These two different ligands exhibit two distinct emission peaks in the photoluminescence (PL) profile of the MOF (Figure 5b). The emission peaks observed at 570 nm and 630 nm correspond to RITC and DBP-Pt ligands, respectively. Now, when the MOF was subjected to exposure with the oxygen gas, the intensity of the emission peak at 630 nm was found to be quenched, while the other maxima at 570 nm remained unchanged. This study explores the fine controlled oxygen gas sensing application of LMOF. In literature, beside oxygen gas sensing, there are libraries of other LMOFs based on various other gas sensing studies which demonstrated the potential of MOF based gas sensors (Liu et al. 2014, Yu et al. 2014, Hao and Yan 2016, Chen et al. 2013a, Nakatsuka et al. 2020).

Explosive sensors

Explosives, a highly reactive substances, are majorly assembled with nitroaromatics, nitroaliphatics and peroxide-based species (Figure 6), which upon sudden release caused explosion with redemption of great potential energy. Explosives containing nitroaromatic compounds (NACs) have become a great threat to the entire environment, especially to the human health because of the ever-increasing uses in the different terrorist activities around the world. These life-threatening high energetic materials have seen a rise in their amounts in the environment, which cause very serious pollution. Therefore, considering the environmental protection prospects along with the national homeland security and anti-terrorism operations, scientific community turns their attention towards selective, sensitive and efficient detection of explosives. Although, in terms of high sensitivity and accuracy, few efficient instrumental techniques have been developed to detect nitro-explosive such as surface-enhanced Raman spectroscopy (SERS), ion mobility spectrometry (IMS), energy dispersive X-ray diffraction (EDXRD), cyclic voltammetry, etc., high cost, low portability, and labor intensiveness grant them a number of limitations or disadvantages over optical sensing-based techniques. Recently, fluorescent materials such as metal-organic frameworks (MOFs) based optical sensing have emerged as a reliable effective way to detect nitro-explosives due to few advantages such as cost effectiveness, easy operational-portability, fast response, low limit of detection, very selective recognition, etc.

High surface area, tuneable porosity, strong luminescent nature, suitable host-guest interaction of LMOFs and facile modification of MOFs with targeted functional moieties promote MOFs as a promising candidate for the fabrication of efficient sensory materials. Nitro-explosives sensing by MOFs based materials have been reported both in vapor and liquid phase in the literature. The very first example of a MOF probe-based sensor for the sensitive detection of trace number of nitro-explosives was reported by Li group (Lan et al. 2009). In 2009, this group used a luminescence MOF (LMOF) known as LMOF-111 for the detection of 2,4-dinitrotoluene (DNT) as well as 2,3-dimethyl-2,3-dinitrotoluene (DMNB) nitro-explosives in vapor phase at room temperature.

The one-dimensional channels based microporous MOF, (LMOF-111) $Zn_2(bpdc)_2(bpee)\cdot2DMF$, where, bpdc is 4,4'-biphenyldicarboxylate and bpee is 1,2-bis(4-pyridyl)ethylene), reveal ligand-based fluorescence emission which was drastically quenched upon addition of two selected nitro-explosive, DNT and DMNB. LMOF-111 exhibits high sensitivity, great quenching efficiencies and fast response time towards nitroaromatic (DNT) and nitroaliphatic (DMNB) with superior reversibility (Figure 7a). The probable reason behind the quenching mechanism was found to be the photoinduced electron transfer process from the LUMO of the MOF to the LUMO of analytes, lying slightly in lower energy, upon excitation.

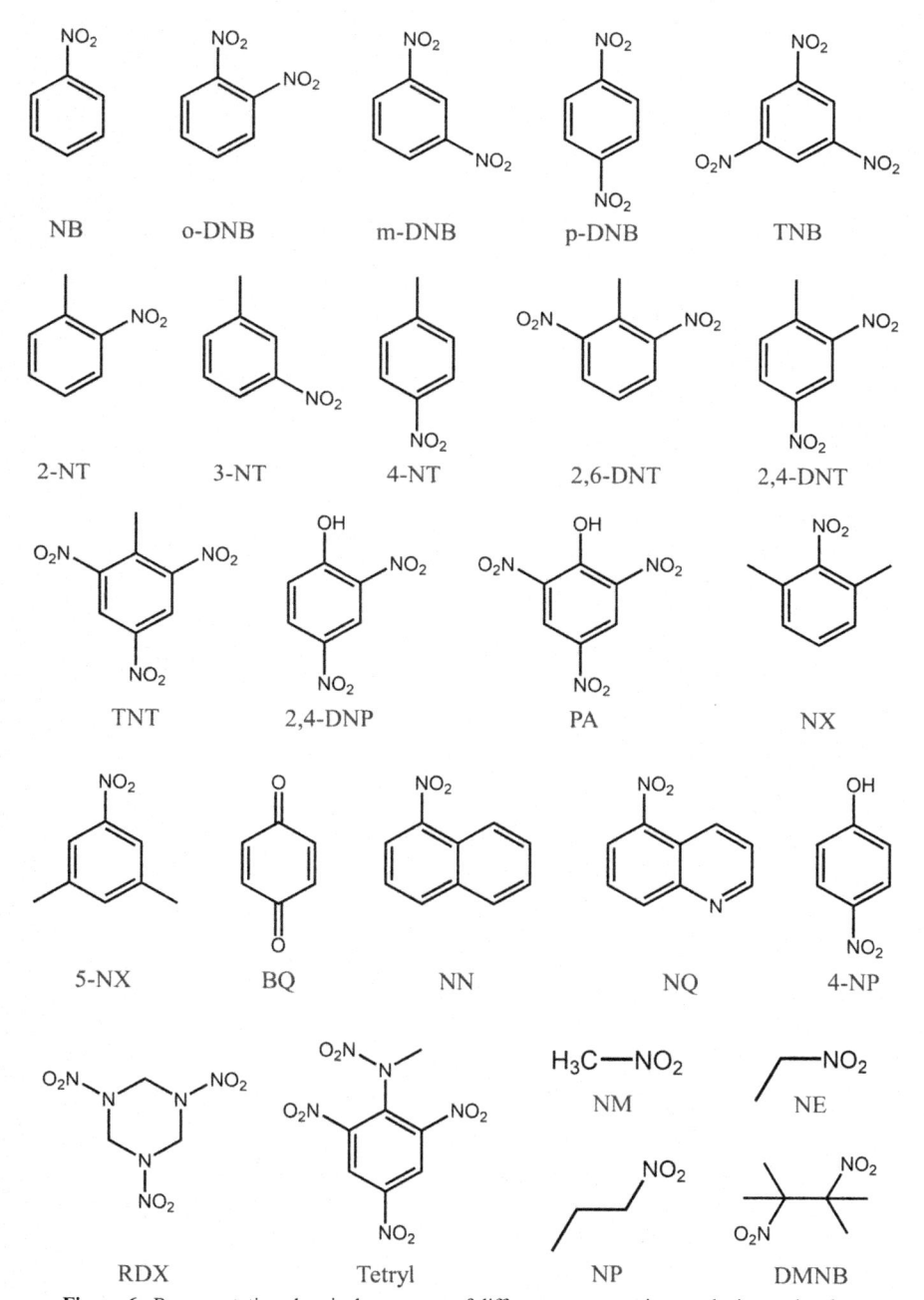

Figure 6. Representative chemical structures of different common nitro-explosive molecules.

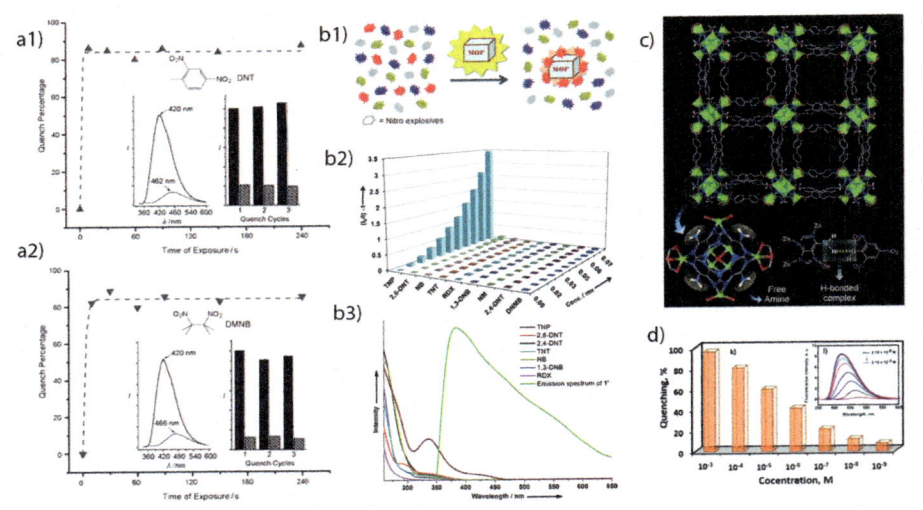

Figure 7. (a1) Time-dependent fluorescence quenching profiles by DNT; (a2) by DMNB. Insets: the corresponding fluorescence spectra before and after exposure to the analyte vapors for 10 s (left) and three consecutive quench/regeneration cycles (right); (b1) [Cd(ndc)0.5(pca)], for the selective detection of nitro explosive; (b2) Stern–Volmer plots for various analytes; (b3) Spectral overlap between the emission spectrum of the LMOF and the absorption spectra of analytes; (c) Crystal structure of bio-MOF-1, with proposed H-bonding interaction between adenine amine group and TNP molecule; (d) Plot of quenching of TNP in water against different concentrations; (inset) the change in fluorescence intensity of the MOF with concentration of TNP. (a1 and a2: Reprinted with permission from Lan et al. (2009). b1 and b2: Reprinted with permission from Nagarkar et al. (2013). C: Reprinted with permission from Joarder et al. (2015). d: Reprinted with permission from Asha et al. (2016)).

The same group demonstrated the vapor phase sensing of nitro functionalized aromatic with a d^{10} metal (Zn^{2+}) based MOF [Zn_2(TCPPE), where the ligand TCPPE is tetrakis[4-(4-carboxyphenyl)phenylethene], (Liu et al. 2015). This one-dimensional channel pore-based 3D MOF shows a strong emission peak at 461 nm, which upon guest removal shifted towards higher wavelength (535 nm) indicating the strong intermolecular electronic coupling. Further, the emission peak at 535 nm was observed to be rapidly quenched upon incremental addition of nitrobenzene as well as 2,4-dinitrotoluene vapors. In a similar systematic way, Ghosh et al. reported a 2D luminescent MOF, [$Zn_{1.5}$(L)(H2O)]·1.5benzene, where L is 4,4',4"-(benzene-1,3,5-triyltris(oxy)tribenzoic acid) for the sensing of both nitro-aromatic and nitro-aliphatic in vapor phase (Chaudhari et al. 2013). After the successful removal of guest molecules (benzene) by mild heating under vacuum which was entrapped inside the 2D pi-stacked layered structure of the MOF, it was subjected to expose towards a number of nitro-explosive vapors. The fluorescence intensity of the MOF was found to be quenched significantly not only in the case of a series of nitro-aromatics but also in the case of nitro-aliphatics. In another study, Parkin and coworker demonstrated the role of nano-space porosity of MOFs in the sensing performance of nitro-compounds (Jurcic et al. 2015). For this report, they synthesized two MOFs, namely, Dy(dcbpy) (DMF)$_2$(NO$_3$) and [Zn(dcbpy)(DMF)], where, dcbpy is [2,2'-bipyridine-4,4'-dicarboxylate] varying two different metal nodes (Zn and Dy) with identical organic linker, which shows similar shape and size but differ in the porosity. Both the MOFs were luminescent in nature. The fluorescence emission of the Zn-MOF was observed to be quenched for all the selected nitro-compounds (NB, 2,4-DNT, DMNB and *p*-NT), whereas, in case of Dy-MOF, only the smaller nitro-aromatic compounds (NB and *p*-NT) were able to selectively decrease the fluorescence intensity of the MOF, as evidenced from the emission profile. The group pointed out that the selectivity of Dy-MOF towards NACs (*p*-NT and NB) originated from the minimal porosity of the MOF. In contrast, as the Zn-MOF was porous enough to incorporate all the analytes into its nano-space, it exhibits fluorescence change behavior for all the NACs selected for this study.

The versatile continuous finding of detection of nitro-explosives by MOFs not only has been done in vapor phase, but also in liquid state. The liquid phase sensing or sensing of NACs in different common organic solvents such as acetonitrile (MeCN), dimethyl acetamide (DMA), dimethyl formamide (DMF), ethanol, etc., with LMOF has also significantly flourished in the last few years. Nagarkar and coworkers reported the sensitive and selective detection of 2,4,6-trinitrophenol or picric acid in acetonitrile with a luminescent 3D Cd-based MOF, termed as $\{Cd(NDC)_{0.5}(PCA)\}$. xG; [NDC = 2,6-napthalenedicarboxylate, PCA = 4-pyridinecaboxylate and G = guest molecules] (Nagarkar et al. 2013). Taking both the considerations of short-ranged electron transfer and long-ranged energy transfer process between the sensory probe and analytes into account, this group demonstrated the fluorescence quenching response of LMOF with the addition of TNP (Figure 7b). The synthesized MOF shows a high fluorescence quenching efficiency value towards TNP in MeCN medium as $K_{sv} = 3.5 \times 10^4$ M^{-1}. On the other hand, the MOF exhibits considerably minimal turn-off response on the fluorescence intensity towards other nitro-compounds.

The most probable quenching mechanism behind such high sensitive and selective detection of TNP with this Cd-MOF was attributed to the strong interaction between the Lewis basic moieties of the PCA linker of the MOF with the acidic phenolic photons of TNP. Since TNP molecules have the lowest LUMO energy and the strongest electron acceptor properties, and as there is a lot of overlap between the absorbance spectrum of the TNP analyte and the emission spectrum of the sensor, the fluorescence quenching of the MOF is much stronger. In 2014, utilizing the similar logical approach of Lewis basic sites interaction with TNP molecules, the same group reported the first example of water stable Zr-based luminescence MOF, that is $Zr_6O_4(OH)_4(L)_6$, (L = 2-phenylpyridine-5,4'-dicarboxylic acid) for the selective detection of nitro-aromatics (TNP) in water (Nagarkar et al. 2014). However, the ability of hydrogen-bonding as well as ionic interactions of the amine group with TNP molecules was significantly higher than the previously used pyridyl Lewis basic interactive sites for TNP recognition, as systematically reported by the same group in 2015. Ghosh et al. demonstrated a greater sensitive detection of TNP molecules with a basic amine group functionalized water stable, highly porous luminescence metal -organic framework, Bio-MOF-1 $[Zn_8(ad)_4(bpdc)_6O_2Me_2NH_2$, ad = adenine, bpdc = 4,4'-biphenyldicarboxylate] (Joarder et al. 2015). The Bio-MOF-1 shows very sensitive (limit of detection (LOD) 12.9 nM per 2.9 ppb) and selective detection of TNP over other NACs. Moreover, the group shows the suitable hydrogen-bonding interaction between the amine group of adenine molecules and phenolic hydrogen of TNP molecules by crystallizing the co-crystal of adenine and TNP molecules (Figure 7c). Utilizing the similar strategy, Mandal et al. reported the selective and very sensitive (ppb level) detection of TNP in water (Figure 7d) (Asha et al. 2016). Shi and coworkers found the gradual decrease of fluorescence intensity of $[Zn_3(tdpat)(H_2O)_3]$ upon incremental addition of nitrobenzene in methanolic solution (Ma et al. 2013). An Eu-based MOF, $Eu_2(bdc)_3(H_2O)_2(H_2O)_2$ (H_2bdc = 1,4-benzenedicarboxylic acid), was employed as an efficient sensory probe for the detection of TNT and DNT in ethanol by Chen and Qian group (Xu et al. 2011). Apart from this, libraries of various nitro-explosive detection in different solvent medium have been reported by other groups also which demonstrated the versatile real time application of LMOFs in the field of sensing of NACs (Lustig et al. 2017, Cui et al. 2012, Hu et al. 2014, Yang et al. 2015, Wang et al. 2016).

Ion sensors

Ions are the charged species, which are broadly classified into cations and anions. These mono or polyatomic ions not only play an important role in the various biological and environmental processes, but also have a significant implication in different fields. Taking into account these considerations, LMOFs have been widely explored in the field of various ions sensing. As in the real-world scenario, most of the ions exist in the water system; aqueous phase detection of ions is one of the elementary needs for LMOFs based ion sensors. On the other hand, considering the real-time application, selective recognition of ions in presence of other interfering ions is another fundamental demand of LMOFs for ions sensing. In this section, few representative literatures of LMOFs based sensing of cations (Cu^{2+}, Hg^{2+}, Fe^{3+}, Al^{3+}, Pb^{2+}, UO_2^{2+}, etc.) and anions (CrO_4^{2-}, $Cr_2O_7^{2-}$, I^-, F^-, PO_4^{3-}, etc.) are discussed.

In 2014, Lin and coworkers reported an eight coordinated UiO (University of Oslo)-type MOF for the metal ions detection study in water (Carboni et al. 2014). The novel MOF was constructed with succinic acid-functionalized terphenyldicarboxylate organic ligand and Zr-oxo cluster as secondary building units (SBUs). The high porosity along with large surface area, water stability, luminescent nature and carboxylate functionalization of the synthesized MOF inspired the authors to perform the aqueous phase metal ion sensing. The MOF shows a strong emission at 390 nm in water, which was found to be quenched with the addition of metal ions with paramagnetic nature such as Fe^{2+}, Mn^{2+}, Co^{2+}, Cu^{2+}, Ni^{2+}. Moreover, in the presence of other metal ions, having no unpaired d-electrons like Mg^{2+}, Zn^{2+}, and Cd^{2+}, the emission intensity of the MOF was not altered. Among the metal ions which have quenching ability, Mn^{2+} and Cu^{2+} exhibits highest quenching efficiency with LOD less than 0.5 ppb. The preconcentration effect of the metal ions into the nanopore of the MOF was found to be the reason behind such high sensitivity, as revealed from the ICP-MS data.

Mercury (Hg) is one of the most toxic elements. Pollution generated due to mercury has a number of detrimental consequences on human health as well as on the entire environment. Therefore, Hg^{2+} ions' detection is always an extremely important topic in the domain of sensing. In literature, few LMOFs have been employed as an efficient sensor for Hg^{2+} ions recognition. Recently, Ghosh and coworker reported aqueous phase Hg^{2+} ions detection with post-synthetically modified UiO-66 MOF (Samanta et al. 2018). The group first synthesized a butyne functionalized isostructural MOF of UiO-66, which was luminescent in nature with an intense emission. The fluorescence intensity of the UiO-66@butyne was observed to be quenched upon additive addition of aqueous mercury solution with a high sensitivity. The MOF exhibits a very selective detection response towards Hg^{2+} ions even in presence of other metal cations. Moreover, oxymercuration reaction-based chemodosimetric approach was considered as the mechanism behind the fluorescence quenching phenomena (Figure 8a). In another report, Li et al. 2016 demonstrated the "turn-on"

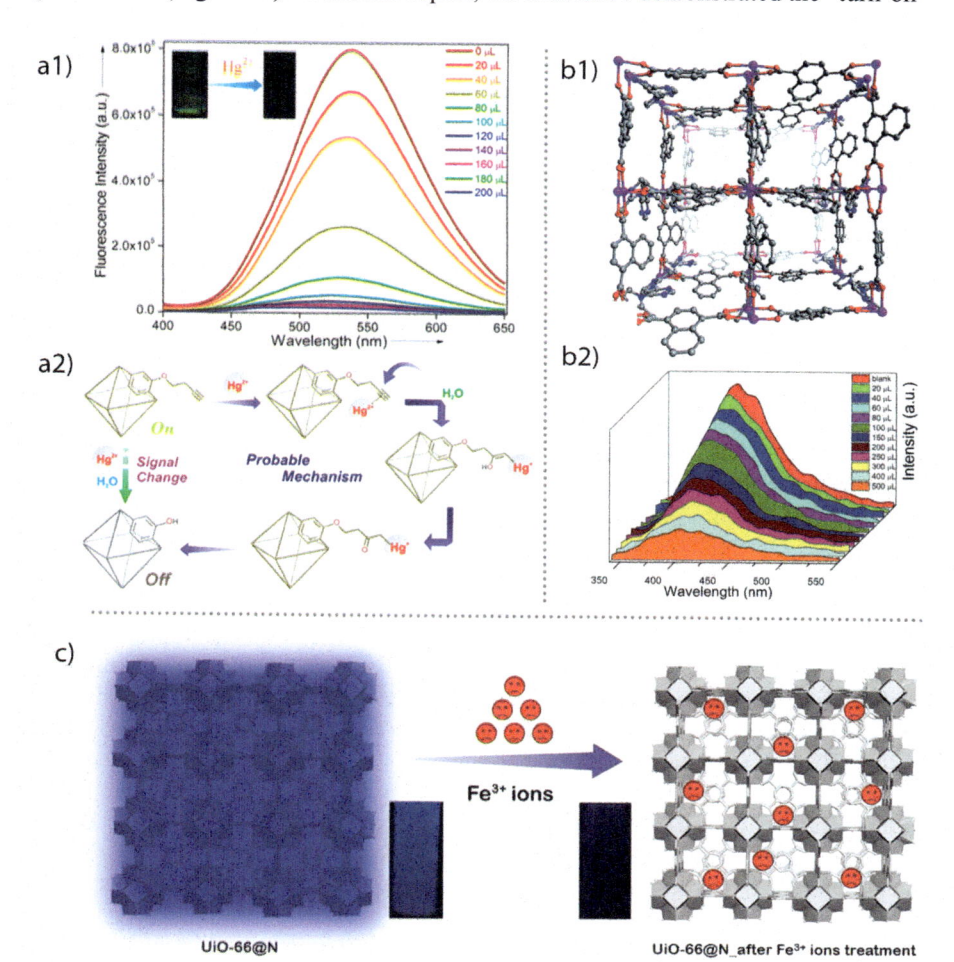

Figure 8. (a1) UV-vis spectra of the MOF shows gradual decrease in the emission intensity upon incremental addition of Hg2+ solution; (a2) Oxymercuration reaction-based chemodosimetric approach for Hg2+ sensing by UiO-66@butyne; (b1) Representation of the single crystal structure of the MOF; (b2) Emission profile of the MOF in DMF : water solution upon the incremental addition of UO22+ ions; (c) Schematic representation of Fe3+ ions detection by UiO-66@N. (a1 and a2: Reprinted with permission from Samanta et al. (2018). b1 and b2: Reprinted with permission from Chen et al. (2017). c: Reprinted with permission from Fajal et al. (2020)).

detection of Hg^{2+} ions with a lanthanide MOF nanoparticle (Li et al. 2016). Wang and co-workers reported a Cd-based MOF, namely, $\{[Cd_{1.5}(C_{18}H_{10}O_{10})]\cdot(H_3O)(H_2O)_3\}n$ (where, EDDA is [ethane-1,2 diylbis(oxy)]diisophthalic acid), for selective sensing of Hg^{2+} ions in aqueous medium (Wu et al. 2015). Apart from these, few other Hg^{2+} sensing by LMOFs is reported in the literature (Zhu et al. 2013, Wen et al. 2015, Rudd et al. 2016).

In line with mercury, lead is considered as another toxic and hazardous metal that can be detected by LMOFs. Liu and coworkers reported the aqueous phase detection of toxic lead (Pb^{2+}) ions by a newly synthesized lanthanide MOF, namely, $[Tb(L)(H_2O)_5]_n$. solvent (H_2L: 3,5-dicarboxyphenol anion) (Ji et al. 2017). The 2D green emissive LMOF exhibits very selective sensing of Pb^{2+} ions in presence of other interfering metal cations. In addition, the high quenching efficiency ($K_{sv} = 1.75 \times 104$ M^{-1}) and very low limit of detection (LOD = 10^{-7} M) towards Pb^{2+} ions in water promote this MOF as an efficient sensor. Moreover, a MOF crystal-based film was fabricated in order to perform the similar sensing study for practical applicability. Along with this report, few other LMOFs based electrochemical sensors have been reported in the literature for Pb^{2+} ions detection (Guo et al. 2016, Cui et al. 2015). Besides lead, cadmium (Cd^{2+}) is another heavy metal which is also considered as a toxic one. Therefore, the sensing of Cd^{2+} ions by LMOFs is well explored in the literature (Hao and Yan 2015, Roushani et al. 2016, Wang et al. 2017). A photoluminescent based "turn-on" Cd^{2+} ions sensing by LMOF was reported by Hao and Yan (Hao and Yan 2015). First, the group synthesized a well-known MOF, $[Zr_6O_4(OH)_4(H_2btec)_6]$ (H_4btec: 1,2,4,5-benzenetetracarboxylic acid) and then post-synthetically incorporated Eu^{3+} ions into the MOF. Thus modified, Eu^{3+} ions loaded MOF-composite exhibit five distinct emission peaks at 578 nm, 591 nm, 614 nm, 650 nm and 695 nm, which correspond to the $^5D_0 \rightarrow {}^7F_0$, $^5D_0 \rightarrow {}^7F_1$, $^5D_0 \rightarrow {}^7F_2$, $^5D_0 \rightarrow {}^7F_3$ and $^5D_0 \rightarrow {}^7F_4$ transitions, respectively. Inspired from the luminescence property of this MOF, the group performed the Cd^{2+} ion sensing in water. An 8-fold enhancement in the intensity of the emission profile of the pristine MOF was observed after addition of selective Cd^{2+} ion aqueous solution. Along with good selectivity, the sensitivity of the MOF towards Cd^{2+} ion was very high as the limit of detection was found to be 0.06 μM. Further, the proposed mechanism behind such high selective and sensitive sensing was attributed to the antenna effect of the MOF, as the interaction between the Cd^{2+} ions and Lewis basic carboxylic group of the ligand facilitates the resonance energy transfer from the organic linker to the Eu^{3+} ions. Besides these heavy toxic metal ions, another radioactive cation is uranium (UO_2^{2+}), which is considered as a critical cation, not only with respect to environmental perspective, but also from energy perspective. The LMOFs based uranium detection is very rarely reported in the literature.

Sun et al. reported $[Co_2(dmimpym)(nda)_2]_n$ (1; dmimpym: 4,6-di(2-methyl-imida-zol-1-yl)-pyrimidine, H_2nda: 1,4-naphthalenedicarboxylic acid), a 3D MOF, for uranium sensing in water-DMF mixture (Chen et al. 2017). The MOF, constructed with a free Lewis basic pyrimidyl site containing organic ligands, offers suitable interaction with upcoming uranium analytes. The fluorescence intensity of the MOF was found to be gradually quenched upon incremental addition of UO_2^{2+} solution with a high quenching coefficient value of $K_{sv} = 1.1 \times 10^4$ M^{-1} (Figure 8b). There are few other metal ions (Fe^{3+}, Al^{3+}, Cu^{2+}, etc.) present in the environment, which are well recognized for their biological as well as environmental importance, although the higher concentration of those metal cations in the aquatic system create serious water pollution.

Therefore, selective sensing of those metal ions by LMOFs has received paramount interest in the last few years. Among them, detection of iron cations (Fe^{3+} and Fe^{2+} ions) by LMOFs has been well-explored in the literature. Recently, Xie et al. reported exclusive detection of Fe^{3+} ions in water by two isostructural Zr-MOFs (Wang et al. 2017). They synthesized two new luminescent MOFs, BUT-14, $[Zr_6O_4(OH)_8(H_2O)_4(L^1)_2]$ and BUT-15, $[Zr_6O_4(OH)_8(H_2O)_4(L^2)_2]$, which show high stability in water as well as in different chemical environment as revealed from the PXRD and gas adsorption data. Both the two MOFs exhibit strong fluorescence emission which was observed to be quenched rapidly upon incremental addition of aqueous Fe^{3+} ions solution. Both BUT-14 and BUT-15 demonstrated very sensitive detection of Fe^{3+} ions with a very low detection limit (212 ppb for BUT-14 MOF and 16 ppb for BUT-15 MOF), as well as very selective Fe^{3+} ions recognition in presence of other metal cations. Moreover, the superior reusability of BUT-15 (functionalized with Lewis basic pyridine sites) inspired the author to perform the Fe^{3+} ions sensing experiment in simulated biological samples also, which showed similar quenching performance. In another report, Ghosh and co-worker demonstrated the efficient detection of Fe^{3+} ions in water-methanolic suspension by a Lewis basic decorated chemically stable MOF (Fajal et al. 2020). The UiO-66@N, an isostructural MOF of UiO-66, constructed with 2,5-pyridine dicarboxylic acid organic ligand was employed as a potential sensor for Fe^{3+} ions detection. UiO-66@N shows excellent Fe^{3+} ions detection with a LOD value of 0.69 ppm and a high quenching constant value of K_{sv}: 6.65×10^3 M^{-1} (Figure 8c). Apart from these, there are few other representative examples which are there which are dedicated towards Fe^{3+} ions sensing by LMOFs (Chen et al. 2018, Lu et al. 2014, Let et al. 2020, Chen et al. 2019).

Another two important metal cations are copper (Cu^{2+}) and aluminum (Al^{3+}), which can be detected by LMOFs (Qiao et al. 2016, Chen et al. 2016, 2013b, Yu et al. 2017). In 2016, for the first time, Chen and co-workers reported the naked-eye disclosure of Cu^{2+} ions by a Cd-MOF, formulated as $\{NH_2(CH_3)_2 \cdot Cd_{2.5}(L)_2(H_2O) \cdot (H_2O)\}_n$ (H$_3$L: tricarboxytriphenylamine) (Qiao et al. 2016). A SC-to-SC transformation of the MOF crystal with a color change from yellow to dark green was observed when the MOF crystals were exposed to 10^{-4} mol L^{-1} concentration of Cu^{2+} ions. These phenomena indicated the efficient recognition of Cu^{2+} ions with a short response time of 9 sec. Moreover, other interfering metal cations did not alter the color change, and pointed out the selective Cu^{2+} ions sensing. In recent years, Al^{3+} ions sensing has also caught paramount interest. Sun et al. reported cation exchange based luminescence change phenomena with Al^{3+} and Fe^{3+} ions by a Ln-MOF, $[H_2N(CH_3)_2][Eu(H_2O)_2(BTMIPA)] \cdot 2H_2O$ (H$_4$BTMIPA: 5,5'-methylenebis (2,4,6-trimethylisophthalic acid)) (Chen et al. 2013). The synthesized anionic MOF contains $[H_2N(CH_3)_2]^+$ cation which was free for exchange in the tubular channels of the MOF. The MOF shows dramatic luminescence enhancement when treated with Al^{3+} ions solution in DMF, while quenching response was observed in case of Fe^{3+} ions.

In the above section, we have discussed the various cations sensing by LMOFs, while in the next, few representative examples of anions recognition are discussed. In literature, the anions, which are found to be exclusively detected by LMOFs, are dichromate ($Cr_2O_7^{2-}$), chromate (CrO_4^{2-}), iodide (I^-), arsenate ($HAsO_4^{2-}$), etc. Among them, dichromate anion is an interesting one due to its extreme toxic nature as an environmental hazard. In 2013, outstanding sensing as well as capture of dichromate ($Cr_2O_7^{2-}$) anions was reported by Ruihu Wang and co-workers (Li et al. 2013a). A cationic luminescence MOF ($[Ag_2\text{-}(btr)_2].2ClO_4.3H_2O$, btr = 4,4'-bis(1,2,4-triazole)) has been synthesized which shows superior stability not only in different organic solvent, but also in various pH (from 0 to 10). High water stability of this LMOF allowed to perform the aqueous medium based $Cr_2O_7^{2-}$ capture study. A SC-to-SC transformation of the MOF was demonstrated when exposed to dichromate exchange with free ClO_4^- anions which was present in the pore of the MOF. Besides capture, the strong fluorescence emission intensity of the LMOF was found to be quenched upon $Cr_2O_7^{2-}$ addition. Moreover, the cationic LMOF was able to efficiently remove very diluted dichromate solution (from 14.7 ppm to 90 ppb). Along with these studies, few other LMOFs based anions detection performance has been demonstrated in the literature (Shi et al. 2015, Li et al. 2020).

Humidity and volatile organic compounds (VOCs) sensors

Humidity sensors

Humidity sensors are used in an extent range of fields like medical, agricultural, environmental and industrial facilities. Metal-organic coordinating framework-based humidity sensors have come into the forefront due to its high sensitivity, superior selectivity, simplicity, easy operation, low cost, etc. Dong et al. reported for the first-time luminescence-based humidity sensors based on dehydration and rehydration (Yu et al. 2012). They have synthesized two three dimensional (3D) networks formulated as $[Ln(L)(C_2H_2O_4)_{0.5}(H_2O)].2H_2O$ (Ln: Tb and Eu) with a binary mixed linker system of pyridyl-4,5-imidazole dicarboxylic acid and oxalic acid. They have used lanthanum metal ions; which resulted into two structure, one using Tb and other using Eu. Both of these MOFs are isostructural in nature. Both the synthesized networks have ellipsoid channels (8.9*7.5 Å) which can accommodate water molecules reversibly. A suitable interaction between the guest water molecules and the O-H group of the ligand plays a significant role in the luminescence tuning behavior of both the MOFs. By controlling such interaction of O-H vibrational oscillator on the Ln MOFs and the guest water molecule, they control the luminescent intensity which is detected by naked eye. Finally, they concluded higher sensitivity at higher humidity. Later on, the same group reported a MOF based naked eye colorimetric humidity sensor (Yu et al. 2014). They synthesized a Cu(I)-MOF formulated as $CH_3CN.MeOH.1.5H_2O$. In presence of 33% humidity, it changes color from bright yellow to red brown which is recognized by naked eye. Besides this colorimetric change, there is quenching of luminescent intensity with encapsulation of water molecules. On the other hand, LMOF based water molecules sensing was reported by Douvali et al. (Douvali et al. 2015). The group developed a 3D flexible MOF, [[Mg(H$_2$dhtp)(H$_2$O)$_2$]DMAc, (H$_4$dhtp = 2,5-dihydroxy-terephthalic acid; DMAc = N,N-dimethylacetamide)] with luminescence Mg(II) metal ions. Thus, synthesized LMOF exhibits superb detection ability along with good reusability towards trace water (0.05–5% v/v), when the sensing experiment has been done in a range of different organic solvents.

Volatile organic compounds (VOCs) sensors

US-Environmental protection agency (US-EPA) has selected VOCs as one of the potential toxic pollutant among the serious hazardous environment pollutants, generally available in the air and water system (McDonald et al. 2018). VOCs are defined as benzene and/or its derivatives aromatic compounds, aliphatic compounds and common organic solvents with the boiling point less than or equal to 250°C at 1atm pressure (Lai et al. 2021). Due to its extreme harmful effect, VOCs cause severe problems to the entire ecosystem, specially to the human health (Yi et al. 2016). Therefore, considering the destructive consequences of VOCs upon environment, selective detection of these toxic species (both in solution and vapor phase) has evolved as a cutting-edge research interest in the domain of LMOFs based sensing study (Takashima et al. 2011, Zhang et al. 2014, Desai et al. 2020, Chen et al. 2007, Zhou and Yan 2016, Liu et al. 2020, Li et al. 2013). LMOFs based VOCs detection study has been further categorized into aromatic, aliphatic, aldehydes, ketones, alcohols and different organic solvents recognitions.

In 2011, Kitagawa group demonstrated for the first time "molecular decoding" based selective sensing of aromatic VOCs utilizing an interpenetrated luminescent coordination polymer (CP), [Zn$_2$(bdc)$_2$(dpNDI)]$_n$ (ligand bdc is 1,4-benzenedicarboxylate and dpNDI is N,N'-di(4-pyridyl)-1,4,5,8-naphthalenediimide) (Takashima et al. 2011). The strategically embedded naphthalene di-imide (NDI) moiety in the CP was selectively interacting with the upcoming guest, VOCs molecules with the unique NDI-VOCs interactions, which delivered an exciplex emission with charge-transfer characteristic. This exciplex emission resulted in the distinguished turn-on, naked eye emissions to each selected guest VOC molecule. In addition, distinct ionization potential, average lifetime, emission wavelength of different aromatic VOCs was observed. Further, the group performed the single-crystal-to-single-crystal transformation study of the CP with toluene VOC, which demonstrated the confinement of VOC into the nanopore of the MOF (Figure 9a1 and 9a2).

Another interesting report by Zhao et al. showed the recognition of aromatic hydrocarbons with a LMOF (Hu et al. 2014). Utilizing the advantage of aggregation-induced emission (AIE) effect, the group successfully disclosed the selective sensing of aromatic VOCs. In this work, the author first constructed a tetraphenylethene (TPE) based LMOF (NSU-1). NSU-1 selectively interacts with the aromatic hydrocarbon, which further exhibited distinguished emission, indicating selective recognition of VOCs. The reason behind this phenomenon was attributed to the conformational altering of the AIE effect in the ligand (Figure 9b and Figure 10a).

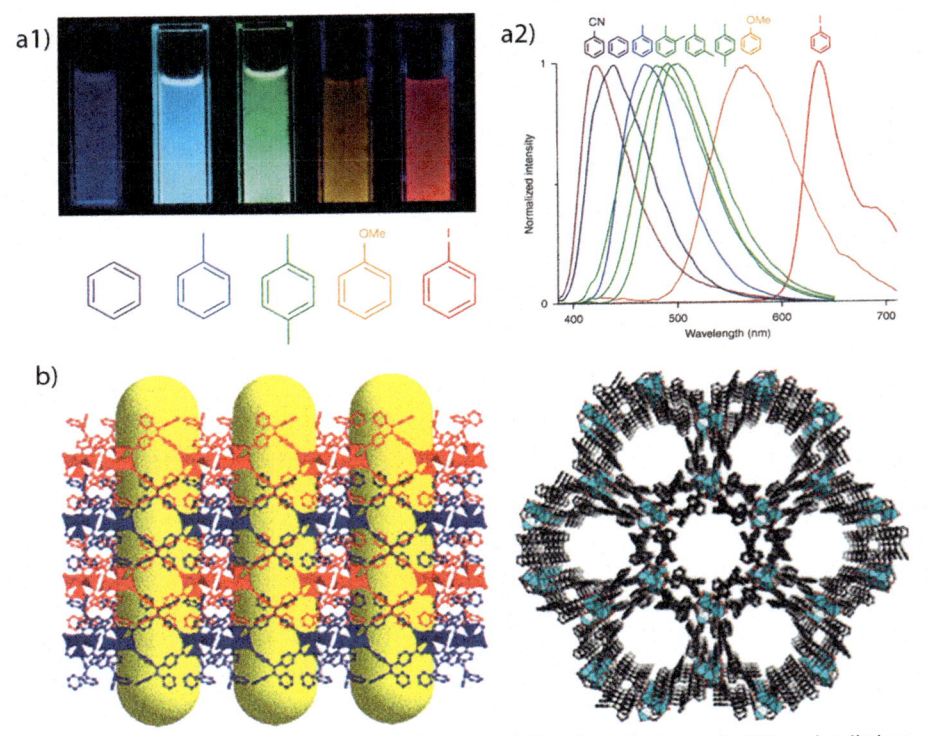

Figure 9. (a1) MOF powders suspended in different organic liquid is indicated, under 365 nm irradiation; (a2) Normalized luminescent spectra of guest-containing MOFs upon excitation at 370 nm. (b) Crystal structure of NUS-1. (a1 and a2: Reprinted with permission from Takashima et al. (2011). B: Reprinted with permission from Zhang et al. (2014)).

Biomolecule sensors

Luminescence MOF based biomolecules sensing is another interesting topic. The first report of a well-known biomolecule, dipicolinic acid (DPA) sensing with a luminescence MOF has been done by Lin and coworker (Rieter et al. 2007). In an another report, Qian, Chen and co-workers pointed out the "turn-on" fluorescence response towards the same biomolecules (DPA) with a LMOF, $[Eu_2(fma)_2(ox)(H_2O)_4]4H_2O$ (fma = fumarate and ox = oxalate) (Xu et al. 2012). The MOF displays superior sensitivity as well as high selective detection of DPA. In 2011, another biomolecule, salicyldehyde sensing was achieved when a luminescence sensitizer (triphenylamine) was incorporated in to the pore of a MOF, Tb–TCA (H_3TCA = tricarboxytriphenylamine) and was subjected to perform the sensing ability (Wu et al. 2011). Talking about biomolecules, amino acid is one of the important and significant biomolecules.

The aqueous phase sensing of amino acid with MOF was performed by Moorthy et al. (Chandrasekhar et al. 2016). Lin et al. demonstrated the highly enantioselective fluorescence-based sensing of chiral amino alcohols by a chiral porous coordination polymer or MOF, $[Cd_2(L)(H_2O)_2]_6.5DMF_3EtOH$ [LH_4 is (R)-2,2'-dihydroxy-1,1'-

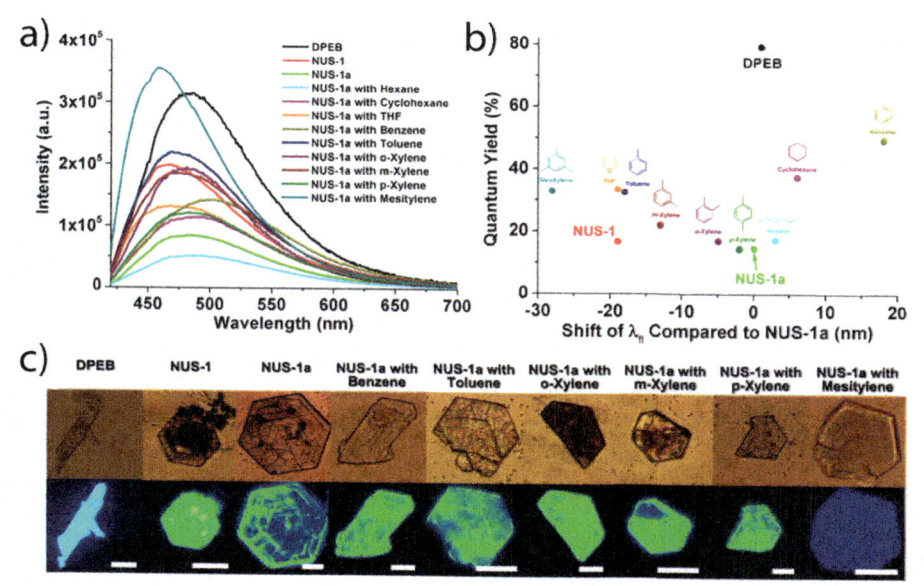

Figure 10. (a) Fluorescent emission spectra of NSU-1; (b) Quantum yield profile of the MOF; (c) Fluorescence microscopy images (first row, optical images; second row, fluorescence images; scale bar, 30 mm). (Reprinted with permission from Zhang et al. 2014).

binaphthyl-4,4',6,6'-tetrakis(4-benzoic acid)] in acetonitrile medium (Wanderley et al. 2012). In 2016, Moorthy and coworker reported enantioselective differentiation between the akin chiral amino acids with configuration (L-(-) and D-(+)-isomers of histidine) by a water stable luminescent 3D MOF, known as Zn–PLA, where the ligand was PLA= pyrene-tetralactic acid (Chandrasekhar et al. 2016). Ruan, Chang, Li and coworkers reported the sensing of DNA strands by luminescent, flexible linker derived two isostructural LMOFs, [Zn(L)(HDMA)$_2$(DMF)(H$_2$O)$_6$] and [Cd(L)(HDMA)$_2$(DMF)(H$_2$O)$_3$] (ligands, H$_4$L = bis-(3,5-dicarboxy-phenyl)terephthalamide and HDMA = protonated dimethyl ammonium cation) (Figure 11a1 and 11a2) (Wang et al. 2014). Apart from these representative examples, a few other studies of biomolecules sensing have explored the potential of LMOFs in this interesting field of research.

pH sensors

LMOFs based sensing application has been explored widely due to its unique ability to recognize the target specific analytes, selectively. The high porosity, large surface area, tuneable functionality, unique host-guest interaction inside the nano-space as well as heterogeneous nature, all these characteristics promote the LMOFs as a productive and economically effective sensory material in the real-time application of chemical sensing. In the previous section, we have discussed the LMOFs based biomolecule sensing, to which pH sensing is critically relevant. In 2014, the Li group reported a luminescence MOF based pH sensing (Deibert and Li 2014). PCN-222, also known as [Zr$_6$(OH)$_8$(tcpp)$_4$, H$_4$tcpp = tetrakis(4-carboxyphenyl)porphyrin], was

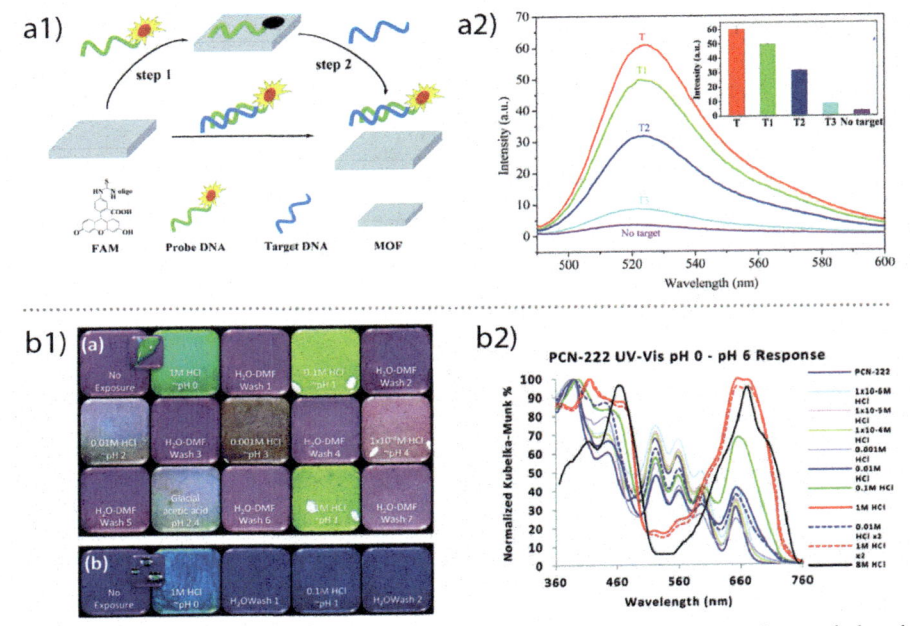

Figure 11. (a1) Schematic representation of nucleic acid sensing by MOF (FAM: fluorescein-based fluorescent dye); (a2) Fluorescence emission spectra of FAM-labeled probe DNA (P) under different conditions; (b1) Images of PCN-222 samples under different condition; (b2) UV-Vis spectra of PCN-222 MOF after being exposed to various concentrations of HCl. (a1 and a2: Reprinted with permission from Wang et al. (2014). B1 and b2: Reprinted with permission from Deibert and Li (2014))

employed as an effective pH sensor, which shows both colorimetric and fluorescent changes in acidic condition. The LMOF exhibits good stability in a wide pH range (pH 11 to 0), especially in strong acidic conditions, as revealed from the PXRD data. In extreme acidic condition, pH 0 to 3, the MOF was found to be a solid color change from purple to deep green (Figure 11b1 and 10b2). This obvious color tuning was observed due to the disruption of the pi system present in the macrocyclic tcpp ligand of the MOF. As mentioned earlier, the pH sensing is directly relevant to bio-sensing; for the real-time, the first intracellular pH detection by LMOF in living cells was reported by Wenbin Lin and co-workers (He et al. 2014). Fluorescein isothionate (FITC) is a powerful fluorescent molecule that shows pH-dependent ratiometric fluorescence. In this work, the group covalently threaded the FITC molecules with UiO MOF, $[Zr_6O_4(OH)_4(amino\text{-}TPDC)_6$, amino-TPDC is 2'-amino-[1,1':4',1''-terphenyl]-4,4''-dicarboxylic acid], through a thiourea linkages to perform the intracellular pH sensing into human non-small cell lung cancer H460 cell endosomes. An amine decorated MOF, that is, UiO-66-NH$_2$, also known as $[Zr_6O_4(OH)_4(bdc\text{-}NH_2)_6$, bdc-NH$_2$ = 2-aminoterepthlate], was reported by Aguilera-Sigalat et al., for pH sensing in 2014 (Aguilera-Sigalat et al. 2014). The MOF shows a fluorescence emission originating from its ligand, containing a free amine group. Upon decreasing the pH, the intensity of the emission profile was found to be decreased, which was attributed to the protonation of the pendant amine group. In another report, a lanthanum-based MOF, UPC-5 or $[H_3O][Eu_3(HBPTC)_2(BPTC)(H_2O)_2]4DMA$ (ligand H_4BPTC is 3,3',5,5'-biphenyltetracarboxylic acid) was demonstrated for pH sensing by Daofeng Sun and co-workers (Meng et al. 2015). The MOF exhibits two distinct types of coordination environments of the BPTC ligand towards Eu^{3+} metal ions. In the first mode, all the four-carboxylate groups of the ligand are in deprotonated manner and coordinate with metal ions, although, in the second mode, one out of the four carboxylates group remains protonated and does not take part in coordination. Now in basic condition, this protonated carboxylate group becomes deprotonated, which causes distribution in the antenna effect, due to which the phenomenon of transferring excitation energy from the ligand to the metal ions has been perturbed, which causes quenching in the fluorescence emission.

Conclusion and Future Outlook

In conclusion, MOF has emerged as one of the most fertile and promising platforms toward efficient recognition and desensitization of several environmentally toxic species as well as relevant inorganic and organic compounds such as VOCs, ionic species, biomolecules, environmental pollutants, toxic molecules, etc. The porous architectures and freedom of tunability of MOFs induce the advantages of sensing and sequestration of relevant environmental pollutants as the targeted analyte molecule can be occluded inside the porous voids and can produce a significant interaction between the recognition center and the aimed analyte to generate strong signals. Moreover, researchers across different disciplines have exploited the liberty of alteration of physical as well as the chemical properties of MOFs by several strategies including changing their connectivity, substitution of the metal nodes and organic likers, post-synthetic functionalization, etc. The quality to offer various kinds of photophysical properties along-with the freedom of appropriate tuning of the pore surfaces and electronic properties for a particularly targeted application has made LMOFs favorite towards their utility for real-time sensing-based applications.

However, there are always few areas of improvement towards which further research of L-MOFs should be focused such as the reversibility and regeneration criteria. Apart from that, development of MOF based composite materials is another crucial aspect, which requires serious scientific attention to manifest the true potential of MOFs towards such sensing applications. In addition, another key aspect to address is the real-time utilization of MOFs as standalone powders. Hence, hybridization of MOFs with different kinds of polymers can be investigated in detail to meet the demand of actual scenarios. Another crucial aspect is the stability and sustainability of functionalized MOFs in the harsh condition of wastewater sludge. Hence, development of new design-principles is required to produce functional MOFs that can have a wide range of chemical as well as hydrolytic stability. In addition, cost-effectiveness of the sensory material is another very important parameter for its long-term applicability. Hence, utilization of widely available and cheap materials should be considered for design and synthesis of L-MOFs. So, with the continuous improvement in rational design and development of LMOF based sensory materials for targeted applications, new avenues in scientific research as well as real-time utility material are expected to open in future.

Conflicts of interest

There are no conflicts to declare.

Acknowledgments

YM and SD are thankful to IISER-Pune for research fellowship; SF acknowledges DST-Inspire fellowship (DST/INSPIRE/03/2016/001694). SKG is thankful to IISER Pune for its research facilities and DST-SERB (Project No. CRG/2019/000906) for financial support.

References

Aguilera-Sigalat, J., and Bradshaw, D. 2014. A colloidal water-stable MOF as a broad-range fluorescent pH sensor via post-synthetic modification. Chem. Comm. 50: 4711–4713.

Allendorf, M. D., Bauer, C. A., Bhakta R. K., and Houk, R. J. T. 2009. Luminescent metal-organic frameworks. Chem. Soc. Rev. 38: 1330–1352.

An, J., Shade, C. M., Chengelis-Czegan, D. A., Petoud, S., and Rosi, N. L. 2011. Zinc-adeninate metal-organic framework for aqueous encapsulation and sensitization of near-infrared and visible emitting lanthanide cations. J. Am. Chem. Soc. 133: 1220–1223.

Asha, K. S., Vaisakhan, G. S., and Mandal, S. 2016. Picogram sensing of trinitrophenol in aqueous medium through a water stable nanoscale coordination polymer. Nanoscale. 8: 11782–11786.

Bennett, M. 2003. TICs, TIMs, and Terrorists. Todays Chemist At Work. 12: 21–26.

Bitzer, J., and Kleist, W. 2019. Synthetic strategies and structural arrangements of isoreticular mixed-component metal-organic frameworks. Chem. Eur. J. 25: 1866–1882.

Bo, Z., Yuan, M., Mao, S., Chen, X., Yan J. H., and Cen, K. F. 2018. Decoration of vertical graphene with tin dioxide nanoparticles for highly sensitive room temperature formaldehyde sensing. Sens. Actuators B-Chem. 256: 1011–1020.

Bobbitt, N. S., Mendonca, M. L., Howarth, A. J., Islamoglu, T., Hupp, J. T., Farha, O. K., and Snurr, R. Q. 2017. Metal-organic frameworks for the removal of toxic industrial chemicals and chemical warfare agents Chem. Soc. Rev. 46: 3357-3385.

Burrows, A. D. 2011. Mixed-component metal-organic frameworks (MC-MOFs): enhancing functionality through solid solution formation and surface modifications. Cryst. Eng. Comm. 13: 3623–3642.

Burtch, N. C., Jasuja, H., and Walton, K. S. 2014. Water stability and adsorption in metal-organic frameworks. Chem. Rev. 114: 10575–10612.

Cable, M. L., Kirby, J. P., Grey, H. B., and Ponce, A. 2013. Enhancement of anion binding in lanthanide optical sensors. Acc. Chem. Res. 46: 2576–2584.

Cai, H., Huang, Y. L., and Li, D. 2019. Biological metal-organic frameworks: Structures, host-guest chemistry and bio-applications. Coord. Chem. Rev. 378: 207–221.

Campbell, M. G., Sheberla, D., Liu, S. F., Swager, T. M., and Dinca, M. 2015. Cu3(hexaiminotriphenylene)2: An Electrically conductive 2D metal-organic framework for chemiresistive sensing. Angew. Chem. Int. Ed. 54: 4349–4352.

Carboni, M., Lin, Z., Abney, C. W., Zhang, T., and Lin, W. 2014. A Metal-organic framework containing unusual eight-connected Zr-Oxo secondary building units and orthogonal carboxylic acids for ultra-sensitive metal detection. Chem. Eur. J. 20: 14965–14970.

Chandrasekhar, P., Mukhopadhyay, A., Savitha, G., and Moorthy, J. N. 2016. Remarkably selective and enantiodifferentiating sensing of histidine by a fluorescent homochiral Zn-MOF based on pyrene-tetralactic acid. Chem. Sci. 7: 3085–3091.

Chaudhari, A. K., Nagarkar, S. S., Joarder, B., and Ghosh, S. K. 2013. A Continuous π-stacked starfish array of two-dimensional luminescent MOF for detection of nitro explosives. Cryst. Growth Des. 13: 3716–3721.

Chen, B., Wang, L., Zapata, F., Qian, G., and Lobkovsky, E. B. 2008. A luminescent microporous metal-organic framework for the recognition and sensing of anions. J. Am. Chem. Soc. 130: 6718–6719.

Chen, B., Wang, L., Xiao, Y., Fronczek, F. R., Xue, M., Cui, Y., and Qian, G. 2009. A luminescent metal-organic framework with Lewis basic pyridyl sites for the sensing of metal ions. Angew. Chem. Int. Ed. 48: 500–503.

Chen, B., Xiang, S., and Qian, G. 2010. Metal-organic frameworks with functional pores for recognition of small molecules. Acc. Chem. Res. 43: 115–1124.

Chen, C. H., Chen, X. S., Li, L., Huang, Y. B., and Cao, R. 2018. Highly selective sensing of Fe3+ by an anionic metal-organic framework containing uncoordinated nitrogen and carboxylate oxygen sites. Dalton Trans. 47: 3452–3458.

Chen, H., Fan, P., Tu, X., Min, H., Yu, X., Li, X., Zeng, J. L., Zhang, S., and Cheng, P. 2019. A bifunctional luminescent metal–organic framework for the sensing of paraquat and Fe3+ ions in water. Chem. Asian J. 14: 3611–3619.

Chen, Q., Chang, Z., Song, W. C., Song, H., Song, H. B., Hu, T. L., and Bu, X. H. 2013a. A controllable gate effect in cobalt(II) organic frameworks by reversible structure transformations. Angew. Chem. Int. Ed. 52: 11550–11553.

Chen, W. M., Meng, X. L., Zhuang, G. L., Wang, Z., Kurmoo, M., Zhao, Q. Q., Wang, X. P., Shan, B., Tung, C. H., and Sun, D. 2017. A superior fluorescent sensor for Al^{3+} and UO_2^{2+} based on a Co(II) metal-organic framework with exposed pyrimidyl Lewis base sites. J. Mater. Chem. A. 5: 13079–13085.

Chen, Y. Z., and Jiang, H. L. 2016. Porphyrinic metal–organic framework catalyzed heck-reaction: Fluorescence "TurnOn" Sensing of Cu(II) Ion. Chem. Mater. 28: 6698–6704.

Chen, Z., Sun, Y., Zhang, L., Sun, D., Liu, F., Meng, Q., Wang, R., and Sun, D. 2013b. A tubular europium-organic framework exhibiting selective sensing of Fe3+ and Al3+ over mixed metal ions. Chem. Comm. 49: 11557–11559.

Cornil, J., dos Santos, D. A., Crispin, X., Silbey, R., and Bre´das, J. L. 1998. Influence of interchain interactions on the absorption and luminescence of conjugated oligomers and polymers: A quantum-chemical characterization. J. Am. Chem. Soc. 120: 1289–1299.

Cui, J., Wong, Y. L., Zeller, M., Hunter, A. D., and Xu, Z. 2014. Pd uptake and H_2S sensing by an amphoteric metal-organic framework with a soft core and rigid side arms. Angew. Chem. Int. Ed. 53: 14438–14442.

Cui, L., Wu, J., Li, J., and Ju, H. 2015. Electrochemical sensor for lead cation sensitized with a DNA functionalized porphyrinic metal-organic framework. Anal. Chem. 87: 10635–10641.

Cui, Y., Yue, Y., Qian, G., and Chen, B. 2012. Luminescent functional metal-organic frameworks. Chem. Rev. 112: 1126–1162.

Dai, J. C., Wu, X. T., Fu, Z. Y., Hu, S. M., Du, W. X., Cui, C. P., Wu, L. M., Zhang, H. H., and Sun, R. Q. 2002. A novel ribbon-candy-like supramolecular architecture of cadmium(II)-terephthalate polymer with giant rhombic channels: Two fold interpenetration of the 3D 8210-a net. Chem. Comm. 2: 12–13.

Dalgliesh, C. E. 1951. Biological degradation of tryptophan. Q. Rev. Chem. Soc. 5: 227–244.

De Cola, L., and Belser, P. 1998. Photoinduced energy and electron transfer processes in rigidly bridged dinuclear Ru:Os complexes. Coord. Chem. Rev. 177: 301–346.

Deibert, B. J., and Li, J. 2014. A distinct reversible colorimetric and fluorescent low pH response on a water-stable zirconium-porphyrin metal-organic framework. Chem. Comm. 50: 9636–9639.

Desai, A. V., Samanta, P., Manna, B., and Ghosh, S. K. 2015. Aqueous phase nitric oxide detection by an amine-decorated metal-organic framework. Chem. Comm. 51: 6111–6114.

Desai, A. V., Sharma, S., Roy, A., Shirolkar, M. M., and Ghosh, S. K. 2020. Specific recognition of toxic allyl alcohol by pore-functionalized metal-organic frameworks. Mol. Syst. Des. Eng. 5: 469–476.

Ding, M., Flaig, R. W., Jiang, H. L., and Yaghi, O. M. 2019. Carbon capture and conversion using metal-organic frameworks and MOF-based materials. Chem. Soc. Rev. 48: 2783–2828.

DMello, M. E., Sundaram, N. G., Singh, A., Singh, A. K., and Kalidindi, S. B. 2019. An amine functionalized zirconium metal–organic framework as an effective chemiresistive sensor for acidic gases. Chem. Comm. 55: 349–352.

Dong, X., Wang, Y. R., Wang, J. Z., Zang, S. Q., and Mak, T. C. W. 2015. Highly selective Fe3+ sensing and proton conduction in a water-stable sulfonate-carboxylate Tb-organic-framework. J. Mater. Chem. A. 3: 641.

Douvali, A., Tsipis, A. C., Eliseeva, S. V., Petoud, S., Papaefstathiou, G. S., and Malliakas, C. D. 2015. Turn-on luminescence sensing and real-time detection of traces of water in organic solvents by a flexible metal-organic framework. Angew. Chem. Int. Ed., 54: 1651–1656.

Dutta, S., Samanta, P., Joarder, B., Let, S., Mahato, D., Babarao, R., and Ghosh, S. K. 2020. A water-stable cationic metal-organic framework with hydrophobic pore surfaces as an efficient scavenger of oxo-anion pollutants from water. ACS Appl. Mater. Interfaces, 37: 41810–41818.

Easun, T. L., Moreau, F., Yan, Y., Yang, S., and Schröder, M. 2017. Structural and dynamic studies of substrate binding in porous metal-organic frameworks. Chem. Soc. Rev. 46: 239–274.

Eliseevaa, S. V., and Bunzli, J. C. G. 2010. Lanthanide luminescence for functional materials and biosciences. Chem. Soc. Rev. 39: 189–227.

Fajal, S., Samanta, P., Dutta, S., and Ghosh, S. K. 2020. Selective and sensitive recognition of Fe^{3+} ion by a Lewis basic functionalized chemically stable metal-organic framework (MOF), Inorg. Chim. Acta. 502: 119359.

Falcaro, P., Ricco, R., Yazdi, A., Imaz, I., Furukawa, S., Maspoch, D., Ameloot, R., Evans, J. D., and Doonan, C. J. 2016. Application of metal and metal oxide nanoparticles@MOFs. Coord. Chem. Rev. 307: 237–254.

Fan, J., Zhu, H. F., Okamura, T. A., Sun, W. Y., Tanga, W. X., and Ueyamab, N. 2003. Three-dimensional photoluminescent pillared metal-organic framework with 4.82 topological channels obtained from the assembly of cadmium(II) acetate and trimellitic salt. New J. Chem. 27: 1409–1411.

Fang, Y., and Wang, E. 2013. Electrochemical biosensors on platforms of graphene. Chem. Commun. 49: 9526–9539.

Fei, H., Cahill, J. F., Prather, K. A., and Cohen, S. M. 2013. Tandem postsynthetic metal ion and ligand exchange in zeolitic imidazolate frameworks. Inorg. Chem. 52: 4011–4016.

Frischa, M., and Cahill, C. L. 2005. Syntheses, structures and fluorescent properties of two novel coordination polymers in the U-Cu-H3pdc system. Dalton Trans. 1518–1523.

Fumanal, M., Corminboeuf, C., Smit, B., and Tavernelli, I. 2020. Optical absorption properties of metal-organic frameworks: solid state versus molecular perspective. Phys. Chem. Chem. Phys. 22: 19512.

Ghosh, S. K. (Ed.). 2019. Metal-organic frameworks (MOFs) for environmental applications, 231–238, Elsevier.

Gunninga, N. S., and Cahill, C. L. 2005. Novel coordination polymers and structural systematics in the hydrothermal M,M' trans-3(-3-pyridyl)acrylic acid system. Dalton Trans. 2788–2792.

Guo, H., Wang, D., Chen, J., Weng, W., Huang, M., and Zheng, Z. 2016. Simple fabrication of flake-like NH2-MIL-53(Cr) and its application as an electrochemical sensor for the detection of Pb2+. Chem. Eng. J. 289: 479–485.

Guo, X., Geng, S., Zhuo, M., Chen, Y., Zaworotko, M. J., Cheng, P., and Zhang, Z. 2019. The utility of the template effect in metal-organic frameworks. Coord. Chem. Rev. 391: 44–68.

Han, L., Yuan, D., Wu, B., Liu, C., and Hong, M. 2006. Syntheses, structures and properties of three novel coordination polymers with a flexible asymmetrical bridging ligand. Inorganica Chim. Acta. 359: 2232–2240.

Hao, J. N., and Yan, B. 2015. A water-stable lanthanide-functionalized MOF as a highly selective and sensitive fluorescent probe for Cd2+, Chem. Comm. 51: 7737–7740.

Hao, J. N., and Yan, B. 2016. A dual-emitting 4d-4f nanocrystalline metal-organic framework as a self-calibrating luminescent sensor for indoor formaldehyde pollution. Nanoscale. 8: 12047–12053.

He, C., Lu, K., and Lin, W. 2014. Nanoscale metal-organic frameworks for real time intracellular pH Sensing in live cells. J. Am. Chem. Soc. 136: 12253–12256.

Heine, J., and Muller-Buschbaum, K. 2013. Engineering metal-based luminescence in coordination polymers and metal-organic frameworks. Chem. Soc. Rev. 42: 9232.

Hu, Z., Deibert, B. J., and Li, J. 2014. Luminescent metal–organic frameworks for chemical sensing and explosive detection. Chem. Soc. Rev., 43: 5815–5840.

Huang, R. W., Wei, Y. S., Dong, X. Y., Wu, X. H., Du, C. X., Zang, S. Q., and Mak, T. C. W. 2017a. Hypersensitive dual-function luminescence switching of a silver-chalcogenolate cluster-based metal-organic framework. Nat. Chem. 9: 689–697.

Huang, Y. B., Liang, J., Wang, X. S., and Cao, R. 2017b. Multifunctional metal-organic framework catalysts: synergistic catalysis and tandem reactions. Chem. Soc. Rev. 46: 126–157.

Huo, P., Chen, T., Hou, J. L., Yu, L., Zhu, Q. Y., and Dai, J. 2016. Ligand-to-ligand charge transfer within metal-organic frameworks based on manganese coordination polymers with tetrathiafulvalene-bicarboxylate and bipyridine ligands. Inorg. Chem. 55: 6496–6503.

Ji, G., Liu, J., Gao, X., Sun, W., Wang, J. Z., Zhao, S., and Liu, Z. 2017. A luminescent lanthanide MOF for selectively and ultra-high sensitively detecting Pb2+ ions in aqueous solution. J. Mater. Chem. A. 5: 10200–10205.

Jiang, M., Weng, Y. G., Zhou, Z. Y., Ge, C. Y., Zhu, Q. Y., and Dai, J. 2020. Cobalt metal–organic frameworks incorporating redox-active tetrathiafulvalene ligand: Structures and effect of LLCT within the MOF on photoelectrochemical properties. Inorg. Chem. 59: 10727–10735.

Joarder, B., Desai, A. V., Samanta, P., Mukherjee, S., and Ghosh, S. K. 2015. Selective and sensitive aqueous-phase detection of 2,4,6-trinitrophenol (TNP) by an amine-functionalized metal-organic framework. Chem. Eur. J. 21: 965–969.

Jurcic, M., Peveler, W. J., Savory, C. N., Scanlon, D. O., Kenyone, A. J., and Parkin, I. P. 2015. Thevapour phase detection of explosive markers and derivatives using two fluorescent metal-organic frameworks. J. Mater. Chem. A. 3: 6351–6359.

Karmakar, A., Desai, A. V., and Ghosh, S. K. 2016. Guest-responsive metal-organic frameworks as scaffolds for separation and sensing applications. Coord. Chem. Rev. 307: 313–341.

Karmakar, A., Samanta, P., Dutta, S., and Ghosh, S. K. 2019. Fluorescent "Turn-on" sensing based on metal-organic frameworks (MOFs). Chem. Asian J. 14: 4506–4519.

Kreno, L. E., Leong, K., Farha, O. K., Allendorf, M., Duyne, R. P. V., and Hupp, J. T. 2012. Metal-organic framework materials as chemical sensors. Chem. Rev. 112: 1105–1125.

Kumar, S., Jain, S., Nehra, M., Dilbaghi, N., Marrazza, G., and Kim, K. H. 2020. Green synthesis of metal-organic frameworks: A state-of-the-art review of potential environmental and medical applications. Coord. Chem. Rev. 420: 213407.

Lai, C., Wang, Z., Qin, L., Fu, Y., Li, B., Zhang, M. et al. 2021. Metal-organic frameworks as burgeoning materials for the capture and sensing of indoor VOCs and radon gases. Coord. Chem. Rev. 427: 213565.

Lan, A., Li, K., Wu, H., Olson, D. H., Emge, T. J., Ki, W. et al. 2009. A luminescent microporous metal-organic framework for the fast and reversible detection of high explosives. Angew. Chem. Int. Ed. 48: 2334–2338.

Let, S., Samanta, P., Dutta, S., and Ghosh, S. K. 2020. A Dye@MOF composite as luminescent sensory material for selective and sensitive recognition of Fe(III) ions in water. Inorg. Chim. Acta. 500: 119205.

Li, C. P., Zhou, H., Chen, J., Wang, J. J., Du, M., and Zhou, W. 2020. A highly efficient coordination polymer for selective trapping and sensing of perrhenate/pertechnetate. ACS Appl. Mater. Interfaces. 12: 15246–15254.

Li, H. Y., Zhao, S. N., Zang, S. Q., and Li, J. 2020a. Functional metal-organic frameworks as effective sensors of gases and volatile compounds. Chem. Soc. Rev. 49: 6364–6401.

Li, J. R., Yu, J., Lu, W., Sun, L. B., Sculley, J., Balbuena, P. B., and Zhou, H. C. 2013b. Porous materials with pre-designed single-molecule traps for CO_2 selective adsorption. Nat. Comm. 4: 1538.

Li, J. R., Kuppler, R. J., and Zhou, H. C. 2009. Selective gas adsorption and separation in metal-organic frameworks. Chem. Soc. Rev. 38: 1477–1504.

Li, Q., Wang, C., Tan, H., Tang, G., Gao, J., and Chen, C. H. 2016. A turn on fluorescent sensor based on lanthanide coordination polymer nanoparticles for the detection of mercury (II) in biological fluids. RSC Adv. 6: 17811–17817.

Li, X., Xu, H., Kong, F., and Wang, R. 2013a. A cationic metal-organic framework consisting of nanoscale cages: Capture, separation, and luminescent probing of $Cr_2O_7^{2-}$ through a single-crystal to single-crystal process. Angew. Chem., Int. Ed. 52: 13769–13773.

Li, Y., Zhang, S., and Song, D. 2013. A luminescent metal-organic framework as a turn-on sensor for DMF vapor. Angew. Chem., Int. Ed. 52: 710–713.

Liu, C. Y., Chen, X. R., Chen, H. X., Niu, Z., Hirao, H., Braunstein, P., and Lang, J. P. 2020. Ultrafast luminescent light-up guest detection based on the lock of the host molecular vibration. J. Am. Chem. Soc. 142: 6690–6697.

Liu, S. Y., Qi, X. L., Lin, R. B., Cheng, X. N., Liao, P. Q., Zhang, J. P., and Chen, X. M. 2014. Porous Cu(I) triazolate framework and derived hybrid membrane with exceptionally high sensing efficiency for gaseous oxygen. Adv. Funct. Mater. 24: 5866–5872.

Liu, X. G., Wang, H., Chen, B., Zou, Y., Gu, Z. G., Zhao, Z. et al. 2015. A luminescent metal-organic framework constructed using a tetraphenylethene-based ligand for sensing volatile organic compounds. Chem. Comm. 51: 1677–1680.

Lu, Y., Yan, B., and Liu, J. L. 2014. Nanoscale metal-organic frameworks as highly sensitive luminescent sensors for Fe2+ in aqueous solution and living cells. Chem. Comm. 50: 9969–9972.

Luo, F., and Batten, S. R. 2010. Metal-organic framework (MOF): lanthanide(III)-doped approach for luminescence modulation and luminescent sensing. Dalton Trans. 39: 4485–4488.

Lustig, W. P., Mukherjee, S., Rudd, N. D., Desai, A. V., Li, J., and Ghosh, S. K. 2017. Metal-organic frameworks: functional luminescent and photonic materials for sensing applications. Chem. Soc. Rev. 46: 3242–3285.

Lustig, W. P., and Li, J. 2018. Luminescent metal-organic frameworks and coordination polymers as alternative phosphors for energy efficient lighting devices. Coord. Chem. Rev. 373: 116–147.

Ma, D., Li, B., Zhou, X., Zhou, Q., Liu, K., Zeng, G., Li, G., Shi, Z., and Feng, S. 2013. A dual functional MOF as a luminescent sensor for quantitatively detecting the concentration of nitrobenzene and temperature. Chem. Comm. 49: 8964–8966.

Mao, S., Wen, Z. H., Ci, S. Q., Guo, X. R., Ostrikov, K., and Chen, J. H. 2015. Perpendicularly oriented MoSe2/graphene nanosheets as advanced electrocatalysts for hydrogen evolution. Small 11: 414–419.

McDonald, B. C., de Gouw, J. A., Gilman, J. B., Jathar, S. H. et al. 2018. Volatile chemical products emerging as largest petrochemical source of urban organic emissions. Science 359: 760–764.

Medishetty, R., Zare, ba, J. K., Mayer, D., Samoc, M., and Fischer, R. A. 2017. Auto-controlled fabrication of a metal-porphyrin framework thin film with tuneable optical limiting effects. Chem. Soc. Rev. 46: 4976–5004.

Meng, Q., Xin, X., Zhang, L., Dai, F., Wang, R., and Sun, D. 2015. A multifunctional Eu MOF as a fluorescent pH sensor and exhibiting highly solvent-dependent adsorption and degradation of rhodamine B. J. Mater. Chem. A. 3: 24016–24021.

Meng, Z., Aykanat, A., and Mirica, K. A. 2019. Welding metallophthalocyanines into bimetallic molecular meshes for ultrasensitive, low-power chemiresistive detection of gases. J. Am. Chem. Soc. 141: 2046–2053.

Meyer, L. V., Schönfeld, F., Zurawski, A., Mai, M., Feldmannb, C., and Müller-Buschbaum, K. 2015. A blue luminescent MOF as a rapid turn-off/turn-on detector for H_2O, O_2 and CH_2Cl_2, MeCN: $^3_\infty$[Ce(Im)3ImH]·ImH. Dalton Trans. 44: 4070.

Moldovan, O., Iniguez, B., Deen, M. J., and Marsal, L. F. 2015. Graphene electronic sensors- review of recent developments and future challenges. IET Circuits Devices Syst. 9: 446–453.

Mukherjee, S., Desai, A. V., More, Y. D., Inamdar, A. I., and Ghosh, S. K. 2016. A bifunctional metal-organic framework: Striking CO_2-selective sorption features along with guest-induced tuning of luminescence. Chem. Plus. Chem. 81: 702–707.

Mukherjee, S., Sensharma, D., Qazvini, O. T., Dutta, S., Macreadie, L. K., Ghosh, S. K., and Babarao, R. 2021. Advances in adsorptive separation of benzene and cyclohexane by metal-organic framework adsorbents. Coord. Chem. Rev. 437: 213852.

Mutihac, L., Lee, J. H., Kim, J. S., and Vicens, J. 2011. Recognition of amino acids by functionalized calixarenes. Chem. Soc. Rev. 40: 2777.

N. Le Floc'h, Le, N., Otten, W., and Merlot, E. 2011. Tryptophan metabolism, from nutrition to potential therapeutic applications. Amino Acids 41: 1195–1205.

Nagarkar, S. S., Joarder, B., Chaudhari, A. K., Mukherjee, S., and Ghosh, S. K. 2013. Highly selective detection of nitro explosives by a luminescent metal-organic framework. Angew. Chem. Int. Ed. 52: 2881–2885.

Nagarkar, S. S., Desai, A. V., and Ghosh, S. K. 2014. A fluorescent metal-organic framework for highly selective detection of nitro explosives in the aqueous phase. Chem. Comm. 50: 8915–8918.

Nakatsuka, S., Watanabe, Y., Kamakura, Y., Horike, S., Tanaka, D., and Hatakeyama, T. 2020. Solvent-vapor-induced reversible single-crystal-to-single-crystal transformation of a triphosphaazatriangulene based metal-organic framework. Angew. Chem. Int. Ed. 59: 1435–1439.

Nickerl, G., Senkovska, I., and Kaskel, S. 2015. Tetrazine functionalized zirconium MOF as an optical sensor for oxidizing gases. Chem. Comm. 51: 2280–2282.

Pamei, M., and Puzari, A. 2019. Luminescent transition metal-organic frameworks: An emerging sensor for detecting biologically essential metal ions. Nano-Struct. Nano-Objects. 19: 100364.

Pramanik, S., Zheng, C., Zhang, X., Emge, T. J., and Li, J. 2011. New microporous metal-organic framework demonstrating unique selectivity for detection of high explosives and aromatic compounds. J. Am. Chem. Soc. 133: 4153–4155.

Qiao, C., Qu, X., Yang, Q., Wei, Q., Xie, G., Chen, S., and Yang, D. 2016. Instant high-selectivity Cd-MOF chemosensor for naked-eye detection of Cu(II) confirmed using in situ microcalorimetr., Green Chem. 18: 951–956.

Qin, J. S., Yuan, S., Wang, Q., Alsalme, A., and Zhou, H. C. 2017. Mixed-linker strategy for the construction of multifunctional metal-organic frameworks. J. Mater. Chem. A. 5: 4280–4291.

Rasheed, T., and Nabeel, F. 2019. Luminescent metal-organic frameworks as potential sensory materials for various environmental toxic agents. Coord. Chem. Rev. 401: 213065.

Razavi, S. A. A., and Morsali, A. 2020. Metal ion detection using luminescent-MOFs: Principles, strategies and roadmap. Coord. Chem. Rev. 415: 213299.

Resolution A/RES/64/292; United Nations General Assembly, July 2010.

Rieter, W. J., Taylor, K. M. L., and Lin, W. 2007. Surface modification and functionalization of nanoscale metal-organic frameworks for controlled release and luminescence sensing. J. Am. Chem. Soc. 129: 9852–9853.

Rocha, J., Carlos, L. D., Paz, F. A. A., and Ananias, D. 2011. Luminescent multifunctional lanthanides-based metal-organic frameworks. Chem. Soc. Rev. 40: 926–940.

Rojas, S., and Horcajada, P. 2020. Metal-organic frameworks for the removal of emerging organic contaminants in water. Chem. Rev. 120: 8378–8415.

Roushani, M., Valipour, A., and Saedi, Z. 2016. Electroanalytical sensing of Cd^{2+} based on metal-organic framework modified carbon paste electrode. Sens. Actuators B. 233: 419–425.

Rubio-Gimenez, V., Almora-Barrios, N., Escorcia-Ariza, G., Galbiati, M., Sessolo, M., Tatay, S., and Martí-Gastaldo, C. 2018. Origin of the chemiresistive response of ultrathin films of conductive metal-organic frameworks. Angew. Chem. Int. Ed. 57: 15086–15090.

Rudd, N. D., Wang, H., Fuentes-Fernandez, E. M. A., Teat, S. J., Chen, F., Hall, G., Chabal, Y. J., and Li, J. 2016. Highly efficient luminescent metal-organic framework for the simultaneous detection and removal of heavy metals from water. ACS Appl. Mater. Interfaces. 8: 30294–30303.

Samanta, P., Desai, A. V., Sharma, S., Chandra, P., and Ghosh, S. K. 2018. Selective recognition of Hg2+ ion in water by a functionalized metal-organic framework (MOF) based chemodosimeter. Inorg. Chem. 57: 2360–2364.

Samanta, P., Let, S., Mandal, W., Dutta, S., and Ghosh, S. K. 2020. Luminescent metal-organic frameworks (LMOFs) as potential probes for the recognition of cationic water pollutants. Inorg. Chem. Front. 7: 1801–1821.

Schulz, M., Gehl, A., Schlenkrich, J., Schulze, H. A., Zimmermann, S., and Schaate, A. 2018. A calixarene-based metal-organic framework for highly selective NO_2 detection. Angew. Chem. Int. Ed. 57: 12961–12965.

Sen, A., Desai, A. V., Samanta, P., Dutta, S., Let, S., and Ghosh, S. K. 2018. Post-synthetically modified metal-organic framework as a scaffold for selective bisulphite recognition in water. Polyhedron. 156: 1–5.

Sharma, S., Let, S., Desai, A. V., Dutta, S., Karuppasamy, G., Shirolkar, M. M., Babarao, R., and Ghosh, S. K. 2021. Rapid, selective capture of toxic oxo-anions of Se(iv), Se(vi) and As(v) from water by an ionic metal-organic framework (iMOF). J. Mater. Chem. A. 9: 6499–6507.

Shi, P. F., Hu, H. C., Zhang, Z. Y., Xiong, G., and Zhao, B. 2015. Heterometal-organic frameworks as highly sensitive and highly selective luminescent probes to detect I– ions in aqueous solutions. Chem. Commun. 51: 3985–3988.

Shustova, N. B., McCarthy, B. D., and Dinca, M. 2011. Turn-on fluorescence in tetraphenylethylene-based metal-organic frameworks: An alternative to aggregation-induced emission. J. Am. Chem. Soc. 133: 20126–20129.

Shustova, N. B., Cozzolino, A. F., Reineke, S., Baldo, M., and Dincă, M. 2013. Selective turn-on ammonia sensing enabled by high-temperature fluorescence in metal-organic frameworks with open metal sites. J. Am. Chem. Soc. 135: 13326–13329.

Smith, M. K., Jensen, K. E., Pivak, P. A., and Mirica, K. A. 2016. Direct self-assembly of conductive nanorods of metal-organic frameworks into chemiresistive devices on shrinkable polymer films. Chem. Mater. 28: 5264–5268.

Smith, M. K., and Mirica, K. A. 2017. Self-organized frameworks on textiles (SOFT): Conductive fabrics for simultaneous sensing, capture, and filtration of gases. J. Am. Chem. Soc. 139: 16759–16767.

Song, T., Zhang, P., Zeng, J., Wang, T., Ali, A., and Zeng, H. 2017. Tunable conduction band energy and metal-to ligand charge transfer for wide-spectrum photocatalytic H_2 evolution and stability from isostructural metal-organic frameworks. Int. J. Hydrog. Energy. 42: 26605–26616.

Stavila, V., Talin, A. A., and Allendorf, M. D. 2014. MOF-based electronic and opto-electronic devices. Chem. Soc. Rev. 43: 5994–6010.

Sun, C. Y., Wang, X. L., Zhang, X., Qin, C., Li, P., Su, Z. M. et al. 2013. Efficient and tuneable white-light emission of metal-organic frameworks by iridium-complex encapsulation. Nat. Comm. 4: 2717.

Takashima, Y., Martinez, V. M., Furukawa, S., Kondo, M., Shimomura, S., Uehara, H., Nakahama, M., Sugimoto, K., and Kitagawa, S. 2011. Molecular decoding using luminescence from an entangled porous framework. Nat. Comm. 2: 168.

Taylor, K. M. L., Rieter, W. J., and Lin, W. 2008. Manganese-based nanoscale metal-organic frameworks for magnetic resonance imaging. J. Am. Chem. Soc. 130: 14358–14359.

Wanderley, M. M., Wang, C., Wu, C. D., and Lin, W. 2012. A chiral porous metal-organic framework for highly sensitive and enantioselective fluorescence sensing of amino alcohols. J. Am. Chem. Soc. 134: 9050–9053.

Wang, B., Lv, X. L., Feng, D., Xie, L. H., Zhang, J., Li, M. et al. 2016. Highly stable Zr(IV)-based metal-organic frameworks for the detection and removal of antibiotics and organic explosives in water. J. Am. Chem. Soc. 138: 6204–6216.

Wang, B., Yang, Q., Guo, C., Sun, Y., Xie, L. H., and Li, J. R. 2017. Stable Zr(IV)-based metal-organic frameworks with predesigned functionalized ligands for highly selective detection of Fe(III) ions in water. ACS Appl. Mater. Interfaces. 9: 10286–10295.

Wang, G. Y., Song, C., Kong, D. M., Ruan, W. J., Chang, Z., and Li, Y. 2014. Two luminescent metal-organic frameworks for the sensing of nitroaromatic explosives and DNA strands. J. Mater Chem. A. 2: 2213–2220.

Wang, H., Lustig, W. P., and Li, J. 2018. Sensing and capture of toxic and hazardous gases and vapors by metal-organic frameworks. Chem. Soc. Rev. 47: 4729–4756.

Wang, J., Jiu, J. T., Araki, T., Nogi, M., Sugahara, T., Nagao, S., Koga, H., He, P., and Suganuma, K. 2015. Silver nanowire electrodes: conductivity improvement without post-treatment and application in capacitive pressure sensors. Nano-Micro Lett. 7: 51–58.

Wang, M., Guo, L., and Cao, D. 2018. Amino-functionalized luminescent metal-organic framework test paper for rapid and selective sensing of SO_2 gas and its derivatives by luminescence turn-on effect. Anal. Chem. 90: 3608–3614.

Wang, P. L., Xie, L. H., Joseph, E. A., Li, J. R., Su, X. O., and Zhou, H. C. 2019. Metal-organic frameworks for food safety. Chem. Rev. 119: 10638–10690.

Wang, Y., Wang, L., Huang, W., Zhang, T., Hu, X., Perman, J. A., and Ma, S. 2017. A metal-organic framework and conducting polymer based electrochemical sensor for high performance cadmium ion detection. J. Mater. Chem. A. 5: 8385–8393.

Wang, Z., and Cohen, S. M. 2009. Postsynthetic modification of metal-organic frameworks. Chem. Soc. Rev. 38: 1315–1329.

Wen, L., Zheng, X., Lv, K., Wang, C., and Xu, X. 2015. Two amino decorated metal organic frameworks for highly selective and quantitatively sensing of HgII and CrVI in aqueous solution. Inorg. Chem. 54: 7133–7135.

World Urbanization Prospects, UN Department of Economic and Social Affairs, 2014.

Wu, P., Wang, J., Li, Y., He, C., Xie, Z., and Duan, C., 2011. Luminescent sensing and catalytic performances of a multifunctional lanthanide-organic framework comprising a triphenylamine moiety. Adv. Funct. Mater. 21: 2788–2794.

Wu, P., Liu, Y., Liu, Y., Wang, J., Li, Y., Liu, W., and Wang, J. 2015. Cadmium-based metal-organic framework as a highly selective and sensitive ratiometric luminescent sensor for mercury (II). Inorg. Chem. 54: 11046–11048.

Wu, S., Min, H., Shi, W., and Cheng, P. 2019. Multicenter metal-organic framework-based ratiometric fluorescent sensors. Adv. Mater. 32: 1805871.

Wu, X., Macreadie, L. K., and Gale, P. A. 2021. Anion binding in metal-organic frameworks. Coord. Chem. Rev. 432: 213708.

Xu, H., Liu, F., Cui, Y., Chen, B., and Qian, G. 2011. A luminescent nanoscale metal-organic framework for sensing of nitroaromatic explosives. Chem. Comm. 47: 3153–3155.

Xu, H., Rao, X., Gao, J., Yu, J., Wang, Z., Dou, Z., Cui, Y., Yang, Y., Chen, B., and Qian, G. 2012. A luminescent nanoscale metal-organic framework with controllable morphologies for spore detection. Chem. Comm. 48: 7377–7379.

Xu, R., Wang, Y., Duan, X., Lu, K., Micheroni, D., Hu, A., and Lin, W. 2016. Nanoscale metal-organic frameworks for ratiometric oxygen sensing in live cells. J. Am. Chem. Soc. 138: 2158–2161.

Xu, Y., Meng, J., Meng, L. X., Dong, Y., Cheng, Y. X., and Zhu, C. J. 2010. A highly selective fluorescence-based polymer sensor incorporating an (R,R)-salen moiety for Zn2+ detection. Chem. Eur. J. 16: 12898–12903.

Yan, B. 2017. Lanthanide-functionalized metal-organic framework hybrid systems to create multiple luminescent centers for chemical sensing. Acc. Chem. Res. 50: 2789–2798.

Yang, J., Wang, Z., Hu, K., Li, Y., Feng, J., Shi, J. et al. 2015. Rapid and specific aqueousphase detection of nitroaromatic explosives with inherent porphyrin recognition sites in metal-organic frameworks. ACS Appl. Mater. Interfaces. 7: 11956–11964.

Yao, M. S., Zheng, J. J., Wu, A. Q., Xu, G., Nagarkar, S. S., Zhang, G., Tsujimoto, M., Sakaki, S., Horike, S., Otake, K., and Kitagawa, S. 2020. A Dual-ligand porous coordination polymer chemiresistor with modulated conductivity and porosity. Angew. Chem. Int. Ed. 59: 172–176.

Yao, M. S., Lv, X. J., Fu, Z. H., Li, W. H., Deng, W. H., Wu, G. D., and Xu, G. 2017. Layer-by-layer assembled conductive metal–organic framework nanofilms for room-temperature chemiresistive sensing. Angew. Chem. Int. Ed. 56: 16510–16514.

Yi, F. Y., Chen, D., Wu, M. K., Han, L., and Jiang, H. L. 2016. Chemical sensors based on metal-organic frameworks. Chem. Plus. Chem. 81: 675–690.

Yu, M. H., Hu, T. L., and Bu, X. H. 2017. A metal-organic framework as a "turn on" fluorescent sensor for aluminum ions. Inorg. Chem. Front. 4: 256–260.

Yu, Y., Ma, J. P., and Dong, Y. B. 2012. Luminescent humidity sensors based on porous Ln3+-MOFs. Cryst. Eng. Comm. 14: 7157–7160.

Yu, Y., Zhang, X. M., Ma, J. P., Liu, Q. K., Wang, P., and Dong, Y. B. 2014. Cu(I)-MOF: naked-eye colorimetric sensor for humidity and formaldehyde in single-crystal-to-single-crystal fashion. Chem. Comm. 50: 1444–1446.

Yuvaraja, S., Surya, S. G., Chernikova, V., Vijjapu, M. T., Shekhah, O., Bhatt, P. M., Chandra, S., Eddaoudi, M., and Salama, K. N. 2020. ACS Appl. Mater. Interfaces. 12: 18748–18760.

Zhang, J., Xia, T., Zhao, D., Cui, Y., Yang, Y., and Qian, G. Sensors. 2018a. Metal-organic framework film for fluorescence turn-on H_2S gas sensing and anti-counterfeiting patterns. Sens. Actuators B. 260: 63–69.

Zhang, J., Yang, W. B., Wu, X. Y., Kuang, X. F., and Lu, C. Z. 2015. Protonation effect on ligands in EuL: a luminescent switcher for fast naked-eye detection of HCl. Dalton Trans. 44: 13586–13591.

Zhang, K. Y., Gao, P., Sun, G., Zhang, T., Li, X., Liu, S., Zhao, Q., Lo, K. K., and Huang, W. 2018b. Phosphorescent iridium(III) complexes capable of imaging and distinguishing between exogenous and endogenous analytes in living cells. J. Am. Chem. Soc. 140: 7827–7834.

Zhang, M., Feng, G., Song, Z., Zhou, Y. P., Chao, H. Y., Yuan, D., Tan, T. T., Guo, Z., Hu, Z., Tang, B. Z., Liu, B., and Zhao, D. 2014. Two-dimensional metal-organic framework with wide channels and responsive turn-on fluorescence for the chemical sensing of volatile organic compounds. J. Am. Chem. Soc. 136: 7241–7244.

Zhang, S., Wang, Z., Zhang, H., Cao, Y., Sun, Y., Chen, Y., Huang, C., and Yu, X. 2007. Self-assembly of two fluorescent supramolecular frameworks constructed from unsymmetrical benzene tricarboxylate and bipyridine. Inorganica Chim. Acta. 360: 2704–2710.

Zhang, W. H., Ma, W., and Long, Y. T. 2016. Redox-mediated indirect fluorescence immunoassay for the detection of disease biomarkers using dopamine-functionalized quantum dots. Anal. Chem. 88: 5131–5136.

Zhou, Y., Wang, S. X., Zhang, K., and Jiang, X. Y. 2008. Visual detection of copper (II) by azide-and alkyne-functionalized gold nanoparticles using click chemistry. Angew. Chem. Int. Ed. 47: 7454–7456.

Zhou, Y., Yan, B., and Lei, F. 2014. Postsynthetic lanthanide functionalization of nanosized metal-organic frameworks for highly sensitive ratiometric luminescent thermometry. Chem. Comm. 50: 15235–15238.

Zhou, Y., and Yan, B. 2016. A responsive MOF nanocomposite for decoding volatile organic compounds. Chem. Comm. 52: 2265–2268.

Zhu, Y. M., Zeng, C. H., Chu, T. S., Wang, H. M., Yang, Y. Y., Tong, Y. X., Su, C. Y., and Wong, W. T. 2013. A novel highly luminescent LnMOF film: a convenient sensor for Hg^{2+} detecting. J. Mater. Chem. A. 1: 11312–11319.

CHAPTER 6

Biomaterials as Chemical Sensors

Benny Ryplida,[1] Ji Hyun Ryu[2] and Sung Young Park[1,3,4,]*

Introduction

Recently, biomaterials have been used in a wide array of applications such as drug delivery, tissue engineering, and therapeutic, they are now being applied even as a chemical sensor for detecting internal and external environment (Sun et al. 2018). A variation of the biosensor can be achieved by modifying the materials, for example, the sensor consists of hyaluronic acid that can be used to detect cancer due to abundant CD44 receptors on the tumor microenvironment (Bhattacharya et al. 2017a). Other solid examples are the use of boronic functional group to quantify the glucose or the use of redox-responsive system that prefers oxidation-reduction precursors such as ROS (H_2O_2), protein (cysteine), or antioxidant (GSH) (Schumacker 2006, Rusin et al. 2004, Jung et al. 2018, Zhou et al. 2018). Every stimulant located inside the system, such as the cellular microenvironment, can be defined as internal stimuli, while the stimuli that are found or originate from outside of the system, can be described as external stimuli. For cancer optical diagnosis using fluorescence nanoparticles, several internal factors have been analyzed in the past 20 years, such as pH, the redox level, the overexpress enzyme or protein, or any other abnormal status that separate healthy condition and tumor. On the other hand, temperature, heat, light, magnetic field, and electronic field are identified as external trigger (Schumacker 2006, Rusin et al. 2004, Jung et al. 2018, Zhou et al. 2018, Mazrad et al. 2018). As the amount of stimuli is massively present, the huge possibility for biomaterials sensing not only

[1] Department of Green Bio Engineering, Korea National University of Transportation, Chungju 380-702, Republic of Korea.

[2] Department of Carbon Convergence Engineering, Wonkwang University, Iksan, Jeonbuk 54538, Republic of Korea.

[3] Department of Chemical & Biological Engineering, Korea National University of Transportation, Chungju 380-702, Republic of Korea.

[4] Department of IT Convergence Engineering, Korea National University of Transportation, Chungju 380-702, Republic of Korea.

* Corresponding author: parkchem@ut.ac.kr

for 2-D but also prospect for 3-D application including prosthetic, actuating, and human motion is also implied.

The biosensor itself is usually comprised of a bio-receptor, transducer, and methods of recognition. Bio-receptor is an active site that interacts and binds with the target analyte, the transducer is an element that converts the interaction into a measurable signal, and methods of recognition is the form of observation either optical, mass-based, or electronic. Nucleic acid, enzyme, antibody, gene, etc., are usually the most popular bio-receptor due to their specific binding. However, biomimetic, or synthetic material that can mimic the biochemical process is widely studied owing to its low-cost and simple method to approach it. Another most important point about biosensors is the presence of base materials such as metal, polymer, glass, paper, composite, or even polymer. The base materials are essential for immobilization of bioreceptor. However, it is also possible to prepare a two-in-one bioreceptor-base material by combining the function of the receptor, usually a biomimetic polymer, and the base material, for example, a hydrogel prepared from polymer (Tavakoli and Tang 2017, Moreira et al. 2018, Bazin et al. 2017, Serban and Enesca 2020). Comprehensively, the principle of biomaterial sensing lies on the specific response during reaction on the interfaces of the sensing materials (receptor) and the target, which can sensitively and selectively perform the recognition step (Higgins and Lowe 1987). During this recognition phase, the target would attach to the sensor either by the host-guest reaction, click-mechanism or directly undergoes acid-base, reduction-oxidation, or complex reaction between the receptor and sample/target analyte (Perumal and Hashim 2014). After attachment or completion of reaction, the sensor properties can be distinguished before and after the binding, in which the response can differentiate in the form of optical or electronic performance.

As the possibility arises, it is also feasible to apply the materials for the 3-D mechanical deformation, such as flexible materials that can withstand stretching, compression, and bending (Sun et al. 2018, Willner 2002). Interestingly, the mechanical properties of the sensor could be simultaneously determined by evaluating the electronic performance of the sensor. During the stimulation, the sensor would capture the signal by transforming the movement or stimuli into an electronic pattern. The electronic response usually comes in the form of resistance, capacitance, current density, or voltage change (Yang and Suo 2018). The obtained signal can be determined as the change in electron movement during the deformation, either by the effect of dimensional change of the hydrogel or the defect of nanoparticles inside the hydrogel (Ryplida et al. 2019).

Biomaterials and Biosensing Technology

In terminology terms, biomaterials are described as natural or synthetic materials that are suitable for introduction of the materials into living tissue, particularly as part of a medical device. A wide interpretation of biomaterials could be defined as a substance that has been engineered to associate with biological systems for diagnosis or therapeutic. Based on both definitions, biomaterials can be categorized as natural or synthetic material whether it is inorganic materials or organic materials that are able to bind with the biological components such as tissue, organ, etc. During the development of biomaterial for diagnosis or detection, a lot of trial and error has been implemented to enhance the efficiency and targeting performance. The improvement of the system, colorimetry for example, is conducted by modifying the particles, surface functionalization, conjugation of colorant, etc. (Yao et al. 2016, Oliviera et al. 2016). On the other hand, designing devices from a hydrogel becomes a choice whether by using some methods such as surface patterning, loading, incorporation, or hydrogel-coating. As the demand for the sensor increased recently, the materials or devices for creating sensors have been sort into three biggest categories that would be explained comprehensively in this chapter (Figure 1).

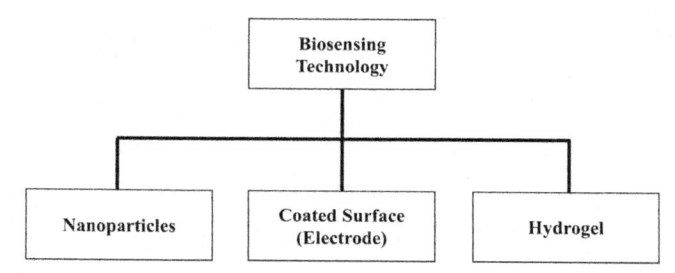

Figure 1. Recent technology of biosensing that has been extensively implemented.

Nanoparticles

Nanoparticles (NPs) are categorized as a material within the size range of 1–100 nm and exhibit unique physical, chemical, optical, and electronic performance, showing different properties than the bulk of materials (Luo et al. 2006). In recent years, NPs are used for sensing applications and mostly are dependent on their fluorescent behavior (Figure 2). In terms of sensing sensitivity towards the biological environment, it demonstrates excellent biocompatibility and is able to avoid nonspecific binding by the cellular biomacromolecules compared to microparticles or molecular probes (Wolfbeis 2015). Moreover, nanoparticles also showed excellent electronic properties for sensing owing to a large specific surface area and high surface free energy, which made it a better choice for detection (Luo et al. 2006). Interestingly, NPs could accommodate different sensing performance even if it is the same particle due to

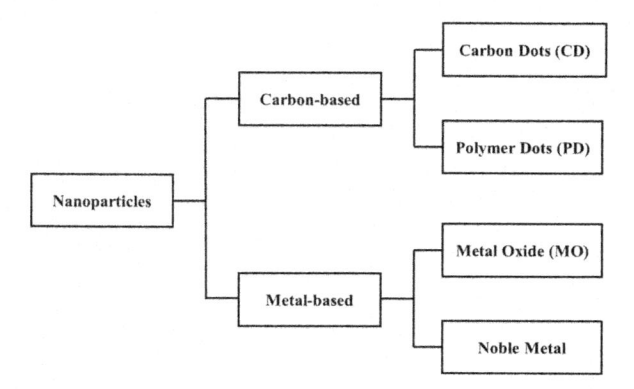

Figure 2. The nanoparticles that are widely used for biosensing applications.

its composition. To make it simpler, complex NPs, and sometimes the same kind of particles, could perform various roles in diverse electrochemical sensing systems, such as enzymatic sensors, immunosensors, and DNA sensors (Luo et al. 2006).

For the nanoparticles-based optical sensor, quantum dots (QD) have emerged as one of the best candidates owing to their biocompatibility to most of the biological system, ease to functionalize, excellent quantum yield, and less cost (Bimberg and Pohl 2011, Kargozar et al. 2020, Bansal et al. 2016). The term QD is a general term for a semiconductor particle, which has a few nanometers in size, and optical and electronic properties that differ from larger particles due to quantum mechanics. The QD mostly consists of inorganic materials such as cadmium sulfide (CdS), lead sulfide (PbS), lead selenide (PbSe), and indium arsenide (InAs) (Bansal et al. 2016, Dong et al. 2019, Zhang et al. 2018a, Franke et al. 2016). However, it is also possible to prepare a QD from organic material, in which the term that is used is different. The QD prepared from organic precursors is usually called carbon quantum dots (CQD), graphene quantum dots (GQD), or carbon dots (CD), and for a fluorescence nanoparticle that is prepared from polymer, polymer dots (PD) would be the suitable term (Zhu et al. 2015, Lim et al. 2015). As for biological application, CD and PD are the best options since most of them are prepared from organic materials that are less toxic compared to inorganic materials.

The advantages of carbon dots (CD) also include their preferable conductivity, outstanding chemical stability, excitation-dependent multicolor emission, and exceptional cellular permeability revealing its wide array of possibilities (Mazrad et al. 2018, Liu et al. 2009). Furthermore, the methods of fabrication of the CD are categorized as simple preparation and are usually separated into two categories, which are top-down techniques and bottom-up strategies which usually describe as carbonization methods as shown in Figure 3 (Li et al. 2017, Xia et al. 2019). For top-down, the CD are prepared by breaking down a larger molecule to create the desired particles *via* known methods such as laser ablation, arc discharge, or chemical/

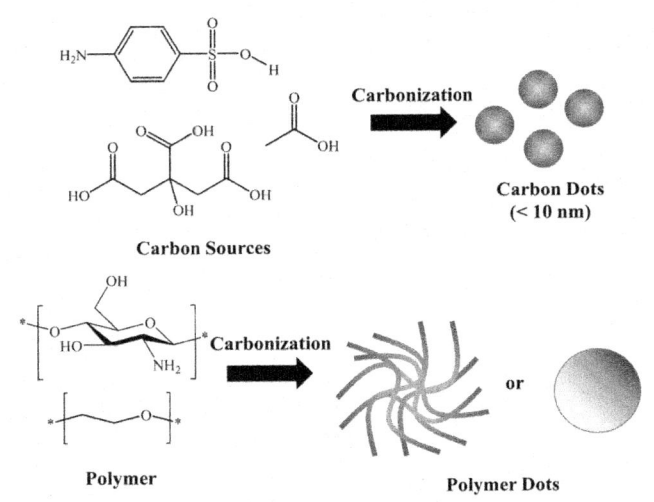

Figure 3. The preparation of carbon dots from carbon particles (e.g., citric acid and acetic acid and polymer dots from polymer (e.g., chitosan and polyethylene glycol) using carbonization step.

electrochemical oxidation (Yu et al. 2016, Su et al. 2014, Zhang and Yu 2016, Liu et al. 2016a). For bottom-up methods, the preparation method is by combining or assembling molecular precursors through different reaction conditions including thermal combustion, microwave-assisted reaction, or other synthetic condition (Bourlinos et al. 2012, Medeiros et al. 2019). In terms of absorption properties, the CD presents a strong optical absorption in the UV region, which extends into the visible light range. Most CD shows an absorption shoulder at a π-π^* transition of C=C bonds, and the n-π^* transition of C=O bonds (Jiang et al. 2020). Another prominent property of a CD is the similar lattice spacing with graphene at 0.21 nm corresponding to the (100) plane of graphite, which would be beneficial for sensing application (Chen et al. 2016). To optimize the optical behavior, several phenomena that are responsible for the change in fluorescent behavior have been introduced and studied, including photo-induced electron transfer (PET), photo-induced charge transfer (PCT), Förster resonance energy transfer (FRET), and inner filter effect (IFE). In PET and PCT, the driving force is the involvement of electron transfer during the process. Moreover, FRET uses energy transfer, in which the energy is transferred between the two chromophores (light-sensitive parts).

Other principles that have been studied are aggregation-induce emission (AIE) and aggregation-caused quenching (ACQ) as shown in Figure 4. In these mechanisms, the state of the nanoparticles becomes the concern. In AIE, the nanoparticles show brighter emission in the solution state rather than in the solid-state, in which the reverse phenomena are ACQ (brighter in solid) (Leduskrasts and Suna 2019, Liu et al. 2016b, Adsetts et al. 2020). Additionally, most of those mechanisms usually happen in organic luminophores (fluorescent dyes).

For the electronic sensor, the adjustment of the CD using dopant (N and B) would promote the electronic performance. Additionally, modifying with a suitable functional group would enhance the selectivity and the sensitivity of the carbon dots sensors, such as a boronic functional group for glucose sensing (Zou et al. 2018) dopamine functional group for metal sensing (Qu et al. 2013) and hyaluronic acid for CD44 of cancer cells sensing (Bhattacharya et al. 2017a). Another material that

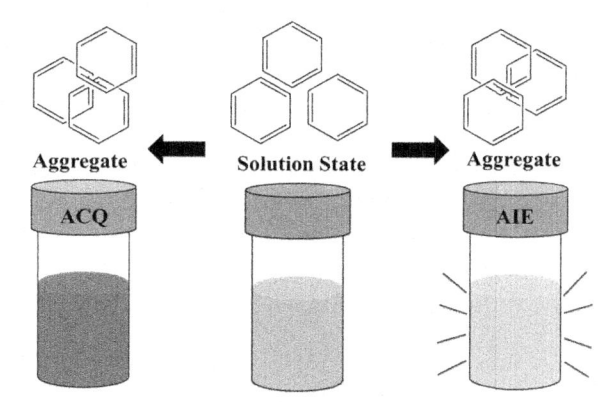

Figure 4. The illustration of nanoparticles during ACQ and AIE phenomena. The fluorescent of the solution becomes brighter during AIE condition, in contrast with ACQ condition that the fluorescent is "OFF".

is extensively used as nanoparticles sensors is metal nanoparticles. The forms of metal that are usually used are metal oxide (MO) and noble metals. For chemical and biological sensing application, various MOs such as zinc oxide (ZnO), iron oxide (Fe_2O_3), titanium oxide (TiO_2), magnesium oxide (MgO), cerium oxide (CeO_2), tin oxide (SnO_2), zirconium oxide (ZrO_2), etc., have been comprehensively investigated (Liu and Liu 2019, Patel et al. 2015, Jana et al. 2019, Hasanzadeh et al. 2015, Shetti et al. 2019, Fallatah et al. 2019). The metal oxide presents several advantages, such as chemical stability, versatile morphology, and the ability to associate with other material (forming composites) (Serban and Enesca 2020). In the case of noble metal, gold (Au), silver (Ag), palladium (Pd), ruthenium (Ru), and platinum (Pt) are largely utilized owing to their low toxicity, stability, and excellent biocompatibility compared to other metal (Wang 2012, Doria et al. 2012). Even though the qualities are superior to others, noble metals are not applied for mass production due to the expensive price during the materials' procurement.

The purpose of metal nanoparticles during the preparation of sensing nanoparticles is for immobilization, catalysis during the reaction, and electron transfer enhancer (Turkmen et al. 2014, Liu et al. 2013, Ronkainen et al. 2010). Especially since most noble metals are electroconductive and MOs are semi-conducting, the electroconductivity would be favored compared to without metal. The detection using metal nanoparticles usually employs the coordination bonds or the complex formation between the metal and the targets. The target would be attached to the metal, creating a bond between the ligand (target) and the metal, in which the process of the data acquisition can be completed. For example, it is stated that gold nanoparticles are able to immobilize protein *via* covalent coordination that forms between the gold atoms and the amine groups of the protein (Liu and Peng 2017, Ferreira et al. 2019, Otten et al. 2013). It has also been reported that TiO_2 and ZrO_2 nanoparticles can immobilize enzymes through the formation of a complex between the metal atom and the enzyme. In short, the electrochemical behavior could be observed after the immobilization by investigating the electron transfer rate or the current density that is produced during the reduction-oxidation reaction. Electrostatic interaction is another potential interaction that occurs during the conjugation of metal and biomolecules (analyte) as shown in Figure 5. It can be

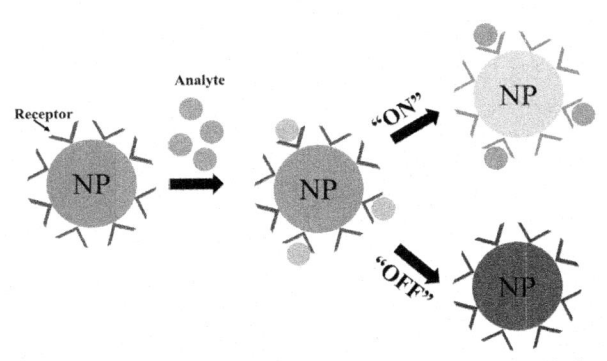

Figure 5. The illustration of nanoparticles detection. The receptor and analyte binding followed by the change in its properties. The nanoparticles' fluorescence will either be "ON" or "OFF" depending on the system.

described as the interaction between positively charged metal nanoparticles (NPs) and negatively charged biomolecules. An example of this mechanism is the detection of specific sequences in genomic DNA by gold NPs, in which DNA represents the negative charges, while gold NP (noble metal) is the positive one (Doria et.al 2012, Li and Rothberg 2004).

Coated surfaces

At the previous sub-point, we have learned about the nanoparticles system for detection or sensing. Now, we will attempt to describe how to improve the detection by applying it to the substrate using coating methods (Figure 6). To cover material on the surface of a substrate, a lot of techniques can be implemented such as vapor deposition (chemical or physical), chemical/electrochemical technique, spraying, roll-to-roll coating (R2R), and physical coating (dip coating and spin coating) (Choy 2003, Fotovvati et al. 2019, Cagnani et al. 2020).

The chemical vapor deposition (CVD) is a method in vacuum condition, in which the substrate is exposed to volatile precursors, then react and decompose on the surface of the substrate producing the desired product. Most of the processes usually create a by-product, which would be eliminated by gas flow through the reaction chamber (Choy 2003). On the other hand, physical vapor deposition (PVD) is portrayed as

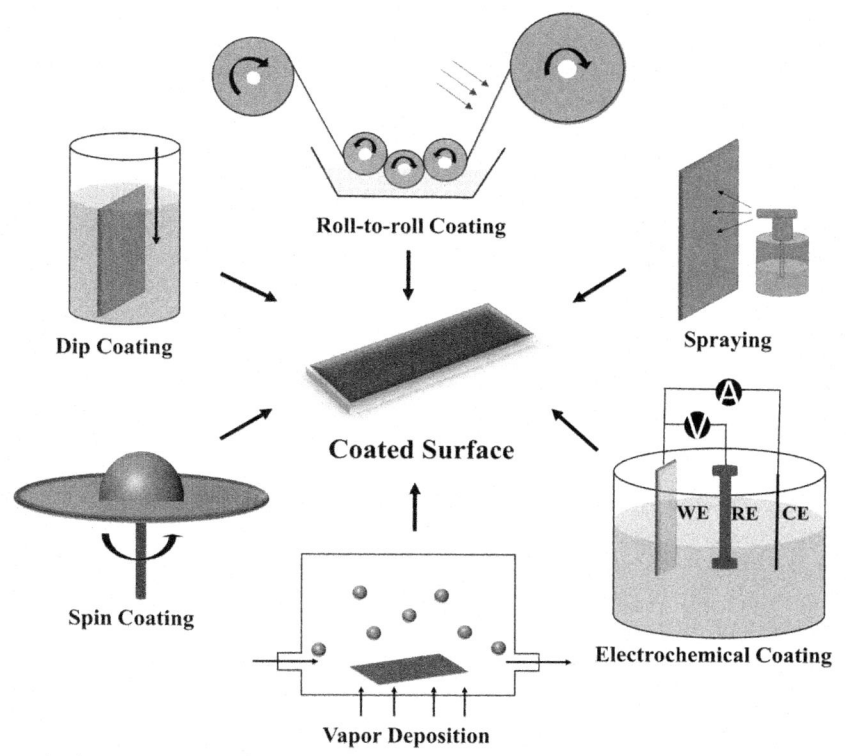

Figure 6. The variation of coating methods (vapor deposition, electrochemical coating, spraying, roll-to-roll coating, dip coating, and spin coating).

a method of coating, in which the materials go from a condensed phase to a vapor phase, then form a thin film condensed phase. Most PVD processes involve sputtering and evaporation (Baptista et al. 2018). Next, a chemical/ electrochemical coating is a process that converts materials to a thin film by using an electrochemical/chemical reaction. The deposition of materials on the surface of substrate can be controlled by varying the condition such as time, voltage, concentration, and pH (Zhang et al. 2018d). For spraying, the method of coating is conducted by spreading the materials onto the surface *via* compressed air, plasma, or thermal spraying (Luo et al. 2018). The principle of spraying is to atomize the particles and homogenously cover the substrate with the particles by using different kinds of forces. Another method is roll-to-roll processing (R2R), in which the process of coating is performed by using pre-made patterned roll from flexible plastic or metal foil. Sometimes, the techniques can be defined as printing (Cagnani et al. 2020). The last method for coating is physical coating such as dip coating and spin coating. The dip-coating approach involves the deposition of a wet liquid film by submerging the substrate into a solution containing hydrolyzable metal compounds (or readily formed particles). The substrate is then removed from the solution and the remaining solvents is eliminated by evaporation, resulting in a homogenous thin film of the substrate.

One of the examples of dip-coating was demonstrated by Sun et al. by preparing a paper sensor for improving detection of complex media. A zwitterionic poly(carboxylbetaine) (pCB) polymer containing four L-3,4-dihydroxyphenylalanine functional groups (DOPA) was grafted to the glucose paper during the dipping process by taking advantages of DOPA adhesive properties. Then, the non-fouling coated paper sensor was further modified to enhance the detection of the bovine serum albumin antibody (anti-BSA) and fibrinogen antibody (anti-Fg) by using 1-ethyl-3-(3-(dimethylamino)propyl)-carbodiimide and N-hydroxysuccinimide (EDC/NHS) coupling reaction (Sun et al. 2017). Moreover, spin coating is a general technique for covering thin films by implementing the centripetal force and the surface tension to a solution of a material and a solvent *via* rotating at high speed to form an even coating. Following the process, the solvent has to be evaporated to create a thin film scaling from a few nanometers to a few microns in thickness. In 2018, Casteleijn et al. prepared a sensor that would be applied for detecting chitin-binding domain (CBD) and protein of *Bacillus* circulans. Chitin in DMA/LiCl solution was used to cover polystyrene-protected gold surface using spin-coating methods achieving a homogenous coated sensor (Casteleijn et al. 2018). Another example of a spin-coating application for biosensor design is the fabrication of ZnO thin film for nitric oxide (NO_2) gas detection. In this process, the zinc acetate dihydrate [$Zn(CH_3COO)_2 2H_2O$] was used to cover a glass substrate using spin-coating methods in open-air followed by drying in at 200°C and annealing at 400°C (Patil et al. 2018).

Recently, the implementation of surface coating was shifted from traditional colorimetry or titration into a more complex electrochemical sensing. Precisely, an electrochemical sensor should involve recognition components that are immobilized on transducers (electrode) and selectively attach to target analytes, producing an electronic signal (Lin and Li 2020). The change of signals can be identified during binding between the electrode and the biofoulant, specifying the result before and after binding in the form of current change, resistance change, or capacitive change. Attachment of target molecules to the electrode disturbed the electron during the reduction-oxidation process or electron transfer, thus influencing the reading of the measurement. Karimi et al. designed a spin-coated ZnO-graphene thin film that can recognize urea *via* immobilization in the form of an impedimetric biosensor. The sensor was fabricated by the spin-coating method and the measurement was conducted using a potentiostat in a three-electrode system with the film as the working electrode, SCE as the reference electrode, and Pt as the counter electrode. The prepared biosensor demonstrated a low detection limit at 3.36 mg/dL and sensitivity up to 0.16 kΩ per mg/dL (Karimi et al. 2018).

Hydrogel

Apart from nanoparticles and coated surface, hydrogel becomes one of the most intriguing platforms for biosensing, owing to its advantageous properties such as exceptional biocompatibility, high water content, and 3-D form stability (Jung et al. 2017, Khajouei et al. 2020).

Hydrogels are water-swellable, three-dimensional (3D) networks of hydrophilic polymers by a physical and/or chemical cross-linking that are comprised of a large amount of water (Hoffman 2012, Ahmed 2015, Mathur et al. 1996). In common, these hydrogels are biocompatible with negligible toxicity and low immune responses that can be utilized for various biomedical applications including tissue adhesives, wound healing materials, and wrinkle fillers. It is also possible to apply it for drug delivery depots, tissue engineering scaffolds, and particularly for biosensors (Hoffman 2012, Ahmed 2015, Mathur et al. 1996, Ghobril and Grinstaff 2015, Li and Mooney 2016, Boateng et al. 2008, Vlierberghe et al. 2011, Kopeček and Yang 2007, Ullah et al. 2015, Richter et al. 2008). For biosensing applications, the detection of the target molecules (i.e., small molecules to macromolecules), diagnosis of diseases such as cancers and diabetes, and monitoring the health conditions are achieved by the observation of changes in hydrogels' behaviors including physicochemical, rheological, mechanical, and electrochemical properties (Tavakoli and Tang 2017, Jung et al. 2017, Kopeček and Yang 2007, Ullah et al. 2015, Richter et al. 2008, Buenger et al. 2012, Russell et al. 1999). Besides, the polymer design and/or chemistry in the hydrogels provide the basis for the detection with high selectivity and sensitivity (Tavakoli and Tang 2017, Jung et al. 2017, Ullah et al. 2015, Richter et al. 2008, Buenger et al. 2012, Russell et al. 1999, Ulijn et al. 2007, Li et al. 2015, Lee et al. 2018, Culver et al. 2017). To develop the hydrogel biosensors, the environmental sensitive polymer, the introduction of stimuli-sensitive functional groups, and conjugation of molecules (i.e., bio-receptor) that specifically react with target molecules are frequently used (Tavakoli and Tang 2017, Jung et al. 2017, Ullah et al. 2015, Richter et al. 2008, Buenger et al. 2012, Russell et al. 1999, Ulijn et al. 2007, Lee et al. 2018, Li et al. 2015, Peppas and Blarcom 2016). Thus, we address the general principle of hydrogel biosensors with preparation and fabrication methods in this section. In addition, we focus on the changes in the hydrogel behaviors when the hydrogels were exposed to the target materials.

During the past few decades, stimuli-sensitive hydrogels have been extensively developed for versatile biomedical applications (Hendrickson and Lyon 2009, Saxena et al. 2012, Peppas and Leobandung 2004, Miyata et al. 2002, He et al. 2008, Qiu and Park 2001, Linden et al. 2004). These stimuli-sensitive hydrogels (Figure 7) exhibit detectable changes in their properties by physical stimuli (temperature, pressure, light, electrical field, magnetic field, ultrasound irradiation), chemical stimuli (pH, ionic strength, and glucose), and biological stimuli (enzyme, antigen, and nucleic acids) (Hendrickson and Lyon 2009, Saxena et al. 2012, Peppas and Leobandung 2004, Miyata et al. 2002, He et al. 2008, Qiu and Park 2001, Linden et al. 2004, Rizwan et al. 2017). Swelling rate, mechanical flexibility, morphology, porosity, and hydrogel opacity are influenced during the stimulation (Chen 2019). For example, Pluronic copolymers composed of poly(ethylene oxide)-poly(propylene oxide)-poly(ethylene oxide) show a thermo-sensitive behavior by forming closely packed individual micelles above low critical solution temperature (LCST), as shown in Figure 8 (Huang et al. 2019, Gioffredi et al. 2016, Yang et al. 2011, Chung et al. 2008). The elastic modulus (G') values of Pluronic above the LCST are far higher than that below the LCST.

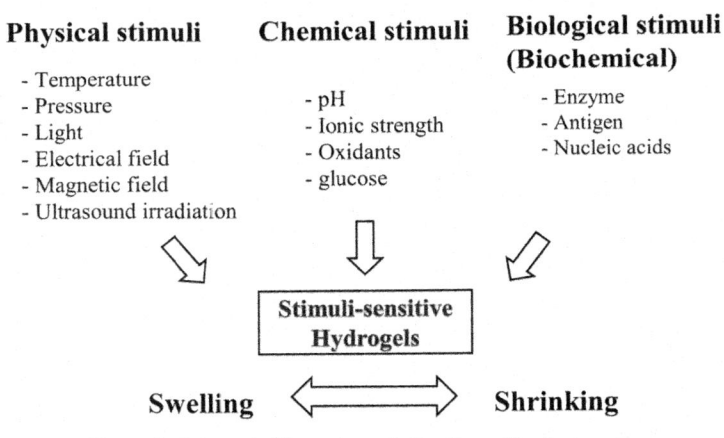

Figure 7. Schematic illustrations of stimuli-sensitive hydrogels.

Pluronic® Temperature-sensitive Behavior

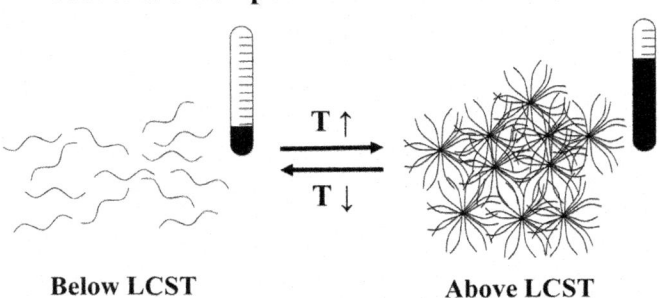

Figure 8. Schematic illustration of temperature-sensitive Pluronic hydrogels.

In addition, the elastic modulus (G') values are high compared to the viscous modulus (G") values above the LCST indicating the formation of hydrogels. It is noteworthy that the temperature can be detected by measuring changes in the rheological properties of hydrogels. Additionally, pH can be observed by monitoring the changes in physical/chemical/mechanical properties and behaviors of hydrogels. By evaluating these phenomena, a hydrogel-based biosensor can be prepared. For instance, chitosan, a representative of amine-rich positively changed polymers, shows different swelling behaviors in different pH solutions because the amine groups are protonated at low pH (Qu et al. 2000, Zhang et al. 2018b, Wang et al. 2004, El-Mahrouk et al. 2016, Qu et al. 1999). To prepare the chitosan-based pH-sensitive hydrogels, the modification or crosslinking with glutaraldehyde, citrate, or tripolyphosphate of chitosan are generally employed. Moreover, the physically crosslinked lactic acid- and/or glycolic acid-grafted chitosan hydrogels exhibit the reversible swelling behaviors between pH 2.2 and pH 7.4 buffer solution. With the increase of pH, the deprotonation of the ionized amino groups in the chitosan-based hydrogels leads to the de-swelling behaviors (Qu et al. 2000). In case of the polymers with negative charges (i.e., carboxylic acid groups), the electrostatic repulsive forces are increased as the pH increased resulting in the swelling of the hydrogel networks (Wang et al. 2015, Kim and Oh 2005, Seeli et al. 2016). Additionally, pH-sensitive colon-targeted hydrocortisone sodium succinate (HSS) delivery system can be achieved by using negatively charged alginate hydrogels crosslinked by calcium ions (Ca^{2+}) (Seeli et al. 2016). The swelling ratio of hydrogels is around 14% at pH 7.4 that is far higher than that at pH 1.2 (~ 1%). Considering the pH of organs, the successful drug delivery to colon tissues can be achieved by pH-sensitive properties. Furthermore, hydrogels can recognize the specific enzymes through the enzymatic reactions (Ebrahimi and Schönherr 2014, Bhattacharya et al. 2017b). Thus, the design and preparation strategy for stimuli-sensitive hydrogels provide the basis for the development of biosensors that can detect the physical, chemical, and biological cues. One of the well-known hydrogel biosensors is a glucose sensor.

As previously reported, monitoring blood glucose levels in diabetic patients are used to manage the disease and to prevent the complication (Jouven et al. 2005, Shaw et al. 1999, Kadowaki et al. 1984). Although there are many devices including portable electronic devices to monitor blood glucose, the hydrogel-based glucose biosensors are important due to the wide range of biomedical applications (Oliver et al. 2009, Klonoff 2005, Rodbard 2016, Chen et al. 2016, Kim et al. 2018). Phenylboronic acid (PBA), boron-containing organic molecules with two hydroxyl groups, can react with the diol groups, particularly *cis*-diol groups, to form the boronate ester (Cambre and Sumerlin 2011, Marco-Duffort and Tibbitt 2019, Ryu et al. 2019). The PBA-diol interactions are highly reversible by adding the stimuli (i.e., pH and/or diol-containing molecules). It indicates that PBA can bind with the glucose including diol groups that can make the different signal. For instance, PBA-functionalized polymer hydrogels lead to volumetric changes when glucose molecules are added into the hydrogel networks by dynamic linkages of boronic acid-diol (Zhang et al. 2013, Elshaarani et al. 2018, Elsherif et al. 2018, Lee et al. 2004). The PBA immobilized in the hydrogels reacts with the diols in glucose forming either 1:1 or 2:1 complex according to the glucose concentrations. When the low concentration of glucose is added into the hydrogels, the 1:1 complex is dominantly formed in the hydrogels resulting in the shrinkage of hydrogels (Figure 9). In contrast, swelling of hydrogel happened when the glucose concentration is increased (the 2:1 complex converts into the two 1:1 complex). The detection of glucose concentration ranges using various PBA derivatives is 0 to 30 mM that is similar to that of clinical interest (2.2–38.9 mM) (Zhang et al. 2013, Updike et al. 2000). In addition, the PBA-containing hydrogels have been established to detect the glucose and deliver the insulin concurrently (Elshaarani et al. 2018). Thus, it is highly desirable to develop the hydrogel-based biosensing materials for multi-purpose biomedical applications.

To fabricate the highly selective and sensitive hydrogel biosensors, the introduction of bioreceptors such as antibody, enzyme, and nucleic acid into the polymeric backbones and/or immobilization of bioreceptors in the hydrogels are performed (Bansal et al. 2016, Mazrad et al. 2018, Updike et al. 2000). Lim et al. reported the antigen- and antibody-immobilized polyacrylic acid hydrogels for the detection of hepatitis B antigens (Lim et al. 2017). The hepatitis B antigens comparatively bind with the antibody-immobilized hydrogels resulting in the swelling of hydrogels. The detection and quantification limits with quartz crystal microgravimetry (QCM) are 0.6 µg/mL and 1.98 µg/mL, respectively. Also, the specific interactions between antibody and antigen allow up to 4 mg/mL of detection of hepatitis B antigens. In addition, the hydrogel biosensors are reusable due to the repeatable and reversible swelling-shrinking behaviors of hydrogels. Yan et al. reported a glucoamylase-entrapped hydrogel crosslinked by aptamers. After target molecules are exposed, the target-aptamer complexes are formed resulting in the disruption of hydrogels. Then, the released glucoamylase catalyzes the hydrolysis reactions of amylose to produce the glucose that can be read by commercially available glucometer. A limit of detection (LOD) is 3.8 µM (1.2 µg/mL) for cocaine and allow 1.6 µM of LOD by prolonging the reaction time (~ 2 hrs). More importantly, these tools can be applied for the various non-glucose biomolecules by inclusion of the aptamer sequences. These enzymatic reactions exhibit different catalytic activities, which depend on the biological conditions and microenvironments indicating that the enzyme-based biosensors can be used to detector molecules in physiological conditions. The nucleic acid is also used to prepare the highly sensitive hydrogel biosensors by hydrogen bond-based pair structures (Baeissa et al. 2010, Helwa et al. 2012). Therefore, bioreceptor-containing polymeric hydrogel biosensors provide selective, sensitive, and accurate information on the detection of biomolecules by highly specific interactions. In some cases, the hydrogel-based sensor can detect biological interaction even without the present bioreceptors. The change of swelling

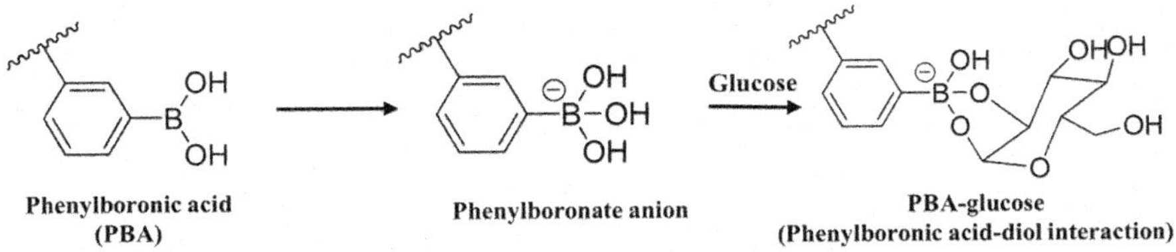

Figure 9. Chemical structures of formation of phenylboronic acid-diol interactions (1:1).

properties, for example, of the hydrogel in response to biological phenomena that change pH is enough for detecting a condition even without the use of bioreceptor. Other than swelling rate, the redox reaction could also be evaluated. For example, with the presence of redox-sensitive functional group the potential difference, resistivity, and capacitance can be obtained (Tavakoli and Tang 2017).

Aforementioned in Section 11.2.1 and 11.2.2, nanoparticles and thin films are used to prepare the hydrogels' biosensors. For instance, metal oxide, noble metal, and nanocarbon (i.e., graphene, carbon nanotube, carbon dots, and polymer dots) materials can be incorporated into the hydrogel networks. Park group reported that carbon dot-incorporated ureidopyrimidinone (UPy)-conjugated gelatin hydrogels show the excellent detection of breast cancers *in vitro* and *in vivo* (Won et al. 2020). UPy group is known to be a quadruple hydrogen bonding unit that could be applied in the preparation of self-healing hydrogels using various polymers (i.e., polyethylene glycol, polyacrylamine, polyurethane, Pluronic, dextran, etc.) (Sijbesma et al. 1997, Beijer et al. 1998, Dankers et al. 2012, Kieltyka et al. 2013, Shi et al. 2013, Teunissen et al. 2014, Hou et al. 2015, Lee-Wang et al. 2010). The conjugation of UPy groups into the gelatin contributes to a stable gelation in the physiological conditions as well as the introduction of self-healing properties to the gelatin (Won et al. 2020). However, no self-healing properties were observed in the Gel-UPy hydrogels when the diselenide bonds-containing carbon dots (dsCD) were incorporated into the Gel-UPy hydrogels. After treatments of glutathione (GSH) or reactive oxygen species (i.e., H_2O_2), the dsCD/Gel-UPy hydrogels have self-healing properties because diselenide bonds are cleaved by GSH or H_2O_2. The self-healing properties of hydrogels contribute to the detection of the GSH and H_2O_2 with high sensitivity. Considering high concentrations of GSH and reactive oxygen species in the cancer cell microenvironments compared to normal cell microenvironments, it is highly beneficial to detect the cancers. Furthermore, the dsCD/Gel-UPy hydrogels exhibit self-healing properties when the hydrogels are implanted onto the tumor sites. Thus, the incorporation of functional nanoparticles into the hydrogels and hydrogel film biosensors are usefully exploited for the detection of various diseases.

Application of the Biomaterials for Sensing

As shown in Section 'Biomaterials and Biosensing Technology', various biomaterials including nanoparticles, coated surface in the form of thin films, and hydrogels are used to prepare the biosensors that can detect the specific biomolecules, cells, and many others. To this end, versatile chemistries to prepare the highly sensitive and selective biosensors by designing the interactions between biosensors and target molecules, immobilizing the receptors on the biosensors, and emphasizing the detection signals are used. This section aims to discuss the design strategy, methodologies (i.e., preparation, fabrication, chemical modification, and detection), and effectiveness for the detection of cancers, bacteria, and enzymes as well as human motions. In addition, perspectives of biosensors for biomedical applications will also be discussed.

Cancer detection

As cancer has become one of the most lethal diseases in the world, the medication and detection for the disease should be precise and accurate. Cancer is also known for its rapid spreading owing to the abnormal cells' cycle behavior, which continuously proliferates cells faster than healthy cells do, making it grow faster than normal/ healthy cells (Collins et al. 1997). For that reason, rapid and real-time diagnosis of cancer/tumor should be developed and one of the methods is using a sensor. Mostly, the detection of cancer depends on tumor tissue or cancer cell marker. Interestingly, many features differentiate normal cells and cancer cells, such as the extracellular pH of the tissue, Glutathione (GSH) level of the cells microenvirontment, and also reactive oxygen species (ROS) which is necessary for cellular apoptotic, protein, or enzyme overabundance in cancer cells (Figure 10).

To make it simple, the extracellular pH of normal tissue is defined at 7.2 ~ 7.4 or usually said as the physiological pH of the organism. However, cancer disease enhances the acid level of the cells, making the pH of the tissue at 6.0 ~ 6.8. In 2019, Phuong et al. manufactured a simultaneous optical and electrochemical sensor-based pH-sensitive polymer-coated surface that can identify the interaction between cells and substrate. According to this study, a zwitterion nanoparticle comprises a quaternary ammonium cation ($-N^+$) and sulfite anion ($-SO_{3^-}$) for a pH-responsive group, and functionalization with a catechol group for surface attachment was prepared initially. To achieve fluorescent and electroconductivity properties, carbon dots were prepared from the functionalized polymer by carbonizing it with oxidative reaction using H_2SO_4 for 1 min. The obtained carbon dots are then coated to various substrates such as silicon wafer (Si wafer), polyethylene terephthalate (PET), polyvinyl chloride (PVC), and polypropylene (PP). The principle of the detection is to induce aggregation by changing the hydrophobicity of the nanoparticles caused by the molecular interaction between the zwitterions. During the acidic and alkaline conditions, the electrostatic interaction would be broken changing the conformation of the surface and alter the affinity from hydrophobic state to a hydrophilic state. The cells, which are practically hydrophobic, would attach to the hydrophobic surface due to the effect of hydrophobic interaction and Van der Waals forces resulting in the change of fluorescence intensity and electronic response in the process (Phuong et al. 2019). Other things that distinguish the cancer cells and healthy cells are the concentration of the GSH and ROS in the cytoplasm of the cell. In cancer conditions, the GSH concentration is four times as high as normal condition, while the ROS such as H_2O_2 is 10-time higher, becoming a major factor during detection (Li et al.

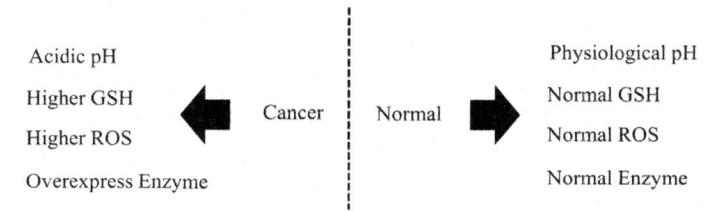

Figure 10. The difference between cancer condition and normal condition.

2020). The GSH, which exists in an organism tissue whether it is a healthy or tumor, can be oxidized to form glutathione disulfide (GSSH) under natural circumstances and it could be converted back to GSH in the presence of NADPH-dependent glutathione reductase preserving homeostasis, a necessary process for cells' growth. Nevertheless, a large quantity of GSH is present in cancer (approximately 2 ~ 20 mM) due to tumor growth and genetic alteration. The reactive oxygen species, on the other hand, consist in the form of superoxide anion (O^{2-}), hydrogen peroxide (H_2O_2), and hydroxyl radicals ($\cdot OH$) in cells. The ROS can be generated endogenously from the NADPH enzyme or mitochondrial metabolism as well as exogenously via photodynamic or non-photodynamic actions. Intracellular superoxide (O^{2-}) is largely produced from NADPH oxidation with the help of NADPH oxidase enzymes (NOXs) or by an electron from the aerobic metabolism of mitochondria. Moreover, the process is extended to the formation of H_2O_2 by O^{2-} and superoxide dismutase (SOD). Low levels of ROS play crucial roles in cellular life cycles, serving as cellular signaling molecules for a protein containing thiol groups, while the irregularity of ROS (100 μM in cancer cells) could cause oxidative stress, resulting in oxidative damage to the cellular constituents (e.g., DNA, lipids, and proteins), apoptosis or necrosis, and probably the promotion of cancer-causing mutations (Tao and He 2018). Therefore, these two factors, GSH and ROS, could open a possibility for a precise detection based on the tumor microenvironment composition. During the detection, the selectivity of the sensor is dependent on the cleavage of the redox-responsive linker such as disulfide (-S-S-), diselenide (-Se-Se-), and sulfide-selenide (-S-Se-) by the GSH of the cells. Normal GSH and ROS concentration would be less impactful to the sensor, while tumor condition influences the sensor even higher (Jia et al. 2019). One recent study introduced a wireless sensor utilizing self-healing hydrogel as the mainframe and redox-responsive functional group for cancer detection. The hydrogel system consists of electroconductive nanoparticles prepared from dopamine functionalized polymer, which would be loaded into self-healing capable matrices. In the report, the hydrogel demonstrated different self-healing behavior according to the environment. Without any treatment, the nanoparticles-loaded hydrogel exhibited no self-healing properties even though the matrix itself has self-healing ability. The self-healing ability was disturbed by the loaded particles preventing hydrogen bonding of the matrices. However, the self-healing properties recovered owing to the new form of hydrogen bonding as the result of H_2O_2 and GSH treatment, enhancing the intermolecular interaction of the hydrogel (Won et al. 2020). Another factor that is usually studied to distinguish between healthy and tumor conditions is the overexpressed biomarker in cancer cells. Xu et al. developed a sensor, which can perform sensitive and selective tumors by investigating the interaction between 2-p-aminophenyl-1, 3, 2-dithiarsenolane (VTA2) based cytosensor and vicinal-dithiol-containing proteins (VDPs), a biomarker for aggressive tumor condition (Xu et al. 2015). The cluster of differentiation 44 (CD44) was one of other protein recognitions that was widely used along with hyaluronic acid (HA) as the ligand. Zhang et al. designed multiwalled carbon nanotubes (MWCNTs) on the indium tin oxide (ITO) electrode to enhance electrochemical sensing and used HA as the molecular probe for ligand-protein interaction with CD44 of the cancer cell (Zhang et al. 2019). With the discovery of these factors, cancer detection showed a promising future and with the hope for precise and rapid detection of the tumor microenvironment.

Bacteria detection

As the previous sub-chapter explains about the detection of cancer cells, we will explain bacteria detection from here onwards. To investigate the bacteria, first, the chemical or physical properties of bacteria have to be studied. The outer membrane of bacteria consists of a high amount of hydroxy functional group (-OH) and phosphate ($-PO_4^{3-}$) of teichoic acids, which enhance the negative charge of the bacteria surface. By taking account of this property, the binding between the sensor and bacteria can be recognized. In 2020, Robby et al. developed a bacteria sensing platform from functionalized carbon dots (FCD) containing a positively charged functional group prepared from quarternized bromoethane and catechol moieties. Both functional groups are essential for the system to provide an sp^2-rich carbon bond, which is required for optical and electrochemical bacterial detection.

Principally, the negative charged bacteria would be attached to the positive charged coated surface via electrostatic interaction (Figure 11). The binding of bacteria to the coated surface influenced the electrochemical reading by increasing the electronic resistance of the system. With the increase in bacteria count (density), the observed resistance would elevate higher. The proposed system displayed a dual property, which is bacteria detection and bacteria-killing. As the FCD responsible for the sensing, the $CsWO_3$ (substrate) is for the bacteria-killing, owing to excellent photo-thermal conversion. The $CsWO_3$-FCD sensor exhibited a preferable result with a limit of detection (LOD) < 10 CFU/mL and almost 100% killing efficiency for *E. coli* and *S. aureus* (Robby et al. 2021). Self-assembled peptide nanostructures (SPNs) is another approach that has been utilized for bacteria detection. Self-assembly is a reversible method, in which the size, function, and morphology of molecular assemblies are managed by the balance of the attractive driving, repulsive opposition, and directional forces. Jeong et al. demonstrated a sensor that can turn

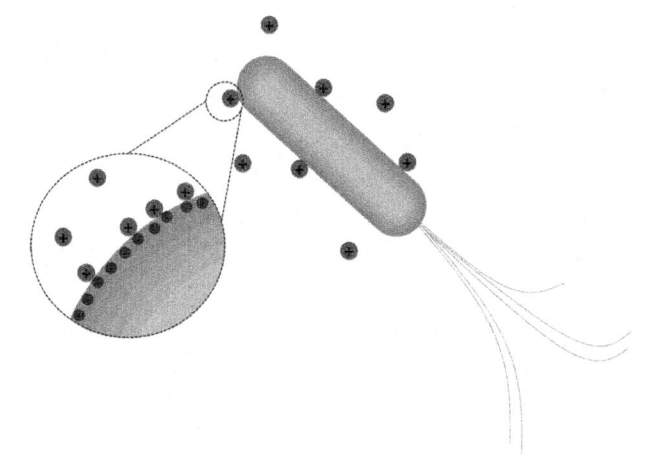

Figure 11. Illustration of the negative charge bacteria interaction with positive charge nanoparticles (bacteria usually has size at the micrometer scale, while nanoparticles should be at nanometer scale).

the fluorescence from "off" to "on" based on the self-assembly of supramolecule and bacteria. During the self-assembly, the coiled-coil chain transforms, becoming a helix-like structure after multivalent binding between the supramolecules and the bacteria. The change of conformation promoted by binding induces different fluorescence behavior to the system, which becomes the principle of the designed sensor (Jeong et al. 2019). Sheikhzadeh et al. invented a sensor utilizing copolymer-aptamer bioconjugation for bacteria sensing using poly[pyrrole-co-3-carboxyl-pyrrole] copolymer, which is commonly used as DNA hybridization biosensors. The polypyrrole-derived nanoparticles were utilized for detection owing to attributes such as ease to synthesize in different solvents at room temperature, excellent chemical stability, and high electrical conductivity. The impedimetric measurement was performed to investigate the intrinsic conjugation of the poly [pyrrole-co-3-carboxyl-pyrrole] copolymer and the aptamer pathogen interaction of *Salmonella Typhimurium*. The prepared sensor showed a linear response in the range of 10^2–10^8 CFU mL^{-1} with a limit of quantification (LOQ) of 10^2 CFU mL^{-1} and a limit of detection (LOD) of 3 CFU mL^{-1} (Sheikhzadeh et al. 2016).

Enzyme detection

In order to maintain homeostasis, the detection and regulation of enzyme activity are very important (Chovatiya and Medzhitov 2014, Rodriguez-Martinez and Ruiz-Torres 1992, Pey 2013, Mu et al. 2008). Also, specific enzymes are overexpressed in various diseases such as cancers and diabetes (Liang and Chan 2007, Park et al. 2018, Kakkar et al. 1995, Kong et al. 2017, Hutter and Maysinger 2013). Thus, it is highly desired to develop enzyme detection for the management of homeostasis and various diseases. Biomaterials (i.e., nanoparticles, thin films, and hydrogels), as mentioned in Section 11.2, are selected to detect the target enzymes. For the detection of enzymes, colorimetric assays, FRET-based assays, and electro-chemical detection methods are used by monitoring the changes of biomaterials due to the catalytic activities of enzymes (Hutter and Maysinger 2013). The enzymes detected in current research include the transferases (i.e., various kinases and methyltransferase), hydrolases (i.e., alkaline phosphatase, matrix metalloproteinase), and Lyases (adenosine deaminase), etc. (Hutter and Maysinger 2013). Alkaline phosphatase (ALP) is one of the most common and important enzymes in clinical chemistry and enzyme immunoassay (Sharma et al. 2014, Moss 1982, Golub and Boesze-Battaglia 2007, Self 1985). In addition, extracellular ALP is expressed by breast cancer cells that have an abnormal level compared with normal cells (Matsumoto et al. 1990, Chang et al. 1994, Sadeghi and Yazdanparast 2003, Choi et al. 2018). These alkaline phosphatases catalyze the reaction of converting ascorbic acid-2-phosphate (AAP) to ascorbic acid (AA), which can be measured by monitoring the concentration of generated AAPs or AAP-mediated reaction products (Nsabimana et al. 2019, Freitas et al. 2019, Balbaied 2019, Jiang et al. 2012). The polymer dot-manganese oxide complexes (PD/MnO$_2$)-based biosensors have been developed for the detection of ALP activity (Sadeghi and Yazdanparast 2003). The formation of AA by ALP promotes the decomposition of MnO$_2$ leading to the recovery of fluorescence of PD and the changes in resistance. The LOD values of the biosensor are 3.98 cells/mL using electrochemical methods that are far higher than that of using fluorescence methods (1,995 cells/mL). Moreover, the monitoring activity of ALP allows the detection of cancers. HAuCl4 is also used to detect the ALP activity in the biosensors. $HAuCl_4$ reacts with ascorbic acid to form a gold nanoparticle that can be observed by scattering peaks at 600 nm that has information of ALP activity (Jiang et al. 2012). Therefore, various enzyme detection based on the design of biomaterials not only analyzes the homeostasis of living organisms but also detects various diseases that will contribute to the survival and improvement in the quality of life of patients in the near future.

Human motion sensing

The comprehensive design of biosensors made it possible to be implemented for 3-D applications such as human motion. Human motion has been categorized as the movement between a part of the body and the space of the displacement. Most of the current biosensor for human motion is divided into stretching, bending, twisting, vibration, or compression (Figure 12). It has widely been utilized in medical applications to investigate body conditions in the form of electromyography (EMG), electroencephalography (EEG), electrocardiography (ECG), electroglottography (EGG), etc.

The principle of the system is converting the mechanical deformation into an identifiable signal such as an electronic response. During the analysis, the electrodes are placed between the designated area, and an electric current is passed through the electrodes. The reading or output result of the measurement would define the condition of the object of observation. For this kind of treatment, hydrogel usually becomes one of the best choices, especially a hydrogel with high biocompatibility. Mostly, polyvinyl alcohol (PVA) has been utilized for such application owing to less-toxic attributes, tissue-mimicking features, and excellent mechanical properties (Jiang et al. 2011). Modification, loading of material, or functionalization has been applied to enhance the properties of PVA into a versatile mechanical biosensor.

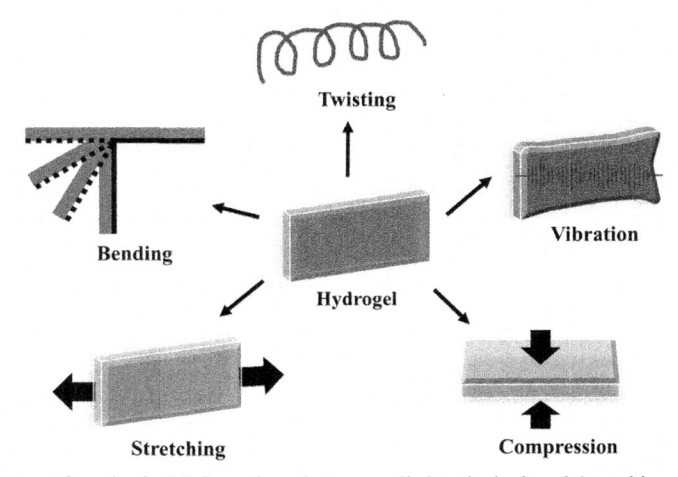

Figure 12. Type of mechanical deformations that are applied to the hydrogel (stretching, compression, vibration, twisting, and bending).

As the biosensor especially for electronic sensor application, the addition of electroconductive materials such as metallic nanoparticles, carbon-based materials, and conductive polymer is essential during the preparation (Ryplida et al. 2020). One of the promising electroconductive materials that has been widely used for biosensor fabrication is PEDOT:PSS polymer due to its high electroconductivity, non-toxic environment, and handling in aqueous solution. However, PEDOT:PSS suffers from its high toxicity to microorganism backstabbing its advantages for biosensor application. To overcome the challenge, PVA comes in handy by not only solving PEDOT: PSS's low biocompatibility, but also the low mechanical strength of common electroconductive hydrogel. Another example is MXenes, which has been extensively introduced as one of the electroconductive materials that are implemented for electrochemical energy storage devices. It is also confirmed to be a potential material for electromechanics sensors owing to sensitivity toward a conformational and deformational shift in the interlayer stack of MXene in response to external mechanical stimuli like force or pressure, making it a suitable material for biosensing (Zhang et al. 2018c). MXene itself is categorized as a 2D-early transition metal carbides/carbonitrides that has great conductivity and abundance surface functional groups (-O-H, -F, -O, etc.) that provide hydrophilicity to the material. A modified hydrogel fabricated from MXene displayed the capability to detect a series of deformation in the form of the electronic signal, even in the slightest movement such as heartbeat, the pulse of the carotid artery, and air gap of larynx or Adam's apple. Both preparation of PEDOT: PSS and MXene into a motion sensor hydrogel is similar in a sort of way.

In general, two approaches are used to develop a flexible, elastic, and stretchable sensor. First is by modification, mixing, blending, or compounding the electroconductive materials (PEDOT: PSS, MXene, metal nanoparticles, etc.) into polymer blends such as PVA. The second approach is by depositing the particles on the surface of an elastomer. Compare to first approach, the second approach exhibits some disadvantages such as the chance of failure owing to poor interfacial interaction between layer and large mechanical mismatch. Therefore, the blending or mixing method is commonly used in experimental research and mass production (Yang et al. 2020). By loading the nanoparticles inside the matrices, not the surface, the capability of both the building polymer and the conductive materials can be fully achieved. For example, the addition of PEDOT: PSS up to 66.7% weight to PVA showed to not only increase the elongation at break but also the tensile strength other than the high conductivity. It can be suggested that PEDOT: PSS/PVA system can be implemented as a motion-sensing (Fan et al. 2019, Peng et al. 2020). MXene-based hydrogel sensor, on the other hand, displays a clear comparison as a mechanical sensor by showing distinct electronic responses during deformation such as tensile and compressive. The different phenomena are influenced by the change of the 3D network structure of the MXene particles embedded in the matrix. Under tensile deformation, the resistance of the system elevates due to an increase in the hydrogel dimension, which causes the spacing between the MXene layer to be separated, reducing the chance of contact during electron transfer. In contrast, compression tends to decrease the resistance as the geometry becomes shorter and the distance between the stack becomes closer, enhancing the chance of contact (Zhang et al. 2018c).

With the rise of the performance of hydrogel-based motion sensors, a sensing application employing wireless connection was introduced in this past decade. By connecting the hydrogel to an electronic circuit consisting of Bluetooth apparatus, real-time human motion can be detected. In instances, the analog signal during the motion (input) is relayed to the processor before transmitted to the smartphone *via* Bluetooth connection. The analog signal that is recorded can be in the form of stretching, bending, or compression (Ryplida et al. 2019). As the acquisition is done in real-time, the change in the morphology and dimension of the hydrogel by the effect of stimuli can be observed directly.

Concluding Remarks

As we have described in this chapter, biosensors can be fabricated either by using nanoparticles, coated surface, and hydrogel. However, all the procedures showed a similar purpose and aim, despite the variety of available methods. For example, nanoparticles, coated surfaces, and hydrogels can be utilized to study the cancer environment or specifically, for cancer cells. Nonetheless, nanoparticles would be the best choice to investigate a tumor tissue, considering the detection can be done by injection or dropping the nanoparticles' solution on the cancer site without any complicated set-up. Despite that, it is also conceivable to use coated surface and hydrogel to observe the condition, though the effectiveness of both methods is still not fully studied and explored. It should be noticed that the nanoparticles are classified as particles that have a size in the range of 1–100 nm. Nanoparticles can be utilized in their form, applied on the surface of a substrate, or loading into a hydrogel matrix, bringing a higher opportunity for the application. Additionally, to implement material such as particles on the surface of substrates, numerous methods of preparation are viable such as vapor deposition, roll-to-roll (R2R) coating, electrochemical, spraying, dip coating, and spin coating. Furthermore, loading nanoparticles into hydrogel has also become one of the methods for sensor preparation.

To summarize, the preparation of biosensors usually consists of bioreceptor material, the form of signal (recognition), and the transducer, in which every part of the sensor has its purpose. Before developing the sensor, another thing to be concerned about is the target analyte. Since the target analyte is the aim of the sensing, identifying the property and the chemical structure of the target would be vital. Without knowing the target, the development of biosensors would be vague and would not be specific. Bioreceptor, the part that is responsible for binding with the target, is usually in the form of biologicals elements or bioselective materials. However, some sensors do not need to have a bioreceptor, especially the one that relies on the reduction-oxidation that happened during the recognition. Nevertheless, the presence of a bioreceptor in the system would enhance the effectivity and the efficiency of the biosensor owing to the targeting behavior and a higher amount of interaction between the bioreceptor and analyte. To fully implement the bioreceptor, it is common to combine it with a base material such as substrate or hydrogel. Substrates that are used for this purpose are glass, metal, paper, and polymer, in which polymer can be processed to fabricate hydrogel, elastomer, plastic, etc. Besides, the design of the sensor based on the mentioned procedure in this chapter is meant to be one of the simple methods of preparation and would open up opportunities for future implementation and a new inventions, especially for bio applications.

References

Adsetts, J. R., Hoesterey, S., Gao, C., Love, D. A., and Ding, Z. 2020. Electrochemiluminescence and Photoluminescence of carbon quantum dots controlled by aggregation-induced emission, aggregation-caused quenching, and interfacial reactions. Langmuir. 36: 14432–14442.

Ahmed, E. M. 2015. Hydrogel: Preparation, characterization, and applications: A review. J. Adv. Res. 6: 105–121.

Baeissa, A., Dave, N., Smith, B. D., and Liu, J. 2010. DNA-Functionalized Monolithic Hydrogels and Gold Nanoparticles for Colorimetric DNA Detection. ACS Appl. Mater. Interfaces. 2: 3594–3600.

Balbaied, Moore. 2019. Overview of Optical and Electrochemical Alkaline Phosphatase (ALP) Biosensors: Recent Approaches in Cells Culture Techniques. Biosensors. 9: 102.

Bansal, A. K., Antolini, F., Zhang, S., Stroea, L., Ortolani, L., Lanzi, M. et al. 2016. Highly luminescent colloidal CdS quantum dots with efficient near-infrared electroluminescence in light-emitting diodes. J. Phys. Chem. C. 120: 1871–1880.

Baptista, A., Silva, F. J. G., Porteiro, J., Míguez, J. L., Pinto, G., and Fernandes, L. 2018. On the Physical Vapour Deposition (PVD): Evolution of magnetron sputtering processes for industrial applications. Procedia Manuf. 17: 746–757.

Bazin, I., Tria, S. A., Hayat, A., and Marty, J. -L. 2017. New biorecognition molecules in biosensors for the detection of toxins. Biosens. Bioelectron. 87: 285–298.

Beijer, F. H., Sijbesma, R. P., Kooijman, H., Spek, A. L. and Meijer, E. W. 1998. Strong dimerization of ureidopyrimidones *via* quadruple hydrogen bonding. J. Am. Chem. Soc. 120: 6761–6769.

Bhattacharya, D. S., Svechkarev, D., Souchek, J. J., Hill, T. K., Taylor, M. A., Natarajan, A. et al. 2017. Impact of structurally modifying hyaluronic acid on CD44 interaction. J. Mater. Chem. B 5: 8183–8192.

Bhattacharya, S., Nandi, S., and Jelinek, R. 2017. Carbon-dot–hydrogel for enzyme-mediated bacterial detection. RSC Adv. 7: 588–594.

Bimberg, D., and Pohl, U. W. 2011. Quantum dots: promises and accomplishments. Mater. Today. 14: 388–397.

Boateng, J. S., Matthews, K. H., Stevens, H. N. E., and Eccleston, G. M. 2008. Wound healing dressings and drug delivery systems: a review. J. Pharm. Sci. 97: 2892–2923.

Bourlinos, A. B., Zbořil, R., Petr, J., Bakandritsos, A., Krysmann, M., and Giannelis, E. P. 2012. Luminescent Surface Quaternized Carbon Dots. Chem. Mater. 24: 6–8.

Buenger, D., Topuz, F. and Groll, J. 2012. Hydrogels in sensing applications. Prog. Polym. Sci. 37: 1678–1719.

Cagnani, G. R., Ibáñez-Redín, G., Tirich, B., Gonçalves, D., Balogh, D. T., and Oliveira, O. N. 2020. Fully-printed electrochemical sensors made with flexible screen-printed electrodes modified by roll-to-roll slot-die coating. Biosens. Bioelectron. 165: 112428.

Cambre, J. N., and Sumerlin, B. S. 2011. Biomedical applications of boronic acid polymers. Polymer (Guildf). 52: 4631–4643.

Casteleijn, M. G., Richardson, D., Parkkila, P., Granqvist, N., Urtti, A., and Viitala, T. 2018. Spin coated chitin films for biosensors and its analysis are dependent on chitin-surface interactions. Colloids Surfaces A Physicochem. Eng. Asp. 539: 261–272.

Chang, T. C., Wang, J. K., Hung, M. W., Chiao, C. H., Tsai, L. C., and Chang, G. G. 1994. Regulation of the expression of alkaline phosphatase in a human breast-cancer cell line. Biochem. J. 303: 199–205.

Chen, C., Xie, Q., Yang, D., Xiao, H., Fu, Y., Tan, Y. et al. 2013. Recent advances in electrochemical glucose biosensors: a review. RSC Adv. 3: 4473.

Chen, D., Wu, W., Yuan, Y., Zhou, Y., Wan, Z., and Huang, P. 2016. Intense multi-state visible absorption and full-color luminescence of nitrogen-doped carbon quantum dots for blue-light-excitable solid-state-lighting. J. Mater. Chem. C. 4: 9027–9035.

Chen, Y. 2019. Hydrogels based on natural polymers, Hydrogels Based on Natural Polymers. Elsevier.

Choi, C. A., Mazrad, Z. A. I., Ryu, J. H., In, I., Lee, K. D., and Park, S. Y. 2018. Membrane and nucleus targeting for highly sensitive cancer cell detection using pyrophosphate and alkaline phosphatase activity-mediated fluorescence switching of functionalized carbon dots. J. Mater. Chem. B. 6: 5992–6001.

Chovatiya, R., and Medzhitov, R. 2014. Stress, inflammation, and defense of homeostasis. Mol. Cell. 54: 281–288.

Choy, K. 2003. Chemical vapour deposition of coatings. Prog. Mater. Sci. 48: 57–170.

Chung, H. J., Lee, Y., and Park, T. G. 2008. Thermo-sensitive and biodegradable hydrogels based on stereocomplexed Pluronic multi-block copolymers for controlled protein delivery. J. Control. Release. 127: 22–30.

Collins, K., Jacks, T., and Pavletich, N. P. 1997. The cell cycle and cancer. Proc. Natl. Acad. Sci. 94: 2776–2778.

Culver, H. R., Clegg, J. R., and Peppas, N. A. 2017. Analyte-responsive hydrogels: intelligent materials for biosensing and drug delivery. Acc. Chem. Res. 50: 170–178.

Dankers, P. Y. W., Hermans, T. M., Baughman, T. W., Kamikawa, Y., Kieltyka, R. E., Bastings, M. M. C. et al. 2012. Hierarchical Formation of supramolecular transient networks in water: A modular injectable delivery system. Adv. Mater. 24: 2703–2709.

de Medeiros, T. V., Manioudakis, J., Noun, F., Macairan, J. -R., Victoria, F., and Naccache, R. 2019. Microwave-assisted synthesis of carbon dots and their applications. J. Mater. Chem. C 7: 7175–7195.

Dong, C., Liu, S., Barange, N., Lee, J., Pardue, T., Yi, X. et al. 2019. Long-wavelength lead sulfide quantum dots sensing up to 2600 nm for short-wavelength infrared photodetectors. ACS Appl. Mater. Interfaces. 11: 44451–44457.

Doria, G., Conde, J., Veigas, B., Giestas, L., Almeida, C., Assunção, M., Rosa, J. et al. 2012. Noble metal nanoparticles for biosensing applications. Sensors. 12: 1657–1687.

El-Mahrouk, G. M., Aboul-Einien, M. H., and Makhlouf, A. I. 2016. Design, optimization, and evaluation of a novel metronidazole-loaded gastro-retentive pH-Sensitive hydrogel. AAPS PharmSciTech. 17: 1285–1297.

Elshaarani, T., Yu, H., Wang, L., Zain-ul-Abdin, Z. -A., Ullah, R. S., Haroon, M. et al. 2018. Synthesis of hydrogel-bearing phenylboronic acid moieties and their applications in glucose sensing and insulin delivery. J. Mater. Chem. B. 6: 3831–3854.

Elsherif, M., Hassan, M. U., Yetisen, A. K., and Butt, H. 2018. Glucose sensing with phenylboronic acid functionalized hydrogel-based optical diffusers. ACS Nano. 12: 2283–2291.

Fallatah, A., Almomtan, M., and Padalkar, S. 2019. Cerium oxide based glucose biosensors: influence of morphology and underlying substrate on biosensor performance. ACS Sustain. Chem. Eng. 7: 8083–8089.

Fan, X., Nie, W., Tsai, H., Wang, N., Huang, H., Cheng, Y. et al. 2019. PEDOT:PSS for Flexible and Stretchable Electronics: Modifications, Strategies, and Applications. Adv. Sci. 6: 1900813.

Ferreira, R. S., Lira, A. L., Torquato, R. J. S., Schuck, P., and Sousa, A. A. 2019. Mechanistic insights into ultrasmall gold nanoparticle–protein interactions through measurement of binding kinetics. J. Phys. Chem. C 123: 28450–28459.

Fotovvati, B., Namdari, N., and Dehghanghadikolaei, A. 2019. On coating techniques for surface protection: a review. J. Manuf. Mater. Process. 3: 28.

Franke, D., Harris, D. K., Chen, O., Bruns, O. T., Carr, J. A., Wilson, M. W. B. et al. 2016. Continuous injection synthesis of indium arsenide quantum dots emissive in the short-wavelength infrared. Nat. Commun. 7: 12749.

Freitas, M., Nouws, H. P. A., and Delerue-Matos, C. 2019. Electrochemical Sensing Platforms for HER2-ECD Breast Cancer Biomarker Detection. Electroanalysis. 31: 121–128.

Ghayedi Karimi, K., Mozaffari, S. A., and Ebrahimi, M. 2018. Spin-coated ZnO-graphene nanostructure thin film as a promising matrix for urease immobilization of impedimetric urea biosensor. J. Chinese Chem. Soc. 65: 1379–1388.

Ghobril, C., and Grinstaff, M. W. 2015. The chemistry and engineering of polymeric hydrogel adhesives for wound closure: a tutorial. Chem. Soc. Rev. 44: 1820–1835.

Gioffredi, E., Boffito, M., Calzone, S., Giannitelli, S. M., Rainer, A., Trombetta, M. et al. 2016. Pluronic F127 hydrogel characterization and biofabrication in cellularized constructs for tissue engineering applications. Procedia CIRP. 49: 125–132.

Golub, E. E., and Boesze-Battaglia, K. 2007. The role of alkaline phosphatase in mineralization. Curr. Opin. Orthop. 18: 444–448.

Hasanzadeh, M., Shadjou, N., and de la Guardia, M. 2015. Iron and iron-oxide magnetic nanoparticles as signal-amplification elements in electrochemical biosensing. TrAC Trends Anal. Chem. 72: 1–9.

He, C., Kim, S. W., and Lee, D. S. 2008. *In situ* gelling stimuli-sensitive block copolymer hydrogels for drug delivery. J. Control. Release. 127: 189–207.

Helwa, Y., Dave, N., Froidevaux, R., Samadi, A., and Liu, J. 2012. Aptamer-functionalized hydrogel microparticles for fast visual detection of mercury(II) and Adenosine. ACS Appl. Mater. Interfaces. 4: 2228–2233.

Hendrickson, G. R., and Andrew Lyon, L. 2009. Bioresponsive hydrogels for sensing applications. Soft Matter. 5: 29–35.

Higgins, I. J., Lowe, and C. R. 1987. Introduction to the principles and applications of biosensors. Philos. Trans. R. Soc. Lond. B. Biol. Sci. 316: 3–11.

Hoffman, A. S. 2012. Hydrogels for biomedical applications. Adv. Drug Deliv. Rev. 64: 18–23.

Hou, S., Wang, X., Park, S., Jin, X., Ma, and P. X. 2015. Rapid self-integrating, injectable hydrogel for tissue complex regeneration. Adv. Healthc. Mater. 4: 1491–1495.

Huang, H., Qi, X., Chen, Y., and Wu, Z. 2019. Thermo-sensitive hydrogels for delivering biotherapeutic molecules: A review. Saudi Pharm. J. 27: 990–999.

Hutter, E., and Maysinger, D. 2013. Gold-nanoparticle-based biosensors for detection of enzyme activity. Trends Pharmacol. Sci. 34: 497–507.

Jana, J., Chung, J. S., and Hur, S. H. 2019. ZnO-associated carbon dot-based fluorescent assay for sensitive and selective dopamine detection. ACS Omega. 4: 17031–17038.

Jeong, W., Choi, S. -H., Lee, H., and Lim, Y. 2019. A fluorescent supramolecular biosensor for bacterial detection *via* binding-induced changes in coiled-coil molecular assembly. Sensors Actuators B Chem. 290: 93–99.

Jia, X., He, J., Shen, L., Chen, J., Wei, Z., Qin, X. et al. 2019. Gradient redox-responsive and two-stage rocket-mimetic drug delivery system for improved tumor accumulation and safe chemotherapy. Nano Lett. 19: 8690–8700.

Jiang, L., Ding, H., Xu, M., Hu, X., Li, S., Zhang, M. et al. 2020. UV–vis–nir full-range responsive carbon dots with large multiphoton absorption cross sections and deep-red fluorescence at nucleoli and *in vivo*. Small. 16: 2000680.

Jiang, S., Liu, S., and Feng, W. 2011. PVA hydrogel properties for biomedical application. J. Mech. Behav. Biomed. Mater. 4: 1228–1233.

Jiang, Z., Wu, M., Liu, G., and Liang, A. 2012. A sensitive enzyme-catalytic nanogold-resonance scattering spectral assay for alkaline phosphate. Bioprocess Biosyst. Eng. 35: 781–787.

Jouven, X., Lemaître, R. N., Rea, T. D., Sotoodehnia, N., Empana, J. -P., and Siscovick, D. S. 2005. Diabetes, glucose level, and risk of sudden cardiac death. Eur. Heart J. 26: 2142–2147.

Jung, I. Y., Kim, J. S., Choi, B. R., Lee, K., and Lee, H. 2017. Hydrogel based biosensors for *in vitro* diagnostics of biochemicals, proteins, and genes. Adv. Healthc. Mater. 6: 1601475.

Jung, J. M., Kim, C., and Harrison, R. G. 2018. A dual sensor selective for Hg^{2+} and cysteine detection. Sensors Actuators B Chem. 255: 2756–2763.

Kadowaki, T., Miyake, Y., Hagura, R., Akanuma, Y., Kajinuma, H., Kuzuya, N. et al. 1984. Risk factors for worsening to diabetes in subjects with impaired glucose tolerance. Diabetologia. 26: 44–49.

Kakkar, R., Kalra, J., Mantha, S. V., and Prasad, K. 1995. Lipid peroxidation and activity of antioxidant enzymes in diabetic rats. Mol. Cell. Biochem. 151: 113–119.

Kargozar, S., Hoseini, S. J., Milan, P. B., Hooshmand, S., Kim, H., and Mozafari, M. 2020. Quantum Dots: A Review from Concept to Clinic. Biotechnol. J. 15: 2000117.

Khajouei, S., Ravan, H., and Ebrahimi, A. 2020. DNA hydrogel-empowered biosensing. Adv. Colloid Interface Sci. 275: 102060.

Kieltyka, R. E., Pape, A. C. H., Albertazzi, L., Nakano, Y., Bastings, M. M. C., Voets, I. K. et al. 2013. Mesoscale modulation of supramolecular ureidopyrimidinone-based Poly(ethylene glycol) transient networks in water. J. Am. Chem. Soc. 135: 11159–11164.

Kim, I. -S., and Oh, I. -J. 2005. Drug release from the enzyme-degradable and pH-sensitive hydrogel composed of glycidyl methacrylate dextran and poly(acrylic acid). Arch. Pharm. Res. 28: 983–987.

Kim, J., Campbell, A. S., and Wang, J. 2018. Wearable non-invasive epidermal glucose sensors: A review. Talanta. 177: 163–170.

Klonoff, D. C. 2005. A Review of continuous glucose monitoring technology. Diabetes Technol. Ther. 7: 770–775.

Kong, W., Wu, D., Xia, L., Chen, X., Li, G., Qiu, N. et al. 2017. Carbon dots for fluorescent detection of α-glucosidase activity using enzyme activated inner filter effect and its application to anti-diabetic drug discovery. Anal. Chim. Acta. 973: 91–99.

Kopeček, J., and Yang, J. 2007. Hydrogels as smart biomaterials. Polym. Int. 56: 1078–1098.

Kushwaha, S., Rai, A., and Saxena, P. 2012. Stimuli sensitive hydrogels for ophthalmic drug delivery: A review. Int. J. Pharm. Investig. 2: 54.

Leduskrasts, K., and Suna, E. 2019. Aggregation induced emission by pyridinium–pyridinium interactions. RSC Adv. 9: 460–465.

Lee-Wang, H. H., Blakey, I., Chirila, T. V., Peng, H., Rasoul, F., Whittaker, A. K. et al. 2010. Novel supramolecular hydrogels as artificial vitreous substitutes. Macromol. Symp. 296: 229–232.

Lee, K., Kim, K., Yoon, H., and Kim, H. 2018. Chemical design of functional polymer structures for biosensors: from nanoscale to macroscale. Polymers. 10: 551.

Lee, Y. -J., Pruzinsky, S. A., and Braun, P. V. 2004. Glucose-sensitive inverse opal hydrogels: analysis of optical diffraction response. Langmuir. 20: 3096–3106.

Li, J., and Mooney, D. J. 2016. Designing hydrogels for controlled drug delivery. Nat. Rev. Mater. 1: 16071.

Li, K., Liu, W., Ni, Y., Li, D., Lin, D., Su, Z. et al. 2017. Technical synthesis and biomedical applications of graphene quantum dots. J. Mater. Chem. B. 5: 4811–4826.

Li, L., Shi, Y., Pan, L., Shi, Y., and Yu, G. 2015a. Rational design and applications of conducting polymer hydrogels as electrochemical biosensors. J. Mater. Chem. B. 3: 2920–2930.

Li, L., Wang, Y., Pan, L., Shi, Ye, Cheng, W., Shi, Yi et al. 2015b. A nanostructured conductive hydrogels-based biosensor platform for human metabolite detection. Nano Lett. 15: 1146–1151.

Li, R., Peng, F., Cai, J., Yang, D., and Zhang, P. 2020. Redox dual-stimuli responsive drug delivery systems for improving tumor-targeting ability and reducing adverse side effects. Asian J. Pharm. Sci. 15: 311–325.

Li, Rothberg, L. J. 2004. Label-free colorimetric detection of specific sequences in genomic DNA amplified by the polymerase chain reaction. J. Am. Chem. Soc. 126: 10958–10961.

Liang, S. -L., and Chan, D. W. 2007. Enzymes and related proteins as cancer biomarkers: A proteomic approach. Clin. Chim. Acta. 381: 93–97.

Lim, S. L., Ooi, C. -W., Tan, W. S., Chan, E. -S., Ho, K. L., and Tey, B. T. 2017. Biosensing of hepatitis B antigen with poly(acrylic acid) hydrogel immobilized with antigens and antibodies. Sensors Actuators B Chem. 252: 409–417.

Lim, S. Y., Shen, W., and Gao, Z. 2015. Carbon quantum dots and their applications. Chem. Soc. Rev. 44: 362–381.

Lin, P. -H., and Li, B. -R. 2020. Antifouling strategies in advanced electrochemical sensors and biosensors. Analyst. 145: 1110–1120.

Liu, B., and Liu, J. 2019. Sensors and biosensors based on metal oxide nanomaterials. TrAC Trends Anal. Chem. 121: 115690.

Liu, H., Xu, S., He, Z., Deng, A., and Zhu, J. -J. 2013. Supersandwich cytosensor for selective and ultrasensitive detection of cancer cells using aptamer-DNA Concatamer-quantum dots probes. Anal. Chem. 85: 3385–3392.

Liu, J., and Peng, Q. 2017. Protein-gold nanoparticle interactions and their possible impact on biomedical applications. Acta Biomater. 55: 13–27.

Liu, M., Xu, Y., Niu, F., Gooding, J. J., and Liu, J. 2016. Carbon quantum dots directly generated from electrochemical oxidation of graphite electrodes in alkaline alcohols and the applications for specific ferric ion detection and cell imaging. Analyst. 141: 2657–2664.

Liu, R., Wu, D., Liu, S., Koynov, K., Knoll, W., and Li, Q. 2009. An aqueous route to multicolor photoluminescent carbon dots using silica spheres as carriers. Angew. Chemie Int. Ed. 48: 4598–4601.

Liu, Z. X., Wu, Z. L., Gao, M. X., Liu, H., and Huang, C. Z. 2016. Carbon dots with aggregation induced emission enhancement for visual permittivity detection. Chem. Commun. 52: 2063–2066.

Luo, S., Wang, Y., Wang, G., Liu, F., Zhai, Y., and Luo, Y. 2018. Hybrid spray-coating, laser-scribing and ink-dispensing of graphene sensors/arrays with tunable piezoresistivity for *in situ* monitoring of composites. Carbon. 139: 437–444.

Luo, X., Morrin, A., Killard, A. J., and Smyth, M. R. 2006. Application of nanoparticles in electrochemical sensors and biosensors. Electroanalysis. 18: 319–326.

Marco-Dufort, B., and Tibbitt, M. W. 2019. Design of moldable hydrogels for biomedical applications using dynamic covalent boronic esters. Mater. Today Chem. 12: 16–33.

Mathur, A. M., Moorjani, S. K., and Scranton, A. B. 1996. Methods for synthesis of hydrogel networks: a review. J. Macromol. Sci. Part C Polym. Rev. 36: 405–430.

Matsumoto, H., Erickson, R. H., Gum, J. R., Yoshioka, M., Gum, E., and Kim, Y. S. 1990. Biosynthesis of alkaline phosphatase during differentiation of the human colon cancer cell line caco-2. Gastroenterology. 98: 1199–1207.

Mazrad, Z. A. I., Lee, K., Chae, A., In, I., Lee, H., and Park, S. Y. 2018. Progress in internal/external stimuli responsive fluorescent carbon nanoparticles for theranostic and sensing applications. J. Mater. Chem. B. 6: 1149–1178.

Miyata, T., Uragami, T., and Nakamae, K. 2002. Biomolecule-sensitive hydrogels. Adv. Drug Deliv. Rev. 54: 79–98.

Moreira, F. T. C., Dutra, R. A. F., Noronha, J. P. C., and Sales, M. G. F. 2013. Electrochemical biosensor based on biomimetic material for myoglobin detection. Electrochim. Acta. 107: 481–487.

Moreira, F. T. C., Truta, L. A. A. N. A., and Sales, M. G. F. 2018. Biomimetic materials assembled on a photovoltaic cell as a novel biosensing approach to cancer biomarker detection. Sci. Rep. 8: 10205.

Moss, D. W. 1982. Alkaline phosphatase isoenzymes. Clin. Chem. 28: 2007–2016.

Mu, T. -W., Fowler, D. M., and Kelly, J. W. 2008. Partial restoration of mutant enzyme homeostasis in three distinct lysosomal storage disease cell lines by altering calcium homeostasis. PLoS Biol. 6: e26.

Nsabimana, A., Lan, Y., Du, F., Wang, C., Zhang, W., and Xu, G. 2019. Alkaline phosphatase-based electrochemical sensors for health applications. Anal. Methods. 11: 1996–2006.

Oliveira, N. M., Zhang, Y. S., Ju, J., Chen, A. -Z., Chen, Y., Sonkusale, S. R. et al. 2016. Hydrophobic hydrogels: toward construction of floating (Bio)microdevices. Chem. Mater. 28: 3641–3648.

Oliver, N. S., Toumazou, C., Cass, A. E. G., and Johnston, D. G. 2009. Glucose sensors: a review of current and emerging technology. Diabet. Med. 26: 197–210.

Otten, L., Richards, S. -J., Fullam, E., Besra, G. S., and Gibson, M. I. 2013. Gold nanoparticle-linked analysis of carbohydrate–protein interactions, and polymeric inhibitors, using unlabelled proteins; easy measurements using a 'simple' digital camera. J. Mater. Chem. B. 1: 2665.

Park, Y., Ryu, Y. -M., Wang, T., Jung, Y., Kim, Sohee, Hwang, S. et al. 2018. Colorectal cancer diagnosis using enzyme-sensitive ratiometric fluorescence dye and antibody-quantum dot conjugates for multiplexed detection. Adv. Funct. Mater. 28: 1703450.

Patel, M. K., Ali, M. A., Krishnan, S., Agrawal, V. V., Al Kheraif, A. A., Fouad, H. et al. 2015. A label-free photoluminescence genosensor using nanostructured magnesium oxide for cholera detection. Sci. Rep. 5: 17384.

Patil, N. B., Nimbalkar, A. R., and Patil, M. G. 2018. ZnO thin film prepared by a sol-gel spin coating technique for NO_2 detection. Mater. Sci. Eng. B. 227: 53–60.

Peng, Q., Chen, J., Wang, T., Peng, X., Liu, J., Wang, X. et al. 2020. Recent advances in designing conductive hydrogels for flexible electronics. InfoMat. 2: 843–865.

Peppas, N. A., and Leobandung, W. 2004. Stimuli-sensitive hydrogels: ideal carriers for chronobiology and chronotherapy. J. Biomater. Sci. Polym. Ed. 15: 125–144.

Peppas, N. A., and Van Blarcom, D. S. 2016. Hydrogel-based biosensors and sensing devices for drug delivery. J. Control. Release. 240: 142–150.

Perumal, V., and Hashim, U. 2014. Advances in biosensors: Principle, architecture and applications. J. Appl. Biomed. 12: 1–15.

Pey, A. L. 2013. Protein homeostasis disorders of key enzymes of amino acids metabolism: mutation-induced protein kinetic destabilization and new therapeutic strategies. Amino Acids. 45: 1331–1341.

Phuong, P. T. M., Ryplida, B., In, I., and Park, S. Y. 2019. High performance of electrochemical and fluorescent probe by interaction of cell and bacteria with pH-sensitive polymer dots coated surfaces. Mater. Sci. Eng. C. 101: 159–168.

Qiu, Y., and Park, K. 2001. Environment-sensitive hydrogels for drug delivery. Adv. Drug Deliv. Rev. 53: 321–339.

Qu, K., Wang, J., Ren, J., and Qu, X. 2013. Carbon dots prepared by hydrothermal treatment of dopamine as an effective fluorescent sensing platform for the label-free detection of Iron(III) ions and dopamine. Chem. - A Eur. J. 19: 7243–7249.

Qu, X., Wirsén, A., and Albertsson, A. -C. 1999. Synthesis and characterization of pH-sensitive hydrogels based on chitosan and D,L-lactic acid. J. Appl. Polym. Sci. 74: 3193–3202.

Qu, X., Wirsén, A., and Albertsson, A. -C. 2000. Novel pH-sensitive chitosan hydrogels: swelling behavior and states of water. Polymer. 41: 4589–4598.

Richter, A., Paschew, G., Klatt, S., Lienig, J., Arndt, K. -F., and Adler, H. -J. 2008. Review on hydrogel-based pH sensors and microsensors. Sensors. 8: 561–581.

Rizwan, M., Yahya, R., Hassan, A., Yar, M., Azzahari, A., Selvanathan, V. et al. 2017. pH sensitive hydrogels in drug delivery: brief history, properties, swelling, and release mechanism, material selection and applications. Polymers (Basel). 9: 137.

Robby, A. I., Kim, S. G., Lee, U. H., In, I., Lee, G., and Park, S. Y. 2021. Wireless electrochemical and luminescent detection of bacteria based on surface-coated $CsWO_3$-immobilized fluorescent carbon dots with photothermal ablation of bacteria. Chem. Eng. J. 403: 126351.

Rodbard, D. 2016. Continuous glucose monitoring: a review of successes, challenges, and opportunities. Diabetes Technol. Ther. 18: S2-3-S2-13.

Rodriguez-Martinez, M. A., and Ruiz-Torres, A. 1992. Homeostasis between lipid peroxidation and antioxidant enzyme activities in healthy human aging. Mech. Ageing Dev. 66: 213–222.

Ronkainen, N. J., Halsall, H. B., and Heineman, W. R. 2010. Electrochemical biosensors. Chem. Soc. Rev. 39: 1747.

Rusin, O., St. Luce, N. N., Agbaria, R. A., Escobedo, J. O., Jiang, S., Warner, I. M. et al. 2004. Visual detection of cysteine and homocysteine. J. Am. Chem. Soc. 126: 438–439.

Russell, R. J., Pishko, M. V., Gefrides, C. C., McShane, M. J., and Coté, G. L. 1999. A fluorescence-based glucose biosensor using concanavalin A and dextran encapsulated in a Poly(ethylene glycol) hydrogel. Anal. Chem. 71: 3126–3132.

Ryplida, B., Lee, K. D., In, I., and Park, S. Y. 2019. Light-induced swelling-responsive conductive, adhesive, and stretchable wireless film hydrogel as electronic artificial skin. Adv. Funct. Mater. 29: 1903209.

Ryplida, B., In, I., and Park, S. Y. 2020. Tunable pressure sensor of f-carbon dot-based conductive hydrogel with electrical, mechanical, and shape recovery for monitoring human motion. ACS Appl. Mater. Interfaces. 12: 51766–51775.

Ryu, J. H., Lee, G. J., Shih, Y. -R. V., Kim, T., and Varghese, S. 2019. Phenylboronic Acid-polymers for biomedical applications. Curr. Med. Chem. 26: 6797–6816.

Sadat Ebrahimi, M. M., and Schönherr, H. 2014. Enzyme-Sensing Chitosan Hydrogels. Langmuir. 30: 7842–7850.

Sadeghi, H., and Yazdanparast, R. 2003. Effect of *Dendrostellera lessertii* on the intracellular alkaline phosphatase activity of four human cancer cell lines. J. Ethnopharmacol. 86: 11–14.

Schumacker, P. T. 2006. Reactive oxygen species in cancer cells: Live by the sword, die by the sword. Cancer Cell. 10: 175–176.

Seeli, D. S., Dhivya, S., Selvamurugan, N., and Prabaharan, M. 2016. Guar gum succinate-sodium alginate beads as a pH-sensitive carrier for colon-specific drug delivery. Int. J. Biol. Macromol. 91: 45–50.

Self, C. H. 1985. Enzyme amplification—A general method applied to provide an immunoassisted assay for placental alkaline phosphatase. J. Immunol. Methods. 76: 389–393.

Şerban, I., and Enesca, A. 2020. Metal Oxides-Based Semiconductors for Biosensors Applications. Front. Chem. 8: 354.

Sharma, U., Pal, D., and Prasad, R. 2014. Alkaline Phosphatase: An Overview. Indian J. Clin. Biochem. 29: 269–278.

Shaw, J. E., Zimmet, P. Z., de Courten, M., Dowse, G. K., Chitson, P., Gareeboo, H. et al. 1999. Impaired fasting glucose or impaired glucose tolerance. What best predicts future diabetes in Mauritius? Diabetes Care. 22: 399–402.

Sheikhzadeh, E., Chamsaz, M., Turner, A. P. F., Jager, E. W. H., and Beni, V. 2016. Label-free impedimetric biosensor for Salmonella Typhimurium detection based on poly [pyrrole-co-3-carboxyl-pyrrole] copolymer supported aptamer. Biosens. Bioelectron. 80: 194–200.

Shetti, N. P., Bukkitgar, S. D., Reddy, K. R., Reddy, C. V., and Aminabhavi, T. M. 2019. Nanostructured titanium oxide hybrids-based electrochemical biosensors for healthcare applications. Colloids Surf. B Biointerfaces. 178: 385–394.

Shi, Z. -M., Wu, C. -F., Zhou, T. -Y., Zhang, D. -W., Zhao, X., and Li, Z. -T. 2013. Foldamer-based chiral supramolecular alternate block copolymers tuned by ion-pair binding. Chem. Commun. 49: 2673.

Sijbesma, R. P., Beijer, F. H., Brunsveld, L., Folmer, B. J. B., Hirschberg, J. H. K. K., Lange, R. F. M. et al. 1997. Reversible Polymers Formed from Self-Complementary Monomers Using Quadruple Hydrogen Bonding. Science. 278: 1601–1604.

Su, Y., Xie, M., Lu, X., Wei, H., Geng, H., Yang, Z. et al. 2014. Facile synthesis and photoelectric properties of carbon dots with upconversion fluorescence using arc-synthesized carbon by-products. RSC Adv. 4: 4839.

Sun, F., Wu, K., Hung, H. -C., Zhang, P., Che, X., Smith, J. et al. 2017. Paper sensor coated with a Poly(carboxybetaine)-multiple DOPA conjugate *via* dip-coating for biosensing in complex media. Anal. Chem. 89: 10999–11004.

Sun, Q., Qian, B., Uto, K., Chen, J., Liu, X., and Minari, T. 2018. Functional biomaterials towards flexible electronics and sensors. Biosens. Bioelectron. 119: 237–251.

Tao, W., and He, Z. 2018. ROS-responsive drug delivery systems for biomedical applications. Asian J. Pharm. Sci. 13: 101–112.

Tavakoli, J., and Tang, Y. 2017. Hydrogel based sensors for biomedical applications: an updated review. Polymersé. 9: 364.

Teunissen, A. J. P., Nieuwenhuizen, M. M. L., Rodríguez-Llansola, F., Palmans, A. R. A., and Meijer, E. W. 2014. Mechanically induced gelation of a kinetically trapped supramolecular polymer. Macromolecules. 47: 8429–8436.

Turkmen, E., Bas, S. Z., Gulce, H., and Yildiz, S. 2014. Glucose biosensor based on immobilization of glucose oxidase in electropolymerized poly(o-phenylenediamine) film on platinum nanoparticles-polyvinylferrocenium modified electrode. Electrochim. Acta. 123: 93–102.

Ulijn, R. V., Bibi, N., Jayawarna, V., Thornton, P. D., Todd, S. J., Mart, R. J. et al. 2007. Bioresponsive hydrogels. Mater. Today. 10: 40–48.

Ullah, F., Othman, M. B. H., Javed, F., Ahmad, Z., and Akil, H. M. 2015. Classification, processing and application of hydrogels: A review. Mater. Sci. Eng. C. 57: 414–433.

Updike, S. J., Shults, M. C., Gilligan, B. J., and Rhodes, R. K. 2000. A subcutaneous glucose sensor with improved longevity, dynamic range, and stability of calibration. Diabetes Care. 23: 208–214.

van der Linden, H. J., Herber, S., Olthuis, W., and Bergveld, P. 2003. Stimulus-sensitive hydrogels and their applications in chemical (micro)analysis. Analyst. 128: 325–331.

Van Vlierberghe, S., Dubruel, P., and Schacht, E. 2011. Biopolymer-based hydrogels As scaffolds for tissue engineering applications: a review. Biomacromolecules. 12: 1387–1408.

Wang, J. 2012. Electrochemical biosensing based on noble metal nanoparticles. Microchim. Acta. 177: 245–270.

Wang, K., Dong, K., Yan, Y., Xu, W., Zhang, L., Zhao, G. et al. 2015. *In vitro* and *in vivo* study of a colon-targeting pH-sensitive hydrocortisone sodium succinate hydrogel. RSC Adv. 5: 80625–80633.

Wang, T., Turhan, M., and Gunasekaran, S. 2004. Selected properties of pH-sensitive, biodegradable chitosan–poly(vinyl alcohol) hydrogel. Polym. Int. 53: 911–918.

Willner, I. 2002. BIOELECTRONICS: Biomaterials for Sensors, Fuel Cells, and Circuitry. Science. 298: 2407–2408.

Wolfbeis, O. S. 2015. An overview of nanoparticles commonly used in fluorescent bioimaging. Chem. Soc. Rev. 44: 4743–4768.

Won, H. J., Ryplida, B., Kim, S. G., Lee, G., Ryu, J. H., and Park, S. Y. 2020. Diselenide-bridged carbon-dot-mediated self-healing, conductive, and adhesive wireless hydrogel sensors for label-free breast cancer detection. ACS Nano. 14: 8409–8420.

Xia, C., Zhu, S., Feng, T., Yang, M., and Yang, B. 2019. Evolution and synthesis of carbon dots: from carbon dots to carbonized polymer dots. Adv. Sci. 6: 1901316.

Xu, Y., Wu, H., Huang, C., Hao, C., Wu, B., Miao, C. et al. 2015. Sensitive detection of tumor cells by a new cytosensor with 3D-MWCNTs array based on vicinal-dithiol-containing proteins (VDPs). Biosens. Bioelectron. 66: 321–326.

Yang, C., and Suo, Z. 2018. Hydrogel ionotronics. Nat. Rev. Mater. 3: 125–142.

Yang, Y., Deng, H., and Fu, Q. 2020. Recent progress on PEDOT:PSS based polymer blends and composites for flexible electronics and thermoelectric devices. Mater. Chem. Front. 4: 3130–3152.

Yang, Z., Nie, S., Hsiao, W. W., and Pam, W. 2011. Thermoreversible Pluronic F127-based hydrogel containing liposomes for the controlled delivery of paclitaxel: *in vitro* drug release, cell cytotoxicity, and uptake studies. Int. J. Nanomedicine. 6: 151–166.

Yao, X., Chen, L., Ju, J., Li, C., Tian, Y., Jiang, L. et al. 2016. Superhydrophobic diffusion barriers for hydrogels *via* confined interfacial modification. Adv. Mater. 28: 7383–7389.

Yu, H., Li, X., Zeng, X., and Lu, Y. 2016. Preparation of carbon dots by non-focusing pulsed laser irradiation in toluene. Chem. Commun. 52: 819–822.

Zhang, C., Losego, M. D., and Braun, P. V. 2013. Hydrogel-based glucose sensors: effects of phenylboronic acid chemical structure on response. Chem. Mater. 25: 3239–3250.

Zhang, J., and Yu, S. -H. 2016. Carbon dots: large-scale synthesis, sensing and bioimaging. Mater. Today. 19: 382–393.

Zhang, R., Zhu, Y., Huang, J., Xu, S., Luo, J. and Liu, X. 2018. Electrochemical sensor coating based on electrophoretic deposition of Au-doped self-assembled nanoparticles. ACS Appl. Mater. Interfaces. 10: 5926–5932.

Zhang, R., Rejeeth, C., Xu, W., Zhu, C., Liu, X., Wan, J. et al. 2019. Label-free electrochemical sensor for CD44 by ligand-protein interaction. Anal. Chem. 91: 7078–7085.

Zhang, W., Jin, X., Li, H., Zhang, R., and Wu, C. 2018. Injectable and body temperature sensitive hydrogels based on chitosan and hyaluronic acid for pH sensitive drug release. Carbohydr. Polym. 186: 82–90.

Zhang, Y. -Z., Lee, K. H., Anjum, D. H., Sougrat, R., Jiang, Q., Kim, H., and Alshareef, H. N. 2018. MXenes stretch hydrogel sensor performance to new limits. Sci. Adv. 4: eaat0098.

Zhang, Y., Wu, G., Ding, C., Liu, F., Yao, Y., Zhou, Y. et al. 2018. Lead selenide colloidal quantum dot solar cells achieving high open-circuit voltage with one-step deposition strategy. J. Phys. Chem. Lett. 9: 3598–3603.

Zhou, H., Tang, J., Lv, L., Sun, N., Zhang, Jie, Chen, B. et al. 2018. Intracellular endogenous glutathione detection and imaging by a simple and sensitive spectroscopic off–on probe. Analyst. 143: 2390–2396.

Zhu, S., Song, Y., Zhao, X., Shao, J., Zhang, J., and Yang, B. 2015. The photoluminescence mechanism in carbon dots (graphene quantum dots, carbon nanodots, and polymer dots): current state and future perspective. Nano Res. 8: 355–381.

Zou, W. -S., Ye, C. -H., Wang, Y. -Q., Li, W. -H., and Huang, X. -H. 2018. A hybrid ratiometric probe for glucose detection based on synchronous responses to fluorescence quenching and resonance light scattering enhancement of boronic acid functionalized carbon dots. Sensors Actuators B Chem. 271: 54–63.

CHAPTER 7

Textile Chemical Sensors

Marta Tessarolo,[1,]* *Isacco Gualandi,*[2] *Luca Possanzini*[1]
and *Federica Mariani*[2]

Introduction

Over the past decade, e-health devices and sensor networks have received the greatest attention among the rising Internet of Things (IoT) technologies. IoT sensors include smart sensors and actuators, and even wearable sensing devices that can be incorporated into daily clothes or worn as accessories or implants. Real-time collection and sharing of information regarding our body status are significant not only for hospitalized situations but also during routine daily activities. For instance, continuous monitoring of vital parameters is crucial to take precautionary measures during sports activity or in the wearer's workplace, especially in the presence of diagnosed, chronic diseases. In this view, wearable sensors and smart textiles can be a mean to realize the so-called personalized-medicine and revolutionize our way to manage healthcare and practice medicine. The main constraints that hinder a straightforward realization of a reliable wearable chemical sensor are related to the low analyte concentration, small sampling volumes and fluid stagnation, mechanical resiliency, signal stability, biofouling and biocompatibility issues, together with the need to operate in physiological conditions, adapt to the human body without causing discomfort and work with low power consumption, thus requiring simple electronic components. In general, the central idea shared by wearable technologies is the wear-and-forget functionality, which automatically excludes all approaches with an intrusive nature that rely, for instance, on blood samples (Windmiller and Wang 2013). Sweat, saliva, interstitial fluids, tears, and breath are biological fluids that contain abundant information about health status, are readily accessible, and can

---

[1] Department of Physics and Astronomy "Augusto Righi", University of Bologna, Viale Berti Pichat 6/2, 40127 Bologna, Italy.
[2] Department of Industrial Chemistry "Toso Montanari", University of Bologna, Viale Risorgimento 4, 40136 Bologna, Italy.
* Corresponding author: marta.tessarolo3@unibo.it

be sampled for non-invasive monitoring. Wound exudate is another biofluid that is currently attracting interest in the field of wearable sensing technologies targeting the realization of smart wound dressings. Upon integration of chemical sensors that are able to non-invasively monitor the wound site, these tools would dramatically impact the current treatment approaches, with the potential to improve significantly the wound management and decrease the healing time.

A distinct class of wearable devices comprises outward-looking sensors, which offer valuable information from the surrounding environment for improved safety and health, particularly for emergency responders or inspectors (Ozanich 2018). Miniaturized dosimeters and belt-worn analytical tools available to the market mainly target real-time monitoring of air pollution, indoor air quality, aerosol exposure, and detection of biothreat agents. Despite portability and robustness, such devices rely on sophisticated detectors that profoundly impact the final cost of the product. On the other hand, a variety of low-cost, wearable chemo-sensors for external monitoring of gaseous threats are at a primary stage of development. Focusing on textile gas sensors that can be conveniently integrated into everyday clothes or work overalls, they have been fabricated on both natural and synthetic fibers, the majority exploiting chemoresistive materials for the detection of volatile organic compounds and ammonia.

Textile Substrates for Wearable Sensors

Among the huge variety of wearable biosensors, the textile-based chemical sensors represent an excellent example for a non-invasive, low cost, comfortable, and lightweight point-of-care platform. From the wide selection of textile materials and fibers, cotton yarns, acrylic fabrics or elastane based textiles are the most common substrate to realize chemical textile sensors. Besides, the choice of the proper textile materials is a relevant aspect in the sensor design since it could affect the sensors' response in terms of sensitivity, stability or washability correlated to the interfacing between textile and sensitive materials (Possanzini et al. 2019). The first step that has to be done to achieve a reliable and reproducible textile sensor platform is to study and investigate the adherence of conductive and sensing compounds with the desired textile substrate. Taking into account the final scope and application of a textile sensor, the proper support to develop and integrate the smart sensor has to be chosen between fibers, yarns and fabrics.

Fibers are the most basic units of textile techonology and the design of flexible, elastic and stretchable fibers that can be knitted or woven is essential to implement the already existing landscape of fiber-based electronics. According to their application, they can be classified into fiber-shaped energy-harvesting as solar cells (Zhang et al. 2014) or triboelectric nanogenerators (Sim et al. 2016); fiber-shaped energy storage devices like supercapacitors (Sun et al. 2016) or lithium-ion batteries (Lu et al. 2017); fiber-shaped electroluminescent devices (Kwon et al. 2018) and fiber-based sensors (Sekar et al. 2019, Wang et al. 2020, Zhou et al. 2017) which generally work via physical or chemical processes. Nowadays, the two relevant strategies to create electronic textiles exploiting the soft, deformable and permeable features of fibers rely on their interaction with textiles through standard textile methods or to weave cathode and anode fiber electrodes in an interlaced configuration to realize the e-textile. However, even if they present real possibilities for the production of skin-like pressure sensors (Li and Wang 2011) or flexible circuits (Liu et al. 2013), their mechanical properties are generally lower than the one required for automatic manufacturing processes and the hand-made realization of this new smart sensors limits their applicability and large-scale production. Conversely, the yarn-based textile sensors are machine sewable and, since the yarns are made by combining different types of fibers, it is possible to produce elastic and conductive yarns (Gil et al. 2019, Guo et al. 2012). Furthermore, many examples of composite structures have been proposed in the last decade based on graphene oxide/nylon yarns (Yun et al. 2013) and functional coatings (Gualandi et al. 2018). Cotton yarns were transformed in e-textile by coating them with a polyelectrolyte-based layer with carbon nanotubes (CNTs) (Shim et al. 2008) or, by a melt-spinning method, elastic silicone rubbers tubes and stainless steel yarns were realized to manufacture strertchable triboelectric yarns (Gong et al. 2019a). As already stated, suitable substrates have a vital role in the production of e-textiles. Different fabrics, such as 100% cotton (woven and knit), 100% polyester and blend substrates (polyester/cotton, cotton/wool, wool/nylon, etc.) have been extensively used. Depending on the sensor type, the less deformable and less porous woven fabric can be chosen instead of the more stretchable and deformable knitted fabrics. Both knitted (Lai et al. 2017) and woven (Choudhary et al. 2015) fabric-based chemical textile sensors are produced using various techniques like dipping (Kim et al. 2018), drop casting (Possanzini et al. 2019), coating (Gualandi et al. 2016, Mariani et al. 2020), pad-dry method (Xu et al. 2018a), screen printing (Gualandi et al. 2016, Takamatsu et al. 2015, 2016) and *in situ* polymerisation (Tunáková et al. 2018). In addition, the different material plays a fundamental role to individuate the textile substrate. For instance, PEDOT:PSS based sensors need a hydrophilic substrate such as cotton fabric (Bihar et al. 2017, Gualandi et al. 2016). On the other hand, Ag/AgCl ink and carbon ink for screen printing require a hydrophobic substrate as GORE-TEX (Windmiller and Wang 2013). Cotton fabrics show high breathability and/or air permeability, which make them preferred for skin-contact sensors (Dąbrowska et al. 2016). PET fabrics can be easily functionalized with metallic layers to increase electrical conductivity and electromagnetic shielding (Liu et al. 2016a). Even 100% polyester (with very low moisture content compared to natural fibers), 100% wool or blend of different yarns to optimize the properties of textile for a specific application have been produced for health monitoring (Du et al. 2016), temperature sensing (Tchafa and Huang 2018), motion monitoring (Tessarolo et al. 2018) and for developing low-cost wearable energy-storage devices (Lv et al. 2018).

Transduction Mechanisms

Most of the textile chemical sensors are electrochemical, in particular amperometric, potentiometric and chemiresistors, transistor-based, or optical sensors.

Potentiometric sensors consist of an indicator electrode and a reference electrode that must be immersed in the solution under investigation. The analyte is usually quantified by measuring the potential difference between the two electrodes in the absence of current flow. Therefore, the read-out electronics should be designed with a high input impedance. The dependence of the potential of the working electrode on the analyte concentration is described by the Nernst equation:

$$E = E^{\ominus} + RT / (nF) \ln (a_i)$$

where E is the potential, $E^{\ominus}$ is the standard potential, R is the gas coefficient, F is the Faraday constant, n is the number of electrons, and a_i is the activity of the principal ion. The Nernst equation shows that there is a linear correlation between the potential and the logarithm of the analyte concentration. The signal can be generated by redox reactions involving the analyte or through a membrane that selectively allows its adsorption or passage. More complex architectures such as the ion-sensitive field-effect transistor are described in the literature, but they have never been used to produce a textile device.

Amperometric sensors measure the current developed by a faradic reaction at a working electrode which is usually biased with respect to a reference electrode with a fixed potential. Both electrodes must be immersed in the analyzed sample. Although the most rigorous measurements are carried out in a three-electrode cell endowed with a counter electrode, the simplest devices use a two-electrode configuration. The applied potential is chosen to consume all the analyte at the electrode surface, thus generating a concentration gradient by diffusion from the bulk of the solution. In this way, the recorded current is directly proportional to the concentration of the analyte in the sample.

Organic electrochemical transistors (OECTs) are composed of a channel made of a conductive polymer and a gate electrode both immersed in an electrolytic solution. When a voltage is applied to the gate electrode, it stimulates electrochemical reactions at the channel that change the charge carrier concentration and thus the current that flows through it. Each substance that changes the current flow in the channel can be detected by these devices. Since OECTs combine a transductor and an amplifier, the signal is enhanced.

Chemiresistors are a class of sensors that can detect and quantify the target compound, thanks to a change in the electrical resistance of the sensing material. The interaction between the target analyte and sensing material can occur, for example, by creating traps, acting as a dopant, imposing resistive interfacial barriers, and changing the existing intermolecular interactions between molecular subunits.

Optical sensors exploit the interaction between the electromagnetic radiation and the chemical system. The detection occurs directly for analytes capable of absorbing the radiation or through the use of reagents that modify their absorption spectrum following the reaction with the analyte. In the simplest systems (for example, the litmus paper for pH estimation), the detection and/or quantification is performed through a visual investigation by the comparison of the resulting colors with a reference scale. More advanced systems use a light source, a photodetector, and a monochromator for quantitative analysis of the absorbed radiation. Recently, signal acquisition has been carried with a standard smartphone camera by images analysis.

Materials

The production of any electrochemical textile sensors (potentiometric, amperometric, transistors and chemoresistors) requires the development of conductive textile components that should be mechanically resistant, flexible, stretchable, lightweight, and of low cost. The development of new materials combined with fabrication techniques plays a key role in the design of these devices. Metals, conductive polymers and carbon-based materials are the main classes of materials that are used independently or combined in the production of conductive textiles.

Metals

Metals are certainly commonly used in standard electrochemical sensors due to their electrochemical and high conductivity properties. The main metallic materials are noble metals, such as Pt, Au, and Ag, because they take advantage of their peculiar electrochemical properties. It is worthy to note that metals (except for Au and Pt) can undergo redox reactions from a thermodynamic point of view, with their corrosion and a possible interference on electrochemical signals. For example, platinum is generally used as a counter electrode in electrochemical sensors, while Ag/AgCl as a reference electrode. Platinum, for example, catalyzes the oxidation of H_2O_2 and is therefore exploited in second-generation glucose biosensors (Bruen et al. 2017, Wang 2008). The integration of these materials into textile sensors is complicated, but there are several strategies for manufacturing them. A common method is based on a fully metallic thread that is sewn into a fabric. Coppede et al. (Coppedè et al. 2014) exploited platinum and silver wires to fabricate the gate electrodes of textile OECT sensors for the determination of adrenaline. An interesting approach to obtain the desired stretchability is proposed by Zhao et al. (Zhao et al. 2019) which covered an elastic fiber with gold nanowires (AuNWs). The innovative idea consisted of covering by dry spinning the elastic wire when it was strained and then released. Thanks to a proper functionalization with glucose oxidase (GOx) and Prussian blue (PB), they developed an electrochemical textile sensor for glucose detection demonstrated under 200% stretching. Moreover, the group of Wang (Bandodkar et al. 2016a) developed a stretchable ink made of Ag and Ag/AgCl that can be easily deposited by stamp transfer, or screen printing on fabric.

Conductive polymers

The mainly used conducting polymers in the field of conductive textile are poly (3,4-ethylene dioxythiophene): polystyrene sulfonate (PEDOT: PSS), polyaniline (PANI) and polypyrrole (PPy). Conductive polymers are often dispersed in solution in order to be deposited by common techniques in the manufacturing industry, such as screen printing. This is a good advantage for the textile wearable sensors, which in theory simplifies their industrialization. These materials are used for the fabrication of the semiconductor channel in the OECTs, or as sensitive materials in chemoresistors, amperometric and potentiometric sensors. For example, PEDOT:PSS has good electrocatalytic properties to detect oxidisable compounds. Gualandi et al. exploited this behavior to develop an all-PEDOT:PSS OECT sensor to detect ascorbic acid, dopamine, and adrenaline. PEDOT:PSS was deposited by screen printing or dip coating directly on cotton fabrics (Gualandi et al. 2016). In alternative, conductive polymers can by electrodeposited on wires. For example, PANI is deposited by on-site polymerisation and, thanks to its acid-base behavior, it is widely used in sensors for the determination of pH and ammonia (Stempien et al. 2017). The main advantages of conductive polymers are flexibility, stretchability, chemical tunability, ion-to-electron conversion, biocompatibility, and reversible redox switching, but the electrical conductivity is not as good as metals. In addition, conductive polymers are overoxidized when they are exposed to a strong oxidant or to very anodic potentials.

Carbon materials

Carbon nanotubes (CNT), Multiwalled Carbon Nanotube (MWCNT), graphene, graphene oxide, and carbon fibers have been widely adopted in textile electronics and chemical sensors, thanks to their large active area, high electrical conductivity, and good electrochemical activity. Besides, these materials are inert toward the redox processes involving water, with a consequent wide potential range that can be exploited for electrochemical measurements. Ropes of carbon fibers are directly produced by floating catalyst deposition (Li et al. 2006, Mendoza et al. 2007, Wang et al. 2020) and they have been exploited as conductive substrates for the fabrication of potentiometric and amperometric sensors (Wang et al. 2018) or as sensitive material in chemoresistors.

Using a suitable surfactant, CNTs (Jeerapan et al. 2016, Wang 2005) can be dispersed in aqueous solutions and thus used as an ink to be deposited on cotton yarns through dip-coating technique. Indeed, CNTs show good ion-to-electron transduction, which has been widely used in potentiometric devices. Alternatively, CNTs have been used in composite materials, for example for humidity sensors (Zhou et al. 2017), where CNTs are dispersed in Polyvinyl-alcohol (PVA) creating an electric conductive path that is altered from water absorption.

While CNTs have a characteristic 1D structure, graphene-based materials are characterized by a 2D structure and high specific surface area. For example, reduced graphene combined with polypyrrole (PPy) nanowires has been used for glucose sensors (Wang et al. 2017).

Sweat

The investigation of the chemical composition of human sweat started in the 19th century, intending to identify a correlation with physiological states, diseases, or the use of drugs. Although large volumes of clinical work showed promising results as far back as the 1940s and 1950s, sweat analysis is routinely employed in cystic fibrosis diagnostics and for the detection of illicit drug use.

The sweat analysis is performed for a specific target molecule by exploiting commercially available sampling equipment that is usually combined with an analytical determination by a dedicated instrument. These studies represent the basis for a quick development of sensors with medical relevance. The analysis of chloride content in human perspiration is performed for the diagnosis of cystic fibrosis and it is the most important application for routine analysis. Doping control is another field wherein sweat tests are employed because they offer some advantages such as less opportunity for sample adulteration and non-invasiveness as compared to blood. The main drugs determined in sweat are opiates, buprenorphine, amphetamines, and design drugs, cocaine and metabolites, and Gamma hydroxybu-tyrates. Finally, the quantification of some toxic metals, such as Cd, Pb, Hg, and As, is important because the rates of their excretion were reported to match or even exceed urinary excretion (Sears et al. 2012). The main issues that hinder a large use of human perspiration as bio-fluid in medical practice are the production of enough sweat for analysis, sample evaporation, lack of appropriate sampling devices, and the need for a normalization of the sampled volume. Consequently, sweat collection and analysis is laborious and requires specialized facilities, when it is performed with conventional approaches.

However, sweat sensors are attracting increasing interest thanks to the development of new wearable technologies that are overcoming the previously-described constraints by achieving a real-time sweat collection and analysis. Several insights favor the use of sweat as a noninvasive biofluid compared to saliva, tears, and urine. Innovative wearable sensors are placed outside of the body but in close proximity of the eccrine sweat glands that produce a very low, but continuous, amount of biofluid. Therefore, sweat is the unique bio-fluid that can continually be sampled without consequence or risks to the human subject.

Although other metabolites (glucose, lactate, creatinine, urea, K^+, Mg^{2+}, Ca^{2+}, NH_4^+, Cu^{2+},...) are not actively involved in the generation of the driving force at the basis of sweating, they diffuse in the secreted liquid from the surrounding cells. Consequently, sweat could provide an overview of the chemical composition of the surrounding cells at the time of sweat secretion.

It is worthy to note that the sweat analysis cannot reach the medical relevance of blood test, because our bodies closely control the blood composition while the sweat can be more variable. Its composition is affected by the body region, gender, age, ethnicity, diet, activity level, and acclimation. In addition, endogenous factors, such as hygiene practices, use of products (lotion, cosmetics, and so on), and seasonality, can vary the sweat composition. However, human perspiration could be real-time monitored by wearable sensors in an easy and non-invasive procedure that cannot be accomplished for blood because of the impossibility of taking samples with this time resolution. Sweat might provide an even deeper picture of what is happening in the body because certain biomarkers in sweat may correlate with disease symptoms, physiological states, or the use of particular drugs. Table 1 shows the typical concentration of metabolites in sweat.

Glucose

Diabetes is a widely spread disease affecting hundreds of millions of people. This health issue has driven technology to develop new glucose sensors for self-testing and monitoring. Commercial portable devices consist of disposable strips with a screen-printed electrochemical sensor and a pocket-size reader (Wang 2008). Even if this technology is common and mass-produced, self-testing of blood glucose still suffers from irregular testing frequency and, moreover, the knowledge of daily changes of glucose concentration would allow to efficiently treat and improve the quality

Table 1. Resume of the typical ions' concentration in human sweat. * weighted average value; ** in μM.

Ion	Concentration range in sweat (mM)	References
Ca^{2+}	0.07–12	(Hirokawa et al. 2007)
Potassium	4–24	(Patterson et al. 2000)
Chloride	10–100	(Sato et al. 1989)
Sodium	10–100	(Sato and Sato 1990)
Ammonium	0.5–8	(Sato et al. 1989)
Lactate	5–60	(Derbyshire et al. 2012)
Urea	14.2–30.2	(Huang et al. 2002)
Glucose	0.01–1	(Xu et al. 2018)
Ethanol	2.25–2.5	(Buono 1999)
Pyruvic acid	0.06–1.6	(Barel et al. 2014)
Amino acids*	303	(Barel et al. 2014)
Ascorbic Acid	1.4–62.5**	(Mickelsen and Keys 1943)

of life for people with diabetes. For this reason, many research groups proposed innovative wearable devices for continuous monitoring exploiting other body fluids in place of blood. Among them, sweat has been by far the most studied. In theory, the combination of a sweat sampling system with a wearable chemical sensor optimized for glucose detection should allow monitoring glucose in a non-invasive way. Even when this is achieved, there is still a main issue to overcome: the significant correlation between glucose concentration variations in sweat and in blood. Despite blood analysis being considered as a reference for medicine, it is important to note that low molecular weight metabolites' concentration can vary significantly among different vessels, for example, from vein blood to capillary blood (Karpova et al. 2019). This is even worse if one considers sweat as a matrix for the analysis. For this reason, in the beginning, many researchers thought that small weight metabolites do not reach sweat, or their concentrations are not correlated with the blood ones.

Many important research groups are working on the development of wearable sweat sensors. Diamond and co-workers (Glennon et al. 2016) developed SwEatch. Wang et al. (Bandodkar et al. 2015) proposed tattoo-based electrochemical sensors for the continuous and non-invasive detection of biomarkers in sweat. Gao et al. reported a wearable glucose-sweat sensor included in multi-sensors that provide a comprehensive profile of sweat composition enabling data cross-comparisons (Gao et al. 2016). Rogers and co-workers (Huang et al. 2014) developed a stretchable sensor for continuous sweat monitoring. Thanks to goals achieved by technology, now we know that the concentration of glucose in sweat ranges between 10 μM and 1 mM (Lee et al. 2017, Zhang et al. 2019), 3 orders of magnitude lower than blood concentration. A specific clinical procedure called "glucose tolerance test" (Karpova et al. 2019) is used to find the direct correlation between the glucose concentration present in blood and sweat. Mimicking a hyperglycemia event in healthy subjects, the concentration of glucose is simultaneously monitored in both fluids and, as a result, a variation in the range between 50 μM and 120 μM represents an alteration from the normal state to a potentially high glucose level.

Recent advances in wearable technology have explored fabric substrates to create smart textiles. In this field, glucose sensors have also been adapted to this innovative approach. Most of the textile glucose sensors are based on amperometric electrochemical sensors, organic electrochemical transistors and, on colorimetric sensors. For all of them, the sensing element is functionalized to detect the oxidation event of glucose. The functionalization has the role to catalyze the oxidation reaction. Two approaches have been used:

i. **Enzymatic sensors** in which GOx is immobilized on the sensing element;

ii. **Non-enzymatic sensors** that exploit the electrocatalytic behavior of a functionalized working electrode.

In the first case, the sensors functionalized with the enzyme are called "biosensors" and the presence of GOx catalyzes the oxidation of glucose to gluconic acid as reported in scheme 1:

Glucose + $O_2 \rightarrow$ *gluconic acid* + H_2O_2

The first-generation (Wang 2008) of glucose biosensors (Figure 1a) dates back to early 1960 to an idea of Clark's (Clark 1965). The detection was an indirect evaluation of oxygen consumption. In textile examples, the sensing element detects the presence of H_2O_2 as a product of oxidation reaction which is directly proportional to the glucose concentration. He et al. reported (He et al. 2019a) an innovative approach based on paper-based colorimetric glucose sensors included in a textile microfluidic device. They functionalized cotton fabric with polyurethane (PU) to produce a thermoresponsive textile microfluidic system. Then, a filter paper was pre-treated with

$$Glucose + O_2 \rightarrow gluconic\ acid + H_2O_2$$

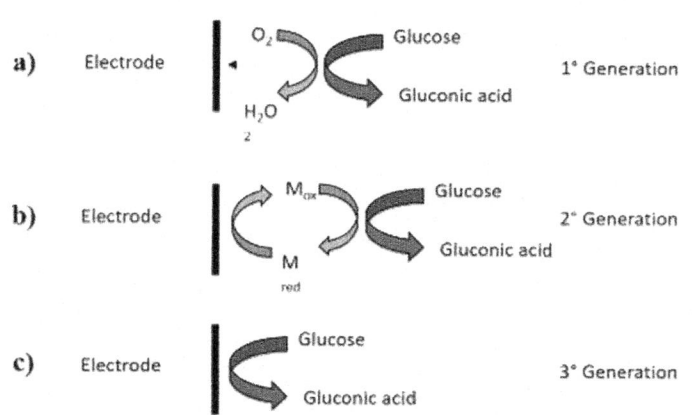

Figure 1. Electrochemical biosensors based on enzymes mechanism. (a) the biosensor detects the presence of a product of the first reaction catalyzed by the enzyme (First generation). (b) redox compound is linked to the electrode and acts as a mediator (Second generation). (c) the electrode surface and the enzyme redox center are electrically in contact (Third generation).

chitosan and then coated with a mixed solution based on GOx, tetramethylbenzidine (TMB), and Horseradish peroxidase (HRP). The chitosan provided a large number of amino groups increasing the density of enzyme immobilization. The detection occured when H_2O_2 oxides methylbenzidine to TMB, in the presence of HRP. As a consequence, paper's color changed from white to blue in function of glucose concentration. The alteration of color was analyzed with a RGB detector and R value was chosen as reference value. The sensors show a linear range from 50 to 600 mM and a limit of detection of 13 mM. Another approach, proposed by Wang et al. (Wang et al. 2017), is based on organic electrochemical transistor. The authors developed a fiber OECT based on polypyrrole (PPy) nanowires and reduced graphene oxide, with GOx immobilized on the surface of the gate electrode. The oxidation event occured close to the gate electrode which in-turn changes the effective gate voltage applied on the transistor and thus affects the channel current (Bernards et al. 2008). In this case, H_2O_2 reacted sequentially with the strong oxidizing lithium perchlorate in the electrolyte, which led to an increase in the conductivity of the OECT channel. The intrinsic amplification of OECT configuration allowed reaching an outstanding sensitivity, up to 0.773 NCR/decade, and LOD down to the nanomolar range.

A recent textile example of first-generation approach is reported by Zhao et al. (Zhao et al. 2019). The sensor employs a three-electrodes configuration and is based on elastic gold fiber. The novelty of this work regards the fabrication method: a solution of nanowires of gold (AuNWs) and styrene-ethylene-butylene-styrene (SEBS) block copolymer is spin-coated on strained elastic fiber and rapidly dried in air forming a long elastic AuNWs/SEBS fiber. Then, the fiber is functionalized with Au film to realize the counter electrode, and Ag/AgCl is deposited on it by cyclic voltammetry to form the reference electrode. In the end, the working electrode has been obtained by electrochemical deposition first of Prussian blue, then GOx, and finally chitosan. Each fiber is wound helically onto a fiber electrode and integrated into fabric. A stretchability up to 200% is demonstrated due to the helical structure. This sensor shows a sensitivity of 11.7 μAmM^{-1} cm^{-2}, a limit of detection of 6 μM, and a linear response up to 500 μM that is relevant for sweat analysis.

In the second generation of biosensors, the sensing element is composed of an additional compound called "mediator" and follows the scheme reported in Figure 1b.

Finally, the third generation of biosensors (Figure 1c) is characterized by the working electrode able to directly detect the GOx oxidation without any redox mediator. This one-step process enhances the sensor sensitivity due to the reduction of electrical losses induced by the mediated component.

A textile example of this kind of sensor was reported by Liu and Lillehoj (Liu and Lillehoj 2016). They fabricated a three-electrode sensor based on polyester thread. The reference electrode was coated with Ag/AgCl ink, while the counter electrode was covered with carbon ink. Then, a working electrode made of a carbon-coated thread was immersed in a GOx solution. Then, each thread has been embroidered into a fabric, with an optimized pattern. The sensor shows a linear range from 0 mM to 50 mM, low interference, and good mechanical properties.

As already mentioned, a different approach consists of non-enzymatic sensors exploiting the electrocatalytic behavior of the functionalized working electrode.

Several functionalizations have been reported but just a few of them have been translated successfully on textiles.

The first approach of non-enzyme sensors uses metals or alloys. An example exploiting the textile substrated was reported by Peng et al. (Peng et al. 2018), which electrochemically deposited gold species on graphene fibers. They used AuNS fiber microelectrode, a fiber of platinum, and a fiber cover with Ag/AgCl as working, counter, and reference electrode, respectively. The high electrocatalytic activity of gold and large-surface-area of graphene fibers allowed detecting H_2O_2 with very low LOD and high sensibility. Exploiting this ability, they demonstrated the detection of glucose in 0,1M NaOH solution.

However, metals and alloys are not electroactive enough to reach high sensibility. Alternative compounds have been investigated. Two groups of materials that own this capability are based on transition metal oxides and sulfides although they suffer from low intrinsic conductivity limiting their application on textile wearable sensors. Therefore, binary transition metal oxides have been created to enhance their electrical properties and electrochemical activities. Among them, the spinel compounds (general composition: AB_2O_4) have been widely investigated for glucose sensing. In parallel, also the binary transition metal sulfides and in particular the compounds of AB_2S_4 show even better performance because, they could maintain structural integrity (Zhu et al. 2015).

Many research demonstrated that $NiCo_2S_4$ shows (Chen et al. 2017), remarkable electrochemical performance in non-enzymatic detection of glucose and in pseudocapacitor applications (Xiong et al. 2015). As an alternative compound, $CuCo_2S_4$ (Liu et al. 2016b) also shows high performance. Recently (Xu et al. 2018), it has been deposited on carbon fiber textiles in a simple one-step method. The as-fabricated $CuCo_2S_4$ carbon fiber is the working electrode, Pt foil the counter electrode and a saturated KCl electrode works as the reference electrode. All measurements have been performed in 0.1 M NaOH solution. Glucose sensor performance was evaluated by chronoamperometry measurement. Sensitivity was calculated to be 3852.7 μAmM^{-1} cm^{-2}. Moreover, it is worth noting that the linear response ranged from 2.5 μM to 3.67 mM. Although over 3.3 mM the amperometric current reaches saturation, these results are compatible with glucose concentration in sweat. However, it is relevant to note that the reported sensors mainly work in an alkaline environment (0,1 M NaOH), not optimal for real-time sweat analysis.

Lactate

The concentration of lactate in blood is correlated with the healthy state of cells. When not enough nutrients and oxygen reach a tissue, cells generate energy through an anaerobic metabolic pathway leading to an increase in the production of lactic acid.

Many reasons can determine this condition, on one hand, an excessive physical exertion during endurance sports such as running, cycling, or boxing. In a persistent condition of anaerobic effort, the accumulation of lactate in the muscles generates muscle fatigue (Promphet et al. 2019). Thus, monitoring lactate concentration can help to improve performance in sports activities. On the other hand, an obstruction of vanes, due to a localized high pressure or in combination with shear and friction, can alter the cell's metabolism. If this condition is maintained for a long time, it can determine an ischemia event with consequent death of cells and necrosis of tissue. For this reason, in standard clinical practice, a high value of lactate concentration in blood is used as an alert for a probable compromised situation. Commercial portable devices for lactate detection are available and are based on standard electrochemical methods using blood samples (Wang et al. 2008). A well-known example is the lactate SCOUT (Senslab). The typical reference level of lactate in sweat is 25 mM but this value is strictly dependent on several factors, such as age, gender, and parts of body perspiration (Jia et al. 2013). However, blood analysis has the same limitations already mentioned in the case of glucose. Wearable textile technology has been developed also to detect lactate continuously. For such kinds of sensors, the functionalization of the sensing element is always enzyme-based. Lactate dehydrogenase (LDH) or lactate oxidase (LOx) has been employed. Both enzymes catalyzed the lactate oxidation in pyruvate. In the cell energy production process (Alam et al. 2018), glucose is converted into adenosine triphosphate (ATP). In one step of this process, LDH catalyzes lactate to pyruvate, through its cofactor, Nicotinamide adenine dinucleotide (NAD+).

$$NAD^+ + Lactate \xrightarrow[LDH]{} NADH + Pyruvate + H^+$$

Alternatively, LOx acts with lactate as GOx acts with glucose, promoting lactate oxidation to pyruvate and releasing hydrogen peroxide (H_2O_2).

$$L - Lactate + LOx \rightarrow pyruvate + LOx_{red}$$

$$LOx_{red} + O_2 \rightarrow LOx_{ox} + H_2O_2$$

Both enzymes have been used for wearable sensors, but the immobilization of LDH with cofactor NAD^+ results a more complex processes and less stable. Thus, in all textile biosensors, LOx has been preferred to exploit both optical and electrochemical transduction mechanism. One example of a colorimetric textile biosensor system has been obtained as a result of the Biotex European project ("Bio-sensing textiles to support health management" 2011). One of the most relevant goals of this sensor is a complex microfluidic fabric system fabricated through a photolithography technique. A sensing colorimetric element is functionalized with a co-immobilization of LOx and horseradish peroxidase. The latter compound changes color in the presence of H_2O_2 which is a product of the lactate oxidation reactions (Scheme 2), and it is directly correlated with its concentration in sweat. As a result, the colorimetric sensor shows a dense purple color when the lactate level is over 5 mM while levels lower than 5 mM lead to a green color (Baysal et al. 2014). More recently, a colorimetric sensor (Promphet et al. 2019) for lactate has also been developed on cotton fabric. Simple fabrication consists of a few steps: first, the cotton substrate was coated with chitosan via padding, then a mixed solution, containing HRP and LOx, was screen printed (Im et al. 2016). As lactate concentration increases, the intensity of the color of the textile sensor visibly increases. Since the reported sensor is specially designed for sport applications, a net color change is observed up to 12.5 mM. This threshold level during sports activities indicates muscle fatigue.

An alternative approach to textile biosensors is the use of electrochemical devices. Commonly, Ag/AgCl is used as a reference electrode, platinum as a counter electrode, and CNT-coated fabric, functionalized with LOx is exploited as the working electrode. Immobilization of the enzyme on fabric is obtained immersing the CNT-coated fabric in a LOx mixed solution (Kai et al. 2018, Liu and Lillehoj 2016), by physical adsorption. In a laboratory perspiration simulation test, with a flow rate of 1 µL cm^{-2} min^{-1}, controlled by a syringe pump, this sensor showed a linear range from 0 to 70 mM.

Ions

Some of the most important components of the human perspiration secreted by the eccrine glands are ions (Na$^+$, Cl$^-$, K$^+$, NH$_4^+$). Several microfluidics models (Sonner et al. 2015) have been developed in the last years to explain the generation of eccrine sweat and its flow, which is strongly coupled with blood-to-sweat biomarkers partition.

On one hand, the level of the electrolytes in sweat strongly depends on the part of the body where they are collected. Even if the Na$^+$, or Cl$^-$ concentrations are very different from the forehead to the lower back (almost a factor of two), it is possible to accurately predict the concentration of the whole-body electrolytes from regional sweat thanks to the high correlation factor (Patterson et al. 2000). Besides this, diet, heat acclimation rate, and genetic predisposition could also affect the concentration of ions in different subjects. On the contrary, aging and sex seem not to influence sweat electrolyte concentrations (Sawka et al. 2007).

On the other hand, numerous clinical and health correlated information can be extracted from the electrolyte's analysis. Although the measurement of Na$^+$ and Cl$^-$ concentration in sweat offers limited correlation with the corresponding values in blood, there is lots of evidence regarding the connection between Cl$^-$ level and medical status. It is used to predict hormonal changes (Lieberman 1966) and to non-invasively diagnose cystic fibrosis when its concentration is abnormally high. Moreover, Na$^+$ and Cl$^-$ concentration can be used to assess hydration level (Bergeron 2003), hyponatremia, and indirectly measure the sweat rate. Conversely, potassium ions' concentration in sweat is directly related to the one in blood and it is independent of sweat rate (Schwartz and Thaysen 1956). Since its concentration in plasma can foretell muscle activity (Medbø and Sejersted 1990) and a vast number of conditions related to hyper- or hypo-kelemia (Newmark and Dluhy 1975), it is very important to achieve a sensing device also able to measure the concentration of this electrolyte.

Among the huge variety of wearable biosensors, the textile-based ions' sensors represent an excellent example for a non-invasive, low cost, comfortable, and lightweight point-of-care platform.

Most of them are based on standard electrochemical techniques, such as potentiometric and amperometric sensors, or exploit the amplification effect of the organic electrochemical transistors (OECT). The most common approach concerns the use of ions-selective membranes (ISM), which exclude all ions except for the one of interest. Although some ISMs are polymeric and do not require other chemical species to sense and immobilize cations, the mostly used ISMs are composites formed by matrix/supporting material, anion or cation excluder component, plasticizer, and ionophore. Figure 2 reports the common chemical materials used to fabricate ISMs for human sweat sensing.

The ability to rapidly and accurately monitor one of the most analyzed electrolytes species in human sweat using a textile device has been presented by Parilla et al. (Parrilla et al. 2016b). They reported a carbon fiber-based sodium ion sensor using commercial carbon fibers (CCF) substrate opportunely modified to realize a potentiometric device in a conventional two-electrode configuration. The working electrode was fabricated dipping the pre-cleaned CCF into the ion-selective membrane cocktail containing: 33 wt.% of Polyvinyl chloride (PVC), 0.2 wt.% potassium tetrakis (4-chlorophenyl)borate (KTClPB), 66 wt.% of 2-nitophenyloctyl ether (NPOE) and 0.7 wt.% sodium ionophore X.

The main solution contained 1 ml of tetrahydrofuran (THF) plus 0.21 wt.% of MWCNT and 5.0 wt.% of polyethylene-co-acrylic acid (PEAA), as also previously reported by Zhu et al. (Zhu et al. 2009). In this case, the matrix material is a high molecular weight polyvinyl chloride (PVC) that presents high strength, low toxicity, redox, and chemical inertness. The KTClPB was used to reduce competitive coordination between the counter ion of the analyte and the ionophore, while the plasticizer 2-nitrophenyloctyl ether (oNPOE) had an important role to dissolve the ionophore (sodium ionophore X) into the polymeric matrix. The ionophore composite was the sensing element of the membrane and it was responsible for ion immobilization producing, in most cases, logarithmic response with a change in analyte-ions activity mediated by Nernst equation. The presence of MWCNTs increased the hydrophobic property and the ion-to-electron transducing capability led to great potential stability and to be unaffected by pH, light, and redox species.

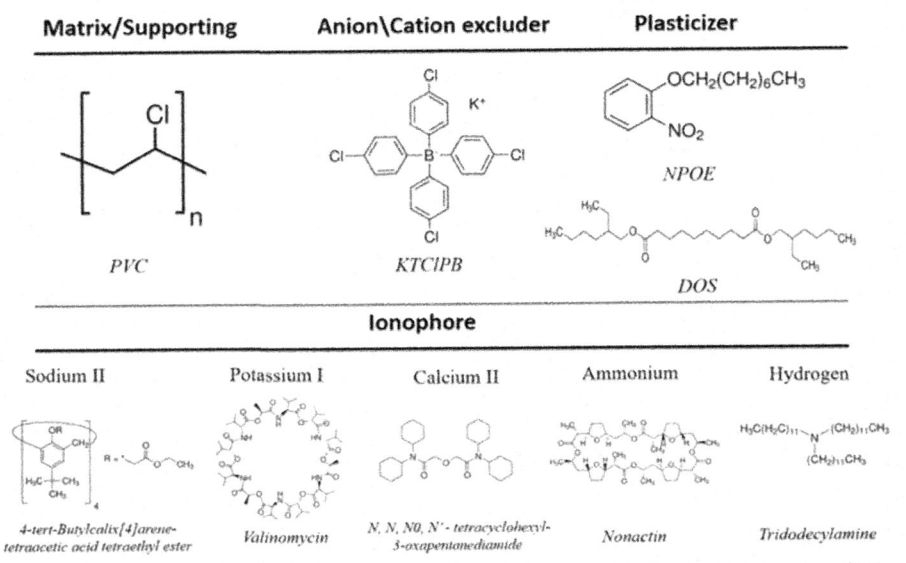

Figure 2. Examples of the chemical structure of the typical components for Ion Selective Membrane (ISMs). The first row shows the common materials used as matrix, anion or cation excluder and plasticizer. The second row collects standard ionophores.

The reference electrode used to extract the electromotive force variation was a solid-state reference membrane made using 10 wt.% of polyvinyl butyral (PVB) in methanol solution containing saturated NaCl as described by Guinovart and co-workers (Guinovart et al. 2013a).

The Na^+- CCF and reference electrodes were integrated into a flexible plastic holder with a proper plastic mask for electrical insulation and a cotton strip for solution sampling. The cotton patch was used to include a microfluidic system that pumps the sweat from the skin to the sensitive region by exploiting the capillary features of the cellulose fibers.

The performance of the sodium ion sensor was evaluated in artificial sweat solution and it presented a Nerstian response of 57.0 mV/decade in a linear dynamic range from 1 mM to 100 mM that is well within the physiologically relevant range for sodium in human perspiration. Moreover, the selectivity coefficients obtained with the fixed interference ion method (FIM) for the Na^+- CCF were similar to conventional membranes using the separate solution method (SSM).

Although it showed long-term stability and a calibration-free approach, a high level of noise was observed during real measurement on an athlete.

On the heels of this idea, Parrilla et al. (Parrilla et al. 2016a) also proposed a variation for the ion-selective membrane cocktail replacing the common PVC matrix with polyurethane (PU) to provide further resistance to mechanical stress and the necessary biocompatibility. These materials can provide exceptional analytical performance when potentiometric techniques are involved and, at the same time, minimize adverse physiological effects like unwanted inflammation or fouling (Yun et al. 1997). Thanks to the attractive mechanical properties of PU, they realized highly stretchable textile-based sensors able to detect the sodium and potassium concentration in human sweat also during extreme mechanical stress. The potassium ion-selective membrane has been realized with 2 wt.% of valinomycin, 32.8 wt.% of PU, 0.5 wt.% KTClPB and 64.7 wt.% of bis (2-Ethylhexyl) sebacate (DOS). The conductive traces (CNT ink and such PU-based membrane) were coupled in a serpentine design to Ecoflex, which is a biocompatible and biodegradable elastic-plastic widely used in the wearable applications. The specific selection of polyurethane as a matrix, coupled with the treated elastic fabric, led to selective and highly stretchable textile sensors that exhibited a Nerstian response of 59.4 mV/ decade and a linear range from 10^{-4} up to 10^{-1} M for sodium ions, and a response of 56.5 mV/decade in the same linear range for the potassium selective electrode. In addition, they performed both crumpling test to assest the mechanical robustness of the sensor and a simulation of a conventional whasing procedure. The calibration plots recorded after these tests led to favorable analytical results.

A breakthrough in the functionalization of textile fibers was taken by Guinovart et al. (Guinovart et al. 2013b) which turned cotton yarns in an ion-selective sensor using carbon nanotubes (CNTs) and ISMs. They exploited the good ion-to-electron transduction features of CNTs as potentiometric devices (Mousavi et al. 2011) coupled with the excellent electrical conductivity. In their work, they proposed a cotton fiber, properly dipped in an ion-selective membrane solution, able to monitor pH as well as potassium (K^+) and ammonium (NH_4) concentration in the liquid. The K^+ ISMs had the same composition presented before, the ammonium ISM contains 32.2 wt.%

of PVC, 1 wt.% of Nonactin and 66.8 wt.% of DOS while the pH sensing membrane contains 33.0 wt.% of PVC, 0.5 wt.% of KTFClPB, 1 wt.% of Tridodecylamine, and 65.5 wt% of DOS. The modified cotton yarns were used as working electrodes in standard potentiometric cells provided with a reference electrode. The promising performance achieved includes stability, quite fast response time, and selectivity. Under optimized condition (e.g., stirred distilled water solution), the calibration curves of the sensors presented an ideal Nernstian behavior with the typical slope of almost 59.2 mV per decade at 25°C. The broad pH linear trend ranged from 3 to 11, while the K^+ and NH_4 sensors had linearity over 4 orders of magnitude.

Besides the great interest inspired by the potentiometric textile sensors, the textile OECT represents a novel class of wearable and portable sensors able to monitor specific analytes in biofluids. During the last decade, different examples have been reported highlighting the promising performance of such textile sensors for the detection of salts (Tarabella et al. 2012), adrenaline (Coppedè et al. 2014), dopamine, ascorbic acid (Gualandi et al. 2016) as well as tyrosine (Battista et al. 2017) in the body fluids employing textile-OECT.

Recently, Coppedè et al. (Coppedè et al. 2020) have exploited the remarkable performance of OECTs, such as the very low operating voltage and absorbed power, coupled with the highest trans-conductance among the transistor-based sensors, to achieve a wearable textile-OECT able to selectively detect the concentration of potassium, and calcium in human sweat. The textile-OECT is composed of a conductive channel made of an acrylic textile covered with the stable, biocompatible, and semiconducting polymer PEDOT:PSS and a gate realized with a silver wire. The textile threads were made conductive, as previously reported by Mattana et al. (Mattana et al. 2011), by immersing the yarns in a semiconducting solution formed by PEDOT:PSS, 10% ethylene glycol and 3% dodecyl benzene sulfonic acid (DBSA) as surfactant, resulting in a linear electrical resistance of 120 Ohm/cm. Even in this case, the ion sensitivity was achieved exploiting ISMs as follows: 32.80 wt.% of high molecular PVC, 65.60 wt.% of DOS, 1 wt.% N, N, N', N' - tetracyclohexyl-3-oxapentanediamide (calcium ionophore II) and 0.6 wt.% of KClBT for calcium membrane while the potassium membrane had the same receipt already reported above. The membrane components were dissolved in Tetrahydrofuran (THF) for proper solubilization and homogenization.

Coppedè and co-workers (Coppedè et al. 2020) studied separately the performance of two different tex-OECT devices recording the channel current versus time upon variation of the gate voltage and analyte concentration. The selectivity feature of the potassium and calcium sensors has been evaluated using sodium ions as interfering species, the most abundant electrolytes in human sweat, since the detection of the other species with respect to sodium is more challenging. Finally, the potassium ISM-based text-OECT reported sensitivity values of 3.49 M^{-1} and 0.71 M^{-1} for K^+ and Na^+, respectively, finding a selectivity factor of 20.34%. The same approach was also followed for the characterization of the device based on a calcium-selective membrane presenting, however, a selectivity factor with respect to the sodium species of 12.19%. Despite the promising results achieved by the transistor configuration based on PEDOT:PSS-functionalized fibers, the requirement of an external and metal gate electrode reduces the flexibility and the wearability of these devices.

To overcome such a problem, Gualandi et al. (Gualandi et al. 2018) reported a novel and breakthrough approach to fabricate textile sensors inspired by the OECT. They synthesized and achieved a newly composed material based on PEDOT:PSS and, Ag/AgCl Nanoparticles integrated into the semiconducting polymer that behave, in such a way, as nanogate electrodes. Consequently, the sensor showed the intrinsic amplification response of a transistor with a simple two-terminal configuration. The textile sensor was made by a cotton thread coated by a mixture of PEDOT:PSS, 1% v/v of (3-glycidyloxypropyl)trimethoxysilane (GOPS), and 20% v/v of ethylene glycol using a roll-to-roll-like technique. The conductive solution homogeneously covers the cotton thread to obtain a conductive fiber with a linear resistance of about 70 Ω/cm. The NPs were deposited on the surface of the polymer layer via a two-step electrodeposition process and they were composed by a core of silver surrounded by a shell of silver chloride. The electronic coupling between the ionic charge and the electrochemically active nanoparticle allows explaining the amplification behavior of the sensors. When the two-terminal textile device was tested in artificial sweat to monitor the chloride ions concentration, the spontaneous and reversible faradaic redox equilibrium $Cl^- + Ag \leftrightarrows AgCl + e^-$ was responsible for the Nernstian relation between the electrochemical potential and chloride concentration. The main analytical signal was the current that flows in the functionalized conductive cotton thread and its variation was linearly proportional to the logarithm of chloride concentration. The sensor characterization in artificial sweat showed a sensitivity of $(51 \pm 9) \ 10^{-3}$ decade^{-1}, short-time response, and selectivity with respect to I^- and Br^- anions. The real-time sweat monitoring using textile sensors technology opens a huge variety of possibilities and new paths to achieve fully-textile multi-sensing platforms for health and sport application as suggested by the possibility to integrate the sensing yarns directly on clothes.

pH

pH variations in human sweat can provide information about the overall body status as well as particular physiological conditions and disorders. For instance, while sweat pH typically ranges from 4.5 to 6.5, a more alkaline sweat (up to pH 9) has been found in patients affected by cystic fibrosis due to impaired bicarbonate reabsorption (Douglas and Mauri 2013). Sweat pH has been correlated to blood glucose levels (Moyer et al. 2012), the intensity of physical exercise, metabolic alkalosis during sport activity (Patterson et al. 2002, Sonner et al. 2015), and skin diseases (Manjakkal et al. 2020).

The route of optical transduction for the realisation of textile pH sensors has been widely explored in literature, with an emphasis on the use of halochromic materials. A halochromic textile changes its color upon pH variations due to the presence of a pH dye, whose reversible acid-base chemistry leads to bathochromic or hypsochromic shifts resulting in visible color changes. While dyeing is one of the most ancient practices where priority is given to color stability, maximisation of color change is desired in textile sensing systems. pH dyes can be combined with the textile matrix by many coloring strategies, including direct application, surface treatments, and direct incorporation during fibers formation (van der Schueren and de Clerck 2012). The pH sensing abilities of dye-functionalised textiles obtained by fibers activation and soaking (Giachet et al. 2017), layer-by-layer method (Park et al. 2011), electrospinning (Van Der Schueren et al. 2010), sol-gel technique (Van Der Schueren et al. 2012), and photo-grafting (Kianfar et al. 2020) have been investigated in relation to the physico-chemical properties of a library of natural and synthetic dyes and fabrics. Among textile optical sensors for pH detection in sweat, one of the first examples of textile platforms for on-body trials was reported by Coyle et al. (Morris et al. 2009), who developed a smart waistband capable of pH and sweat rate monitoring comprising a sensing area with immobilised Bromocresol Purple and a fabric fluidic channel for passive sweat pumping. The wearable platform was coupled with an external control electronics with a LED optical detector and a Bluetooth communication module. More recently, Rosace's group has thoroughly reported on the encapsulation of Litmus (Caldara et al. 2016, 2012a) and Methyl Red (Caldara et al. 2012b) into cotton fabrics by sol-gel method and equipped the halochromic fabric with miniaturized readout electronics including LED optical detector and wireless connectivity. In particular, the authors performed preliminary on-body trials with the wearable module inserted in an elastic waistband, achieving 0.2 pH resolution and an agreement within 0.5 pH with respect to a reference skin pH-meter (Caldara et al. 2016).

Alternatively to optical devices, the realisation of textile pH sensors for sweat analysis based on potentiometric transduction has been introduced only in recent years. The first example was reported by Zamora et al. with the electrodeposition of an iridium oxide film on nylon-based conducting fabrics and stainless-steel mesh fabric (Zamora et al. 2018). The authors reported on the preliminary integration of the textile sensor into a sport belt and recorded a 4% error in the wearable sensor response with respect to a commercial pH test strip during on-body trials. In the last two years, only one potentiometric sensor for sweat pH detection has been realised by printing both the sensing electrode, i.e., thick film graphite-polyurethane composite, and the reference electrode on a cellulose-polyester blend cloth showing sub-Nernstian performances (Manjakkal et al. 2019), while most works in literature are based on PANI electrodeposition to impart pH sensitivity to the working electrode. Among them, PEDOT:PSS wet-spun fibers (Reid et al. 2019) and elastomeric gold fibers woven into a textile matrix (Wang et al. 2020b) have been used as textile substrates for PANI functionalisation and approached the ideal Nernstian sensitivity. Worth mentioning is the work by Yoon et al. where carbon fiber threads extracted from a carbon cloth were modified with PANi and Ag/AgCl and finished with a self-healing polymeric coating (Yoon et al. 2020). The fibers were knitted into a wearable headband with flexible electronics for preliminary on-body measurements. Analogous with the work of Gualandi et al. 2018, a two-terminal pH sensor was recently realised exploiting a similar working principle (Mariani et al. 2020). In fact, a potentiometric pH transducer made of PEDOT doped with a pH dye was electrodeposited on top of a PEDOT:PSS film, the latter acting as a charge transport layer. A spontaneous electrochemical gating takes place, as the current flowing across the PEDOT:PSS film due to the application of a small voltage can be reversibly modulated by the pH dependent electromotive force generated at the interface with the pH transducing layer. The versatile and simple sensing architecture was successfully transferred on a bioceramic fabric by screen printing, thus demonstrating the possibility to integrate a chemical sensor within a bioactive and elastic textile substrate.

Multi-sensing platforms

As we have already stated, the analysis and monitoring of human sweat is an emerging and important tool to perform increasingly accurate medical diagnoses or specific training programs.

Several efforts have been recently made by the research community to deeply study and develop novel wearable sensors to monitor and investigate multiple tailored biomarkers from human perspiration.

The possibility to measure and analyse in real-time and non-invasively important health-related parameters opens new prospective and scenarios in order to obtain a flexible, selective, and high sensitive multi-sensing platform. Several studies, as reported in the previous paragraphs, have been carried out employing non-conventional substrate like fibers and fabrics to achieve flexible and wearable single biosensors. In this section, we report some of the latest examples of textile platform for the detection of multiple analytes and biomarkers in human perspiration. As mentioned above, several biomarkers present in sweat can be used to estimate cystic fibrosis, hydration status, bone mineral loss, and physical stress. In this vein, in 2016 Liu and Lillehoj (Liu and Lillehoj 2016) presented a dual electrochemical sensor for simultaneous detection of glucose and lactate in buffer solution and in whole blood samples. The conductive polyester threads, already cited in paragraph 'Glucose', were functionalized with glucose oxidase or lactate oxidase and embroidered in extended fabric showing linear behavior and no interference effects during multiplex operation.

In addition, Promphet et al. (Promphet et al. 2019) proposed a non-invasive textile-based colorimetric sensing platform for the parallel detection of lactate and pH from the human sweat. The pH sensor pad was realized by screen-printing a mixed solution based on bromocresol green (BCG), methyl orange (MO), cetyltrimethylammonium bromide (CTAB) and, sodium carboxymethylcellulose (NaCMC) onto a chitosan-coated cotton fabric. After an appropriate calibration, based on the mean color intensity, this platform can be used to estimate the lactate level (0–25 mM) and the sweat pH (1–14).

By incorporating further sensors in the same textile platform, it will be possible to carry out simultaneous measurements of numerous analytes from a single sample. One of the first example in the wearable sensing technologies that exploit new and general strategy to realize electrochemical fabric from single sensing fibers was reported by Wang and co-workers in 2018 (Wang et al. 2018). Each sensing unit was able to detect efficiently specific physiological signals directly from human perspiration such as glucose, pH, Na^+, Ca^{2+}, and K^+.

The electrochemical fabric, thanks to its high flexibility and structural integrity, is a promising wearable platform for real-time health monitoring by sewing several types of sensing threads as the building blocks. This approach opens the way to a sensing textile-device able to integrate into itself a great amount of different sensory abilities. In the work here mentioned, the coaxial sensing fibers were realized by coating sensing materials onto carbon nanotube (CNT) fibers previously synthesized via floating catalyst vapor deposition with ferrocene and thiophene as catalysts, and hydrogen and argon as gases. The glucose-sensing fiber was fabricated by drop-casting a chitosan/SWCNT/glucose oxidase mixed solution directly on the CNT nanotube previously coated by electrochemical deposition with Prussian Blue in order to improve the sensitivity. The same electrodeposition technique was also employed to fabricate the pH-sensing fiber by coating the CNT with the Polyaniline (PANI). Regarding the ion-sensing fiber, PEDOT:PSS and ion-selective ionophore were used as an ion-to-electron transducer and a specific ion detector, respectively. The conductive polymer was deposited on the CNT by galvanostatic electrochemical polymerization and, subsequently, the ion-selective membrane was drop-cast on the fiber. The Na^+ membrane contained Na-TFPB, high-molecular-weight PVC, DOS and, sodium ionophore X (weight ratios of 0.55/33/65.45/1). The K^+ membrane wss a mixture of NaTPB, PVC, DOS, and potassium ionophore (0.5/32.75/64.75/2), while the Ca^{2+} membrane contained Na-TFPB, PVC, DOS and, calcium ionophore II (0.55/33/65.45/1). The potentiometric ions and pH sensors, formed by the reference and the sensing electrodes twisted together, showed a linear relationship in the physiologically relevant range. Moreover, the amperometric glucose-sensing fiber operated linearly in the typical glucose concentration range in sweat. As proof of a real application, they performed *in situ* analysis integrating the sensing fabric into a garment to sample and directly analyse human sweat. The work presented by Wang et al. offers a promising and an over-all strategy to increase the impact of textile platforms and, as they showed for the first time, a combined sensing fabric that can simultaneously detect a variety of physiological signals.

Recently, Possanzini et al. (Possanzini et al. 2020) have reported an easy and effective strategy to realize a textile multi sensor platform that behave as an electrochemically gated device, merging the robustness of potentiometic-like transduction mechanism with the highly simple and feasible geometry without the need of a reference electrode. Their multi-thread biosensing platform can detect different bioanalytes simultaneously without interference, and they proposed it for testing chloride ions (Cl^-) concentration and pH level in human sweat. They used different commercial threads, as cotton, silk and polyester, coated with the conducting polymer PEDOT:PSS and properly functionalized with either a nano-composite material (Gualandi et al. 2018) or a chemical sensitive dye (Bromothymol Blue – BTB) (Mariani et al. 2020) to obtain Cl^- and pH selective sensing functionality, respectively.

Another significant example was reported in 2019 by He and co-workers (He et al. 2019) and consists of a flexible sweat analysis patch based on a silk fabric-derived carbon textile (silkNCT) for monitoring six different health-related biomarkers. The silk woven fabric, based on the uniform and mechanically robust silk fibers, can be transformed through thermal treatment in a flexible, nitrogen-doped, and highly conductive carbon textile. The carbonization of silk fabric required a specific heat program in a tube furnace with a mixed atmosphere of argon and hydrogen. The silkNCT fabric was efficiently used to realize an electrochemical sensors array capable of simultaneous and selective detection of Na[+], K[+], lactate, glucose, uric acid, and ascorbic acid. Round patches of silkNCT fabrics were placed on a flexible PET substrate and directly used or combined with other compounds to act as the working electrode of the electrochemical sensors. Enzyme-based amperometric sensors were used for glucose and lactate with the oxidase solution formed by PBS, chitosan, and glucose or lactate oxidase, respectively. The ion-selective sensors for Na[+] and K[+] were fabricated by depositing the PEDOT:PSS and the transparent ion-selective membrane (with the formulations already reported) on SilkNCT fabric. Thanks to its good electrical conductivity, pristine SilkNCT was directly used as the working electrode for the AA and UA sensors. The amperometric sensors relied on the three-electrode configuration with counter and reference electrodes made of carbonized SilkNCT and Ag/AgCl ink-modified conductive tapes, respectively, while the ion-selective sensors exploited the traditional two-electrode potentiometric system. The glucose and lactate sensing patch had a sensitivity of 6.3 nA/μM and 174 nA/μM with a limit of detection of 500 μM and 1000 μM, respectively. The sensitivities for the electrolytes sensors were sub-Nerstian with a linear range in the physiologically relevant human range. The ascorbic acid and uric acid sensors showed promising results, with a sensitivity of 22.7 nA/μM and 196.6 nA/μM, respectively. As a proof of concept, the authors fixed the integrated patch in a wristband to perform real-time sweat analysis in a volunteer's arm and displayed the results on a smartphone.

Wound

Wound healing is the essential physiological process by which damaged tissues repair themselves. It involves four dynamic stages, i.e., vascular response, inflammatory response, proliferation, and remodeling (maturation) (Flanagan 2000, Frykberg and Banks 2015, Mehmood et al. 2014). If progression through the normal healing phases is delayed or incomplete, the wound enters a state of pathologic inflammation characterised by impaired healing and even eventual chronicity (Menke et al. 2007, Qin et al. 2019, Tonnesen et al. 2000). Chronic wounds greatly affect patients' quality of life, are associated with high treatment costs, and are recognized as a major source of mortality in bed-ridden and diabetic patients (Frykberg and Banks 2015, Ochoa et al. 2014, Qin et al. 2019), thus leading to a large social, economic, and medical burden. In order to adopt the most appropriate therapeutic strategy, wound management implicates both wound assessment and evaluation of the healing process.

On one hand, traditional methods for wound assessment are often based on visual inspection or qualitative systems that may lead to subjective interpretations and biased results (Cukjati et al. 2001, Lazarus et al. 1994). Moreover, a number of mutually related factors have been associated with the impaired healing process, including body status (age, nutritional defects, etc.), mechanical stress, presence of infections, as well as pathological disorders like diabetes or autoimmune diseases (Brown et al. 2018, Eming et al. 2014), thus complicating the identification of the actual wound aetiology and prospective healing process. On the other hand, despite several steps having been taken towards the understanding of normal tissue repair stages and influence of physicochemical parameters, the evaluation of the healing process, which is essential to maximise the clinical outcomes of the therapy, remains challenging due to its dynamism and complexity (Lazarus et al. 1994), as well as the lack of firmly established and quantitative tools for wound monitoring over time.

Therapeutic strategies have primarily been targeting the local wound environment as a means to provide the optimal healing milieu, regardless of the underlying condition that causes impaired healing (Eming et al. 2014). In this regard, the choice of the right dressing is crucial to provide moisture, absorb excess wound fluid, and protect the wound site from further trauma, thus accelerating tissue regeneration (Brown et al. 2018, Dabiri et al. 2016, Fonder et al. 2008, Gianino et al. 2018, Mehmood et al. 2014). Dressings can be tailored to the specific healing case and are classified as passive, interactive, advance, and bioactive, the latter comprising biologically active films or even drug delivery systems (Brown et al. 2018). However, the need to periodically visualise the wound site for evaluation of the wound status and therapy efficacy imposes a frequent dressing renewal that is often done unnecessarily (Dargaville et al. 2013, Milne et al. 2016), with a risk of disturbing or interrupting the wound healing process, provoking a physiological stress response or even causing a second injury (Parvaneh et al. 2014, Qin et al. 2019). In this scenario, point-of-care (POC) and wearable sensing technologies are attracting growing interest with potential to revolutionise wound care practice. In particular, the realisation of smart dressings comprising chemical and physical sensors, which are specially designed to detect relevant parameters of wound biochemistry, may provide quantitative and real-time information with both diagnostic and theranostic value, thus allowing timely intervention, improved wound management, and decreased healing time (Brown et al. 2018, Dargaville et al. 2013, Gianino et al. 2018, Mehmood et al. 2014, Milne et al. 2016, Ochoa et al. 2014). Among the assayable biomarkers, the crucial role of pH, exudate composition, moisture, and temperature in wound healing is largely documented and they are commonly identified as potential indicators of healing rate and wound bed state (Cutting 2003, Dargaville et al. 2013, Nocke et al. 2012, Power et al. 2017, Schneider et al. 2007).

Exudate is the fluid secreted by the wound and it has the function of a healing agent in delivering essential nutrients to the wound bed and providing the moist environment essential to the healing process (Cutting 2003, Flanagan 2000). Exudate biomarkers of relevance for wound monitoring are reported in Table 2.

As it can be non-invasively sampled and its composition depends on both healing stage and wound type, wound exudate is an interesting biofluid for continuous wound monitoring by means of smart dressings. Among acute exudate constituents, inorganic salts, urea, uric acid, and creatinine levels are similar to serum. Other components include glucose, lactate, cytokines, matrix metalloproteinases, enzymes, and microorganisms (Cutting 2003, Trengove et al. 1996). Impaired glucose levels, as typical in diabetic wounds, as well as high lactate concentration are known to hinder the healing process (Brem and Tomic-canic 2007, Brown et al. 2018), while uric acid has been related to wound severity, oxidative stress, and bacterial infection (Kassal et al. 2015, Liu and Lillehoj 2017, McLister et al. 2014). Further parameters crucial to the healing process are moisture and temperature. Monitoring the level of hydration can help to identify whether redressing is due and assess if the optimal healing milieu is provided at the wound bed, as too little moisture hinders tissue repair and fluid

Table 2. Resume of exudate biomarkers relevant for wound monitoring.

Biomarker	Relevant range	References
pH	4–6 (healing) 7–9 (non-healing)	(Brown et al. 2018, Dargaville et al. 2013, Schneider et al. 2007, Shukla et al. 2007)
Na^+, K^+, Cl^-, urea, creatinine	Similar to serum	(Cutting, 2003, Trengove et al. 1996)
Glucose	1.1–5.9 mM (healing) 0.6–3.7 mM (non-healing)	(Jankowska et al. 2017, Trengove et al. 1996)
Lactate	5.4–16.7 mM above ~ 10 mM (possible inflammation) above ~ 30 mM (possible infection)	(Loffler et al. 2011, Trengove et al. 1996)
Uric acid	221–751 µM	(Trengove et al. 1996)
Interleukin-6	1.3 pg/µg protein (inflammation)	(Beidler et al. 2009, Brown et al. 2018)
Oxygen	5–20 mm Hg (persistent inflammation)	(Brown et al. 2018, Schreml et al. 2010)

excess leads to skin maceration (Mccoll et al. n.d., Milne et al. 2016, Scott et al. 2020). As for wound temperature, overall higher values were found in non-healing, worsening, or acute wounds due to local vasodilation, while temperature decreases with healing progression (Kruse et al. 2015, Power et al. 2017).

A major challenge in wound exudate analysis is the small sample volume due to limited exuding rate, whose typical value does not exceed 0.9 g cm^{-2} day^{-1} (1.3 mL cm^{-2} day^{-1} reported for pressure-treated wounds) (Dealey et al. 2006) and varies depending on the healing stage and wound type. Moreover, probably due to the complexity and variability of exudate composition and viscosity, no standardised medium has been established to date to simulate the wound fluid in laboratory practice. As a result, quite heterogeneous formulations for simulated wound exudate are reported in literature to assess sensors' performance in real-life conditions, including the industrial standard for the evaluation of fluid uptake by the wound dressings.

pH on wound

A slightly acidic mantle (pH 4 – 6) provides the normal barrier function of healthy skin. In the case of injured skin, the underlying tissue is exposed and this acidic milieu is disturbed as the body's internal pH is 7.4. In acute wounds, pH of the wound bed varies from acidic values during an initial inflammation stage to a more basic granulation step before reepithelization, where the normal skin pH is restored. Conversely, the pH of chronic wounds follows more complex pathways, with oscillations in the range 7 – 9 due to impaired healing (Dargaville et al. 2013, Schneider et al. 2007, Shukla et al. 2007). In general, pH at the wound site has been found to vary according to the wound healing stages, while recovery of a slightly acidic environment is desired as it regulates collagen formation, increases fibroblast activity, and decreases bacterial viability (Brown et al. 2018, Gethin 2007, Lambers et al. 2006). Therefore, non-invasive systems capable of detecting the narrow pH fluctuations occurring at the wound bed could be useful to assess the wound state and the efficacy of the therapeutic strategy.

Thanks to their superior accuracy, potentiometric methods such as the glass surface electrode have been acknowledged as a preferential choice with respect to colorimetric methods (Power et al. 2017). However, a similar approach excludes the integration in textile smart dressings and pH measurement can only be performed at a specific location of the wound. It is worth noting that by skin pH is meant an apparent pH value resulting from the extraction of water-soluble components at the interface between the skin surface and the measuring probe (Schmid-Wendtner and Korting 2006) and in fact the tip of the surface glass electrode is normally moistened with distilled water prior to contact with the wound site (Gethin et al. 2008). This suggests that the pH of the wound bed should be treated in analogy to the wound exudate pH. Several examples of textile pH sensors have been proposed in literature as either raw materials or fully integrated platforms for wound healing monitoring. According to the transduction mechanism, they can be classified into two main categories, i.e., optical and electrochemical sensors.

One of the most straightforward methods for the preparation of halochromic textiles is electrospinning. pH-sensitive dyes can be directly loaded into the electrospinning solution to generate electrospun nanofibrous materials combining the high porosity and permeability of ultrathin polymer fibers with the pH color responsiveness of the dye molecule. Interestingly, halochromic nanofibrous composites have been obtained by mixing hydrophobic and hydrophilic polymers (Gorji et al. 2019), as well as natural and synthetic polymers (Pakolpakçil et al. 2019) leading to smart and highly tailored textiles. A great deal of attention is being devoted to the lifetime, reusability, and photostability of halochormic textiles, with the aim to avoid dye leaching and skin contamination, as well as to allow prolonged wound monitoring. In this regard, the strategy of covalent dye immobilisation has been thoroughly investigated. De Clerk's group has reported on electrospun nanofibrous blends where different pH dyes were covalently immobilised on a synthetic copolymer (Steyaert et al. 2015) and chitosan (Schoolaert et al. 2016) showing remarkable stability after leaching, migration and water fastness tests. Recently, it has been reported that dispersion of chitosan microparticles in a halochromic sponge (Tu et al. 2019) and functionalisation of calcium alginate halochromic fibers with a chitosan derivative (Cui et al. 2020) not only improve the overall dyeing property of the textile matrix but also impart some antibacterial activity. Focusing on covalent immobilisation of dyes, Mohr's group has significantly contributed to the field by studying the functionalisation of cellulose-based textiles. In particular, covalent immobilisation of synthetic azo-dyes was demonstrated in textiles and non-wovens, including T-shirts, facecloths (Mohr and Müller 2015), and cotton swabs (Schaude et al. 2017), achieving remarkable stability to washing cycles and sterilization procedures. The use of fluorescent dyes has been also reported for the preparation of halochromic textiles, including synthetic dyes for covalent functionalisation of cellulose (Mohr 2018), dyes validated for *in vitro* and *in vivo* sensing incorporated in electrospun fibers (Del Mercato et al. 2015) and a fluorescent hydrogel coating for wound pads functionalisation

(Jankowska et al. 2017). Among optical textile sensors, real-time on-site pH monitoring using a smartphone camera has been demonstrated for halochromic hydrogel microfibers obtained through microfluidic spinning (Tamayol et al. 2016) and halochromic polycaprolactone electrospun fibers (Pan et al. 2019). Moreover, Kassal et al. (Kassal et al. 2017) recently reported on a commercial wound dressing modified with a pH-sensitive hydrogel incorporating dye-functionalised cellulose particles and equipped with non-contact, flexible electronics enabling real-time monitoring using international wireless standards RFID and NFC.

Electrochemical textile sensors for pH detection in the wound environment mainly rely on potentiometric sensing. Thanks to the combination of lightweight, flexibility, and pH sensing properties, the conjugated polymer polyaniline (PANI) currently stands out as the gold standard in the realisation of textile, solid-state potentiometric probes. The first work describing a potentiometric pH sensor incorporated into medical textile was reported by Guinovart et al. Screen-printing and electrochemical deposition were used to fabricate reference and working electrodes directly on a commercial adhesive bandage, obtaining Nernstian sensitivity over the 4.35–8.00 pH range and excellent resiliency to bending tests (Guinovart et al. 2014). Remarkably, the authors highlight the importance of achieving maintenance-free and calibration-free wearable chemical sensors. More recently, the electrochemical deposition of PANI has been exploited for the functionalisation of PEDOT:PSS/ MWCNTs coated cotton fibers showing antibacterial properties and Nernstian response over a wide pH range (2 – 12) in complex media (Smith et al. 2019). As an alternative to electrodeposition, textiles can be modified by a simple dip-and-dry process using PANI commercial inks. In this regard, both cotton and polyester threads have been coated with PANI to act as working electrodes and the textile potentiometric systems have been sewn into commercial bandages. In particular, Punjya et al. (Punjiya et al. 2017) have realised a pH sensing smart bandage for chronic wound monitoring, which integrates a C/PANI and Ag/AgCl coated cotton threads and is equipped with a wireless data acquisition and transmission setup using a CMOS potentiostat readout IC, Arduino Nano, and Bluetooth module. Also, based on C/PANI coated cotton threads, a smart wound dressing with an Arduino wireless interface and wireless data transmission has been reported that detects bacterial infections by monitoring the wound pH and allows the typical delivery of antibiotics to the wound bed (Karperien et al. 2019). Finally, a pH-mapping bandage has been developed by sewing functionalised polyester threads in a commercial dressing, where readout electronics and Bluetooth wireless transmission module was encased in a 3D printed and reusable button (Lyu et al. 2018).

Humidity

Wound moisture is a key parameter to the healing process. Indeed, too little moisture can desiccate the wound while too much will lead to its maceration (McColl et al. 2007). Maintenance of an optimum level of moisture allows good healing of the wound in a shorter time. Commonly, clinicians observe the moisture status by removing the dressing, thus disturbing the healing process. Significantly helpful would be to monitor the humidity at the wound bed without removing the dressing, so as to change it only when necessary. Several examples of wearable technologies have been developed for this purpose. A commercial example is the "Wound Sense" (Milne et al. 2016) sensor which sits directly on the wound and detects the moisture status. The sensor is based on two wires of silver chloride. The wound exudate is an electrolyte which affects the impedance between the two electrodes. At a low level of moisture, the impedance is high, vice versa, a wet wound generates low impedance. An intermediate value between two extremes indicates the optimal environment for the healing process. However, the system is just a quality detection, lacking reversibility and autonomous alert or reporting. Other wearable humidity sensors are reported in literature, most of them are fabricated on fabric due to their intrinsic hydrophilicity that helps to sample humidity.

Configuration, materials, and working principles exploit different solutions (Scott et al. 2020). The most common and easiest examples are resistive (Zhu et al. 2011), impeditive (Kutzner et al. 2013, Weremczuk et al. 2012) or capacitive (Zhao et al. 2017) sensors.

A recent relevant example (Ma et al. 2019) of capacitive sensors is based on polyamide fiber cover with copper threads. Polymer works as dielectric materials and copper as the conductive electrode. The humidity alters the whole capacitance of the yarn shaped sensor showing a 3.5 s response and 4 s of recovery. The main advantage regards the connection of humidity sensors with a fiber inductor coil which allows a RFID monitoring.

Another approach enabling wireless connection is humidity colorimetric (Gong et al. 2019) sensors. Gong et al. reported a sensor based on structural color materials which change color due to alteration of the periodic crystalline structure. In their study, they deposited poly (styrene–butyl acrylate–acrylic acid) core-shell colloidal microspheres on fabric. When the sensor is dry, the sensing structural color material is transparent, while when polar solvents, such as water, wet the sensing part, the microspheres change their structural configuration becoming colored. This approach allows a semi-quantitative analysis.

Despite the presence of several examples of textile humidity sensors in literature, just a few of them have been applied to wound monitoring. A relevant example is reported by Zhou et al. (Zhou et al. 2017). They fabricated fibers based on SWCNT dispersed in a matrix of PVA Poly (vinyl alcohol) (PVA). CNTs network was used to create a good electrical conductive path, while PVA is a hygroscopic polymer that swells via water absorption. This implies that, in the presence of water molecules, PVA molecular chain can move and modify the conductive network of CNTs with a consequent increase of the fiber resistivity.

A similar approach is demonstrated by Devaux et al. (Devaux et al. 2011) who substituted the PVA matrix with PLA, another polymer with similar hygroscopic characteristics. Also in this case, PLA molecules partially swelled in the presence of water, and the whole fiber resistivity increased.

Just in the last years, the wound environment has become relevant for wearable technologies, few clinical studies have been conducted and few sensors have been developed for this specific application. The main reason is the biocompatibility requirement to be in contact with damaged skin. Moreover, the high complexity of the wound exudate, and the wide range of wound morphologies limited the application of the most common, rigid humidity sensors. However, those preliminary results open the way for a new important research field.

Surrounding Environment

Textile gas sensors could be non-invasively embedded in clothing for the protection and safety of personnel operating under hazardous conditions such as firefighters, soldiers, and chemists. All these devices operate thanks to a variation of electrical resistance that is proportional to the concentration of target compounds, as happens in a common chemiresistor. The device architecture is simple and can easily be applied to textiles. Indeed, the sensing element is inserted between two electrodes of highly conductive materials. The main sensing materials are conductive polymers and carbon nanomaterials such as carbon nanotubes and graphene oxide. Since the harmful concentrations of toxic gases are known, the collected data can be directly employed to identify dangerous situations without any further studies that are instead mandatory to contextualize the information obtained from biofluids. Nevertheless, transferring the gas sensing technology on textile is more complex than liquid sensing because, for example, common chemosensors based on metal oxide usually require both high temperatures and high voltage to operate. These conditions are not suitable for wearable devices that must work in close contact to the human body. Besides this, the analyte must be adsorbed into the sensitive material with mass transfer processes between at least two phases, with a significant effect on response and recovery times, i.e., the times to obtain a stable signal after exposing and removing the analytes, respectively.

Collins et al. first proposed in 1996 about e-textile sensors for the measurements of NO_2 and NH_3, which remains nowadays the most-studied analytes due to their redox behavior (Collins and Buckley 1996). Researchers have improved the sensing performance by designing new e-textiles with different functional materials or morphologies, but the selectivity remains an open issue. As happens for non-textile chemiresistors, the fabrication of sensor arrays coupled to an advanced statistical analysis is exploited for the fabrication of electronic noses able to identify different substances or odors.

Ammonia

Ammonia is a colorless gas with a characteristic pungent smell. It is among the 10 most-produced chemicals with a lot of applications in several fields and a worldwide annual production of 175 109 Kg. Over half of the synthetic ammonia is used for the production of chemicals for agriculture such as fertilizers, but it is a very important feedstock for several industrial sectors such as textiles, war, mining, and metal manufacturing industries. Ammonia is irritant for eyes, skin, and respiratory tract when its concentration is higher than 50 ppm. In addition, acute ammonia expositions (concentration > 500 ppm) are fatal due to the insurgency of chemical bronchitis, fluid accumulation in the lungs, and chemical burns of the skin. Therefore, non-invasive personal protective equipment able to detect abnormal NH_3 concentrations would improve the occupational safety in several sectors.

All the examples of ammonia textile sensors reported in literature are chemiresistor, based on carbon nanomaterials or polyaniline.

Carbon nanomaterials can be deposited on a textile (Han et al. 2013, Su and Liao 2019) or in the form of carbon nanotubes (CNTs) that can be exploited to directly produce yarns or ropes with a very good mechanical resistance (Li et al. 2006) (textile strength = 210 MPa; Young module = 2.2 GPa). When the electron-rich NH_3 is adsorbed on carbon nanotubes, it exerts a chemical gating that shifts the valence bands and decreases the electrical conductivity (Kong et al. 2000). The use of simple carbon nanomaterials allows the detection in the range of about 10 – 500 ppm, but the sensor performance can be improved by the use of composite materials. The combination of polyelectrolytes and carbon nanomaterials allows better NH_3 adsorption with enhanced performance (Su and Liao 2019, Seesaard et al. 2014). Su et al. describes a flexible single-yarn sensor fabricated by layer-by-layer self-assembly of graphene oxide and poly (allylamine hydrochloride). The limit of detection of such a sensor is 1.5 ppm with an enhancement of sensing performance. Randeniya et al. (Randeniya et al. 2011) have deposited nanocrystalline gold on CNTs ropes to increase the NH_3 adsorption and thus the amount of charge carrier extracted from the CNTs with a significant effect on the performance. This e-textiles displays the best sensitivity among carbon-based devices, and the lowest LOD value (0.5 ppm).

Polyaniline (PANI) is the most-used material for the fabrication of NH_3 textile sensors thanks to low cost, high stability, and easy preparation that takes advantage of high polymerization conversion and simple synthetic approach. In fact, the modification of the fabric is always performed by *in situ* polymerization of aniline that can be also performed by depositing the reagents by inject printing technology (Stempien et al. 2017). The transduction occurs because the reaction between ammonia and the conductive emeraldine salt of PANI converts the conductive polymer into the semiconducting emeraldine base (EB) in accordance with Figure 3.

Since this reaction leads to the deprotonation of PANI nitrogen atoms, polarons are consumed and, therefore, the electrical resistance increases. The starting signal

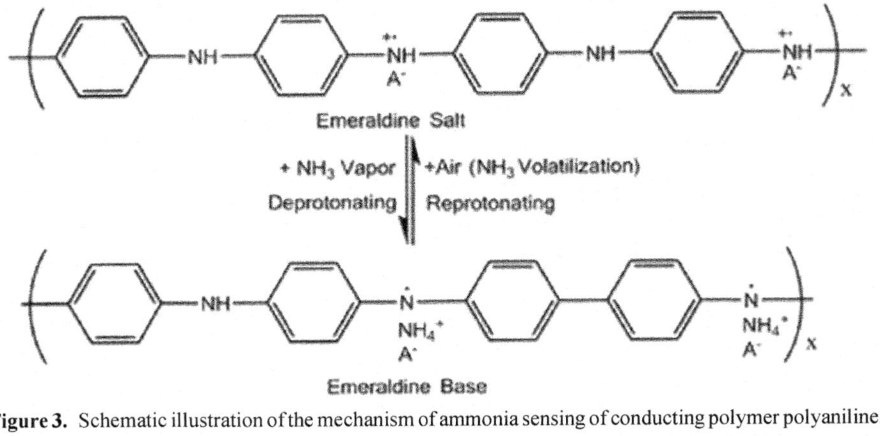

Emeraldine Salt

+ NH_3 Vapor ↕ +Air (NH_3 Volatilization)

Deprotonating ‖ Reprotonating

Emeraldine Base

Figure 3. Schematic illustration of the mechanism of ammonia sensing of conducting polymer polyaniline.

is restored because this reaction is reversible and NH_3 can be desorbed from the polymer. Since conductive polymers are usually p-type conductive materials, they detect NH_3, which is a strong reductant, even if an acid-base group, such as PANI nitrogen atom, is absent. However, PANI outruns other conductive polymers, such as polypyrrole, as NH_3 sensing material for textile devices (Collins and Buckley 1996). The textile sensors for NH_3, NO_2, and warfare agent have been developed since 1996 and the highest NH_3 response was obtained by yarns modified through a layer-by-layer structure made of naphthalenedisulfonic acid and polyaniline. Very high performance can be reached by optimizing the *in situ* PANI polymerization on fabrics by varying the reagent concentrations (aniline, oxidant, H_2SO_4) and the number of fabrication cycles (Qi et al. 2014). The high sensitivity allows for detecting ammonia in the ppb range, while a good selectivity is demonstrated as other volatile compounds show a significantly lower response. The sensing performance of textile sensors based on PANI can be enhanced by varying the PANI dopants. The highest performance is reached by doping PANI with carboxylic acids (formic, acrylic, and trifluor acetic acid). Also, fiber morphology can affect sensing. E-textiles performance can be boosted by depositing PANI on a polyacrylonitrile nanofiber template that ensures a high surface area-to-volume ratio and good mechanical proprieties (Wu et al. 2017). The nanostructured sensor displays a very quick response and very short recovery time in the ppm range. Finally, Polyaniline and carbon nanotubes can be combined to obtain a composite material with enhanced performance (polypropylene) (Maity and Kumar 2018). The sensor exhibits a stable response under mechanical deformation and good selectivity. In addition, the authors studied the effect of temperature showing that response increases slowly up to about 50°C, a limit above which the worst ammonia adsorption occurs with the consequent decrease of sensitivity.

Nitrogen dioxide

Nitrogen dioxide is a brown gas that is classified as hazardous substance (US EPA 1986). It is widely used in the chemical industry as an intermediate in the production of nitric acid, explosives, and as a polymerization inhibitor for acrylates. In the troposphere, NO_2 plays a role in ozone formation by absorbing sunlight. Direct NO_2 exposure (> 100 ppm) can cause irritations and burns. Levels above 1000 ppm can cause death due to asphyxiation from fluid in the lungs. The main danger is the absence of symptoms at the time of exposure other than transient cough, fatigue, or nausea, but it causes edema after an inflammation in the lungs that takes place for hour.

NO_2 has been identified as a target analyte for textile gas sensors due to its oxidizing power that allows for injecting charge carriers in p-type conductors. Although the first attempt of NO_2 detection was performed with yarns modified with conductive polymers (PANI or PPy) (Collins and Buckley 1996), carbon nanomaterials have been mostly used in the following works (Mendoza et al. 2007, Park et al. 2018, Wu et al. 2017, Yun et al. 2018, 2017). The sensors are always chemiresistors. Similar to NH_3, CNTs ropes with good mechanical resistance have been directly exploited for the NO_2 detection (Mendoza et al. 2007). In the last years,

Lee's group has thoroughly investigated the sensing ability of textiles chemically modified with graphene oxide that is then reduced to enhance the conductivity (Lee et al. 2017). Textiles modified with reduced graphene oxide (RGO) took advantage from the high specific surface area, high conductivity, and environmental stability to produce sensors with low signal to noise ratio and stable response to target analytes. All these sensors work in the ppm range, the same request for application in the field of occupational safety and health, and have a stable response after repeated 1000 bending cycles (Yun et al. 2015). Sensing performances can be enhanced by the functionalization with molybdenum disulfide, which transfers holes to RGO after the NO_2 adsorption and thus increases the sensitivity of 4 times with respect to simple RGO-based textiles (Wu et al. 2017). Finally, a highly stretchable textile sensor is obtained using a pre-strain strategy to modify commercial elastic yarns with RGO. The gas sensors showed high NO_2 sensitivity under 200% strain and outstanding mechanical stability as suggested by the maintenance of electrically and sensing performance after 5000 cycles of strain with an elongation of 400%.

Other analytes and electronic noses

Beyond nitrogen dioxide and ammonia, literature reports some examples of conductive textiles that can detect toxic gases, warfare agents, and ethanol. In addition, a sensor array can be exploited to realize a textile electronic nose. Warfare agents detection is important for the fabrication of smart textiles for military use and was already explored by Collins and Buckley (Collins and Buckley 1996). Recently, Maresová et al. (Marešová et al. 2018) designed textile chemiresistors composed of a polymeric ionic liquid deposited between two interdigitated electrodes and graphite. This architecture can be exploited with different sensitive materials and is able to detect toxic gases and warfare agents. The response depends on the molecular weight and dipole moment of the investigated compounds. Ethanol detection can be performed by exploiting V_2O_5 ropes obtained by nano yarns with a process that is similar to drawing a thread from a silk cocoon (Jin et al. 2015). Metal oxides are typical sensing materials for chemiresistors, but they work at high temperatures. In fact, V_2O_5 exhibits the best performance at 330°C with some difficulties in its use in real applications.

A very interesting application is the fabrication of electronic noses because the responses of unselective sensors can be combined to identify specific compounds. Their production requires the fabrication of sensors' arrays; therefore, the sensor architecture is replicated by varying the sensing materials. Kinkeldei et al. (Kinkeldei et al. 2012) proposed carbon black/polymer gas sensors fabricated on a flexible polymer substrate, that can be cut in strips like yarns and woven inside a textile. The sensing performances are changed by varying the polymer mixed with carbon black. The electronic nose detects acetone, IPA, toluene, and methanol by the analysis of the principal components. In addition, the smart textile discriminates against the variation of resistance due to device bending from the contribution that stems from the action of the target compounds. Finally, a textile electronic nose has been fabricated by drop coating and embroidery (Seesaard et al. 2015). The signals are extracted from two interdigitated electrodes that are embroidered on cotton fabrics.

The sensing layers are obtained by deposing a mixture of oxidized carbon nanotubes and a polymer (PVC, cumene-PSMA, PSE and PVP PVC, cumene-PSMA, PSE, and PVP). The electronic nose detects several compounds such as ammonia, ethanol, pyridine, triethylamine, methanol, and acetone and discriminates between different body odors.

Challenges

Stretchability and physical robustness

The main advantage of textile sensors is the ability to be integrated into clothes. This requires two main properties: wearability and washability. Optimization of the fabrication procedure helps to respect these constrains. For example, a pre-treatment of substrates could help to improve the robustness of the sensor to washing. For instance, textile sensors of Joseph Wang's research group are fabricated with a stretchable ink (Jeerapan et al. 2016) of CNT and Ag/AgCl printed on fabric. However, those inks require a hydrophilic substrate, and for this reason, the fabric substrate is pre-treated with an Ecoflex layer. Thanks to this optimization, the final sensors have a good washability (Parrilla et al. 2016a). Also, the group of Yun at al. (Yun et al. 2018) describes NO_2 yarns sensors with a pre-treatment with BSA, which allows maintaining good sensor performance after 10 washing cycles (Yun et al. 2015).

It is relevant to note that the procedure to demonstrate washability is often limited to a laboratory stage, where the sensor is immersed in a beaker with cold water and left under stirring for a certain time, around 40 minutes, for no more than 10 cycles (Gualandi et al. 2016, Parrilla et al. 2016a).

However, in a more realistic case, a laundry machine cycle would be a stronger treatment. To this regard, Mohr's group (Mohr and Müller 2015) has worked on covalent immobilisation of dyes in cellulose-based textiles and demonstrated remarkable stability to washing cycles (performed in a Bauknecht Dynamic Sense WAK7778 laundry machine using 40° and 60°C programs).

On the other hand, conformability is an essential requirement for wearable and on-body sensing devices to achieve the so-called *wear and forget* functionality. Despite being equally important, the resilience to mechanical stress is not always investigated. Examples of mechanical studies on textiles chemical sensors that have been reported in literature include bending tests (Guinovart et al. 2014, Maity and Kumar 2018, Manjakkal et al. 2019, Park et al. 2018, Yang et al. 2018), and stretchability studies (Su and Liao 2019, Wang et al. 2020b). Yun et al. (Yoon et al. 2020) have produced a highly stretchable, mechanically stable, and wearable reduced graphene oxide yarn by exploiting a pre-strain strategy. The yarn was previously covered with BSA (Bovine Serum Albumin). After that, the GO was deposited onto the previously pre-strain yarn. The authors claim that the e-textile exhibits high stability (5000 cycles with a strain of 400%). Moreover, the sensing was also carried out with a strain of 200% without affecting the analytical performances and the electrical features remained unaffected after 1000 bending cycles (Yun et al. 2015). Parilla et al. (Parrilla et al. 2016a) realized highly stretchable textile-based potentiometric sensors by combining polyurethane (PU)-based ion-selective membranes and inks with a serpentine sensor pattern electrodes to increase the mechanical properties.

A recent textile biosensor reported by Zhao et al. is based on elastic gold fiber (Zhao et al. 2019). AuNWs were spin-coated and rapidly dried in air directly on the strained wire forming a long AuNWs/SEBS fiber. Each fiber was wound helically onto a fiber electrode and integrated into a fabric. A stretchability up to 200% has been demonstrated due to the helical structure.

Electrical connection, readout electronics, and power supply

Many efforts have been made to develop innovative and high performing textile chemical sensors. However, a complete textile system includes not only the textile sensor but also connections, readout electronics, and power supply. A common way to connect sensors with electronics imply soldering of metallic wires. Buechley et al. (Buechley and Eisenberg 2009) tried to create solder joints as connectors on a metallic fabric circuit. However, this process is limited by the high temperature required (up to 280°C) which damages most of the commercial fabrics. An alternative is a mechanical gripping between sensor connectors and rigid connectors or metallic wires of circuits. For example, Leśnikowski (Leśnikowski 2016) integrated rigid poppers on fabric as electronic connectors. Li and Tao (Li and Tao 2014) explored unique structures to wrap the conductive track around a stainless-steel needle with a diameter of about 1 mm for forming a circular helix and achieve a helical connection between knitted interconnect and wire. As opposed to soldering, mechanical gripping could be applied to almost all conductive textiles, but the main drawback is that rigid connections cannot keep stability and may break under big deformation. A similar, but more flexible, solution could be conductive adhesive. Siegel et al. (Siegel et al. 2010) applied a commercially conductive adhesive of metallic wires on fiber-based substrate and bonded electronic devices to the metallic pathways attaching sensor electrodes on the adhesive.

Taking into account the readout board, several examples of non-rigid readout system (Han et al. 2013) have been developed to achieve an efficient and small readout system directly located onto the sensor site. Some examples are reported for pH sensing: a bandage (Lyu et al. 2018) with a reader encased in a 3D printed and reusable button, or smart waistbands (Caldara et al. 2016, 2012a, 2012b, Coyle et al. 2009), which include an LED optical detector and wireless connectivity.

Despite several solutions that have been reported, long term durability and multi-plug and unplug repetitions are not yet demonstrated. The same issue remains also for proper insulation of connections in order to avoid cross-talking or short circuit, in a multi-sensing system.

A valid alternative to overcome connection problems could be RFID systems. Kassal et al. (Kassal et al. 2017) recently reported on a pH sensing wound dressing equipped with flexible electronics enabling real-time monitoring using international wireless standards RFID and NFC. Ma et al. (Ma et al. 2019) reported a yarn-shaped humidity sensor with a copper fiber inductor coil for LC wireless monitoring. However, RFID systems need to be interfaced with chemical sensors in which a strong variation of impedance alters the coil signal, thus limiting their applicability.

An innovative approach to integrate whole sensing systems on a cloth is the textile biofuel cell, which provides the required power supply exploiting biomarkers present in body fluids acting as self-powered sensors. Researchers in this field have reached interesting goals thanks to the pioneering work of Katz et al. (Katz et al. 2001). The working principle is the same as in conventional fuel cells: catalysts separate electrons from a parent molecule and force it to reach the cathode thought an electrolyte to generate electric current. In the case of biofuel cells exploited for wearable sensors, the catalyst is an enzyme immobilized on the anode, which catalyzes the oxidation event of metabolites present in body fluid.

In the field of textiles, relevant results have been obtained by the group of Joseph Wang. Their approach is based on stretchable inks (Bandodkar et al. 2016b) printed on fabric. They functionalized the printed CNTs anode with GOx or LOx and NQ as mediator, while the cathode consists of Ag_2O/Ag. Upon adding a biofuel (e.g., glucose or lactate), the biochemical fuel is oxidized by the presence of the enzyme on the anode. The oxidation reaction releases electrons, which are accepted by the silver oxide cathode to complete the power circuit. The great potential of these biofuel cells has been demonstrated by connecting in series six cells functionalized with LOx directly powering 6 LEDs. Based on a similar approach, Yin et al. (Yin et al. 2019) developed a biofuel cell based on CNTs fibers modified with glucose dehydrogenase (GDH) and PolyMG as a mediator, while the gas-diffusion cathode consists of CNR fibers modified with BOD enzyme. It catalyzes the reduction of oxygen in water (Nishizawa 2009). Combining four biofuel cells, they were able to light up a red LED, generating a voltage up to 1.6V when a 100 mM glucose solution wets the cells. Despite the promising preliminary results, application to real life is limited by variations of the environment pH, which influences the biofuel cells' power harvesting ability. Recent results of Sempionatto et al. (Sempionatto et al. 2020) demonstrated that encapsulating an enzyme anode within a hydrophobic carbon-paste helps to maintain high stability under dynamic change of pH conditions. Protected biofuels' cells can recover energy harvesting ability even after long-time exposure to harsh pHs.

Analytical procedure

A critical aspect of wearable textile sensors is data validation, which still limits their commercialization. Main challenges (Wang et al. 2020) are sampling, calibration, selectivity, reliability, and correlation with standard blood analysis. While in outward-looking sensing systems, such as ambient-air gas sensors, the natural diffusion of the analyte at the sensor/gas interface is sufficient to deliver the target gaseous molecule at the sensing site, inward-looking, non-invasive sensors typically have to deal with intermittent and small sample volumes, sample evaporation/stagnation, as well as biofluid rate-dependent analyte concentration. In such cases, the direct contact of the sensing area with skin could compromise the reliability of the detection, while an efficient sampling system should significantly enhance the quality of the analysis. A significant example regards wound exudate analysis. Sample volume is typically small due to limited exuding rate, whose typical value does not exceed 0.9 g cm^{-2} day^{-1} (1.3 mL cm^{-2} day^{-1} reported for pressure-treated wounds) (Dealey et al. 2006) and varies depending on the healing stage and wound type. Textile-based platforms for wound monitoring can profit from the great variety of commercially available dressings, with absorbing capacity, conformability, and drainage features that are tailored to meet the specific wound physiology and healing stage.

Another issue is the correlation of data obtained from external body fluids and blood analysis, that is considered a clinical reference. As discussed for glucose sensors, the knowledge in this field has grown significantly in the last few years, and is attracting increasing attention. Nowadays, they represent a class of innovative wearable sensors that can potentially provide clinically relevant information from body fluids' composition. For example, simultaneous analyses of sweat and blood samples are now possible. It is a new field of research that will open the way to a new kind of personalized medicine.

To reach this goal, an important issue is the need of a well calibrated wearable chemical sensor. Common practices in chemical sensing, such as frequent calibration to correct signal drift and sensor-to-sensor fabrication variations, as well as incubation in conditioning solution or repeated pre-treatment of the sensor and/ or the sample solution, are incompatible with wearable technology and on-body applications (Bandodkar et al. 2016a). In order to address this issue, there are two possible approaches. The first is given by disposable sensors that can be calibrated for a single utilization, as in the case of non-reusable smart wound dressings (Bakker 2016). The second consists in considering a relative value with respect to an initial stable signal. As the body-fluid composition is strictly dependent on the specific condition of each person, an absolute value is often not significant. For example, in sports activities, it is important to monitor the variation of sweat composition from the normal state at rest with respect to the state during the training.

In all cases, the reliability of chemical sensors has to be guaranteed with respect to the interference species. As previously reported, textile chemical sensors for ions in sweat exploit selective membranes. Chemical sensors for glucose and lactate are functionalized with an enzyme which catalyses selectively the desired reactions. For other biomolecules, a promising approach for the detection of other analytes is proposed by Gualandi et al. (Gualandi et al. 2016). They reported an all PEDOT:PSS OECT used for the selective detection of dopamine in the presence of interfering compounds (ascorbic acid, and uric acid). The selective response has been implemented using a potentiodynamic approach, by varying the operating gate voltage and the scan rate.

References

Alam, F., RoyChoudhury, S., Jalal, A. H., Umasankar, Y., Forouzanfar, S., Akter, N., Bhansali, S. and Pala, N. 2018. Lactate biosensing: The emerging point-of-care and personal health monitoring. Biosens. Bioelectron. 117: 818–829.

Bakker, E. 2016. Can Calibration-Free Sensors Be Realized? ACS Sens. 1: 838–841.

Bandodkar, A. J., Jia, W., Yardımcı, C., Wang, X., Ramirez, J., and Wang, J. 2015. Tattoo-based noninvasive glucose monitoring: a proof-of-concept study. Anal. Chem. 87: 394–8.

Bandodkar, A. J., Jeerapan, I., and Wang, J. 2016a. Wearable chemical sensors: present challenges and future prospects. ACS Sens. 1: 464–482.

Bandodkar, A. J., Jeerapan, I., You, J. M., Nuñez-Flores, R., and Wang, J. 2016b. Highly stretchable fully-printed CNT-Based electrochemical sensors and biofuel cells: combining intrinsic and design-induced stretchability. Nano Lett. 16: 721–727.

Barel, A. O., Paye, M., Maibach, H. I., Paye, M., and Maibach, H. I. 2014. Handbook of Cosmetic Science and Technology. CRC Press.

Battista, E., Lettera, V., Villani, M., Calestani, D., Gentile, F., Netti, P. A., Iannotta, S., Zappettini, A., and Coppedè, N. 2017. Enzymatic sensing with laccase-functionalized textile organic biosensors. Org. Electron. 40: 51–57.

Baysal, G., Önder, S., Göcek, İ., Trabzon, L., Kızıl, H., Kök, F. N., and Kayaoğlu, B. K. 2014. Microfluidic device on a nonwoven fabric: A potential biosensor for lactate detection. Text. Res. J. 84: 1729–1741.

Beidler, S. K., Douillet, C. D., Berndt, D. F., Keagy, B. A., Rich, P. B., and Marston, W. A. 2009. Inflammatory cytokine levels in chronic venous insufficiency ulcer tissue before and after compression therapy. J. Vasc. Surg. 49: 1013–1020.

Bergeron, M. F. 2003. Heat cramps: Fluid and electrolyte challenges during tennis in the heat. J. Sci. Med. Sport. 6: 19–27.

Bernards, D. A., MacAya, D. J., Nikolou, M., Defranco, J. A., Takamatsu, S., and Malliaras, G. G. 2008. Enzymatic sensing with organic electrochemical transistors. J. Mater. Chem. 18: 116–120.

Bihar, E., Roberts, T., Ismailova, E., Saadaoui, M., Isik, M., Sanchez-Sanchez, A., Mecerreyes, D., Hervé, T., De Graaf, J. B., and Malliaras, G. G. 2017. Fully printed electrodes on stretchable textiles for long-term electrophysiology. Adv. Mater. Technol. 2: 1600251.

Bio-sensing textiles to support health management [WWW Document]. 2011. https://cordis.europa.eu/project/id/16789/it.

Brem, H., and Tomic-Canic, M. 2007. Cellular and molecular basis of wound healing in diabetes. J. Clin. Investig. 117: 1219–1222.

Brown, M. S., Ashley, B., and Koh, A. 2018. Wearable technology for chronic wound monitoring: Current dressings, advancements, and future prospects. Front. Bioeng. Biotechnol. 6: 47.

Bruen, D., Delaney, C., Florea, L., and Diamond, D. 2017. Glucose sensing for diabetes monitoring: recent developments. Sensors 17: 1866.

Buechley, L., and Eisenberg, M. 2009. Fabric PCBs, electronic sequins, and socket buttons: techniques for e-textile craft. Pers. Ubiquitous Comput. 13: 133–150.

Buono, M. J. 1999. Sweat ethanol concentrations are highly correlated with co-existing blood values in humans. Exp. Physiol. 84: 401–404.

Caldara, M., Colleoni, C., Galizzi, M., Guido, E., Re, V., Rosace, G., and Vitali, A. 2012a. Low power textile-based wearable sensor platform for pH and temperature monitoring with wireless battery recharge. Proc. IEEE Sens. 6–9.

Caldara, Michele, Colleoni, C., Guido, E., Re, V., and Rosace, G. 2012b. Development of a textile-optoelectronic pH meter based on hybrid xerogel doped with Methyl Red. Sens. Actuat. B Chem. 171–172: 1013–1021.

Caldara, M., Colleoni, C., Guido, E., Re, V., and Rosace, G. 2016. Optical monitoring of sweat pH by a textile fabric wearable sensor based on covalently bonded litmus-3-glycidoxypropyltrimethoxysilane coating. Sens. Actuat. B. Chem. 222: 213–220.

Chen, D., Wang, H., and Yang, M. 2017. A novel ball-in-ball hollow NiCo2S4 sphere based sensitive and selective nonenzymatic glucose sensor. Anal. Methods. 9: 4718–4725.

Choudhary, T., Rajamanickam, G. P., and Dendukuri, D. 2015. Woven electrochemical fabric-based test sensors (WEFTS): A new class of multiplexed electrochemical sensors. Lab Chip 15: 2064–2072.

Clark, L. C. 1965. Membrane polarographic electrode system and method with elecxtrochemical compensation. US3539455D.

Collins, G. E., and Buckley, L. J. 1996. Conductive polymer-coated fabrics for chemical sensing. Synth. Met. 78: 93–101.

Coppedè, N., Tarabella, G., Villani, M., Calestani, D., Iannotta, S., and Zappettini, A. 2014. Human stress monitoring through an organic cotton-fiber biosensor. J. Mater. Chem. B. 2: 5620–5626.

Coppedè, N., Giannetto, M., Villani, M., Lucchini, V., Battista, E., Careri, M., and Zappettini, A. 2020. Ion selective textile organic electrochemical transistor for wearable sweat monitoring. Org. Electron. 78: 105579.

Coyle, S., Morris, D., Lau, K. -T., Diamond, D., and Moyna, N. 2009. Textile-based wearable sensors for assisting sports performance, in: 2009 Body Sensor Networks. Institute of Electrical and Electronics Engineers (IEEE), pp. 307–311.

Cui, L., Hu, J. jing, Wang, W., Yan, C., Guo, Y., and Tu, C. 2020. Smart pH response flexible sensor based on calcium alginate fibers incorporated with natural dye for wound healing monitoring. Cellulose. 27: 6367–6381.

Cukjati, D., Reberšek, S., and Miklavčič, D. 2001. A reliable method of determining wound healing rate. Med. Biol. Eng. Comput. 39: 263–271.

Cutting, K. F. 2003. Wound exudate: composition and functions. Br. J. Community Nurs. 8: S4–S9.

Dabiri, G., Damstetter, E., and Phillips, T. 2016. Choosing a wound dressing based on common wound characteristics. Adv. Wound Care. 5: 32–41.

Dąbrowska, A. K., Rotaru, G. -M., Derler, S., Spano, F., Camenzind, M., Annaheim, S., Stämpfli, R., Schmid, M., and Rossi, R. M. 2016. Materials used to simulate physical properties of human skin. Ski. Res. Technol. 22: 3–14

Dargaville, T. R., Farrugia, B. L., Broadbent, J. A., Pace, S., Upton, Z., and Voelcker, N. H. 2013. Sensors and imaging for wound healing: A review, Biosensors and Bioelectronics. Elsevier.

Dealey, C., Cameron, J., and Arrowsmith, M. 2006. A study comparing two objective methods of quantifying the production of wound exudate. J. Wound Care. 15: 149–153.

Del Mercato, L. L., Moffa, M., Rinaldi, R., and Pisignano, D. 2015. Ratiometric Organic Fibers for Localized and Reversible Ion Sensing with Micrometer-Scale Spatial Resolution. Small. 11: 6417–6424.

Derbyshire, P. J., Barr, H., Davis, F., and Higson, S. P. J. 2012. Lactate in human sweat: A critical review of research to the present day. J. Physiol. Sci. 62: 429–440.

Devaux, E., Aubry, C., Campagne, C., and Rochery, M. 2011. PLA/carbon nanotubes multifilament yarns for relative humidity Textile sensor. J. Eng. Fiber. Fabr. 6: 13–24.

Douglas, B. L., and Mauri, E. K. 2013. Cystic Fibrosis: Does CFTR Malfunction Alter pH Regulation?, in: Genetic Disorders. InTech.

Du, D., Li, P., and Ouyang, J. 2016. Graphene coated nonwoven fabrics as wearable sensors. J. Mater. Chem. C. 4: 3224–3230.

Eming, S. A., Martin, P., and Tomic-Canic, M. 2014. Wound repair and regeneration mechanisms. Science TransL. Med. 322: 265sr6.

Flanagan, M. 2000. The physiology of wound healing. J. Wound Care.

Fonder, M. A., Lazarus, G. S., Cowan, D. A., Aronson-Cook, B., Kohli, A. R., and Mamelak, A. J. 2008. Treating the chronic wound: A practical approach to the care of nonhealing wounds and wound care dressings. J. Am. Acad. Dermatol. 58: 185–206.

Frykberg, R. G., and Banks, J. 2015. Challenges in the treatment of chronic wounds. Adv. Wound Care. 4: 560–582.

Gao, W., Emaminejad, S., Nyein, H. Y. Y., Challa, S., Chen, K., Peck, A., Fahad, H. M., Ota, H., Shiraki, H., Kiriya, D., Lien, D. -H. H., Brooks, G. A., Davis, R. W., and Javey, A. 2016. Fully integrated wearable sensor arrays for multiplexed in situ perspiration analysis. Nature. 529: 509–514.

Gethin, G. 2007. The significance of surface pH in chronic wounds. Wounds UK. 3: 52–56.

Gethin, G. T., Cowman, S., and Conroy, R. M. 2008. The impact of Manuka honey dressings on the surface pH of chronic wounds. Int. Wound J. 5: 185–194.

Giachet, F. T., Vineis, C., Sanchez Ramirez, D. O., Carletto, R. A., Varesano, A., and Mazzuchetti, G. 2017. Reversible and washing resistant textile-based optical pH sensors by dyeing fabrics with curcuma. Fiber. Polym. 18: 720–730.

Gianino, E., Miller, C., and Gilmore, J. 2018. Smart wound dressings for diabetic chronic wounds. Bioengineering 5: 51.

Gil, I., Fernández-García, R., and Tornero, J. A. 2019. Embroidery manufacturing techniques for textile dipole antenna applied to wireless body area network. Text. Res. J. 89: 1573–1581.

Glennon, T., O'Quigley, C., McCaul, M., Matzeu, G., Beirne, S., Wallace, G. G., Stroiescu, F., O'Mahoney, N., White, P., and Diamond, D. 2016. 'SWEATCH': A Wearable Platform for Harvesting and Analysing Sweat Sodium Content. Electroanalysis 28: 1283–1289.

Gong, W., Hou, C., Zhou, J., Guo, Y., Zhang, W., Li, Y., Zhang, Q., and Wang, H. 2019a. Continuous and scalable manufacture of amphibicus energy yarns and textiles. Nat. Commun. 10: 1–8.

Gong, X., Hou, C., Zhang, Q., Li, Y., and Wang, H. 2019b. Solvatochromic structural color fabrics with favorable wearability properties. J. Mater. Chem. C. 7: 4855–4862.

Gorji, M., Sadeghianmaryan, A., Rajabinejad, H., Nasherolahkam, S., and Chen, X. 2019. Development of highly pH-sensitive hybrid membranes by simultaneous electrospinning of amphiphilic nanofibers reinforced with graphene oxide. J. Funct. Biomater. 10: 23.

Gualandi, I., Marzocchi, M., Achilli, A., Cavedale, D., Bonfiglio, A., and Fraboni, B. 2016a. Textile organic electrochemical transistors as a platform for wearable biosensors. Sci. Rep. 6: 1–10.

Gualandi, I., Tonelli, D., Mariani, F., Scavetta, E., Marzocchi, M., and Fraboni, B. 2016b. Selective detection of dopamine with an all PEDOT:PSS Organic Electrochemical Transistor. Sci. Rep. 6: 1–10.

Gualandi, I., Tessarolo, M., Mariani, F., Cramer, T., Tonelli, D., Scavetta, E., and Fraboni, B. 2018. Nanoparticle gated semiconducting polymer for a new generation of electrochemical sensors. Sensors Actuators B Chem. 273: 834–841.

Guinovart, T., Bandodkar, A. J., Windmiller, J. R., Andrade, F. J., and Wang, J. 2013a. A potentiometric tattoo sensor for monitoring ammonium in sweat. Analyst. 138: 7031–7038.

Guinovart, T., Parrilla, M., Crespo, G. A., Rius, F. X., and Andrade, F. J. 2013b. Potentiometric sensors using cotton yarns, carbon nanotubes and polymeric membranes. Analyst 138: 5208.

Guinovart, T., Valdés-Ramírez, G., Windmiller, J. R., Andrade, F. J., and Wang, J. 2014. Bandage-based wearable potentiometric sensor for monitoring wound pH. Electroanalysis 26: 1345–1353.

Guo, L., Berglin, L., and Mattila, H. 2012. Improvement of electro-mechanical properties of strain sensors made of elastic-conductive hybrid yarns. Text. Res. J. 82: 1937–1947.

Han, J. W., Kim, B., Li, J., and Meyyappan, M. 2013. A carbon nanotube based ammonia sensor on cotton textile. Appl. Phys. Lett. 102: 193104.

He, J., Xiao, G., Chen, X., Qiao, Y., Xu, D., and Lu, Z. 2019a. A thermoresponsive microfluidic system integrating a shape memory polymer-modified textile and a paper-based colorimetric sensor for the detection of glucose in human sweat. RSC Adv. 9: 23957–23963.

He, W., Wang, C., Wang, H., Jian, M., Lu, W., Liang, X., Zhang, X., Yang, F., and Zhang, Y. 2019b. Integrated textile sensor patch for real-time and multiplex sweat analysis. Sci. Adv. 5: 1–8.

Hirokawa, T., Okamoto, H., Gosyo, Y., Tsuda, T., and Timerbaev, A. R. 2007. Simultaneous monitoring of inorganic cations, amines and amino acids in human sweat by capillary electrophoresis. Anal. Chim. Acta 581: 83–88.

Huang, C. T., Chen, M. L., Huang, L. L., and Mao, I. F. 2002. Uric acid and urea in human sweat. Chin. J. Physiol. 45: 109–115.

Huang, X., Liu, Y., Chen, K., Shin, W. J., Lu, C. J., Kong, G. W., Patnaik, D., Lee, S. H., Cortes, J. F., and Rogers, J. A. 2014. Stretchable, wireless sensors and functional substrates for epidermal characterization of sweat. Small 10: 3083–3090.

Im, S. H., Kim, K. R., Park, Y. M., Yoon, J. H., Hong, J. W., and Yoon, H. C. 2016. An animal cell culture monitoring system using a smartphone-mountable paper-based analytical device. Sens. Actuat. B Chem. 229: 166–173.

Jankowska, D. A., Bannwarth, M. B., Schulenburg, C., Faccio, G., Maniura-Weber, K., Rossi, R. M., Scherer, L., Richter, M., and Boesel, L. F. 2017. Simultaneous detection of pH value and glucose concentrations for wound monitoring applications. Biosens. Bioelectron. 87: 312–319.

Jeerapan, I., Sempionatto, J. R., Pavinatto, A., You, J. -M. M., and Wang, J. 2016. Stretchable biofuel cells as wearable textile-based self-powered sensors. J. Mater. Chem. A. 4: 18342–18353.

Jia, W., Bandodkar, A. J., Valdés-Ramírez, G., Windmiller, J. R., Yang, Z., Ramírez, J., Chan, G., and Wang, J. 2013. Electrochemical tattoo biosensors for real-time noninvasive lactate monitoring in human perspiration. Anal. Chem. 85: 6553–6560.

Jin, W., Yan, S., An, L., Chen, W., Yang, S., Zhao, C., and Dai, Y. 2015. Enhancement of ethanol gas sensing response based on ordered V_2O_5 nanowire microyarns. Sens. Actuat. B Chem. 206: 284–290.

Kai, H., Kato, Y., Toyosato, R., and Nishizawa, M. 2018. Fluid-permeable enzymatic lactate sensors for micro-volume specimen. Analyst. 143: 5545–5551.

Karperien, L., Dabiri, S. M. H., Hadisi, Z., Hamdi, D., Samiei, E., and Akbari, M. 2019. Smart thread based pH sensitive antimicrobial wound dressing. 2019 IEEE Int. Flex. Electron. Technol. Conf. IFETC. 2019. 8–12.

Karpova, E. V., Shcherbacheva, E. V., Galushin, A. A., Vokhmyanina, D. V., Karyakina, E. E., and Karyakin, A. A. 2019. Noninvasive diabetes monitoring through continuous analysis of sweat using flow-through glucose biosensor. Anal. Chem. 91: 3778–3783.

Kassal, P., Kim, J., Kumar, R., De Araujo, W. R., Steinberg, I. M., Steinberg, M. D., and Wang, J. 2015. Smart bandage with wireless connectivity for uric acid biosensing as an indicator of wound status. Electrochem. Commun. 56: 6–10.

Kassal, P., Zubak, M., Scheipl, G., Mohr, G. J., Steinberg, M. D., and Murković Steinberg, I. 2017. Smart bandage with wireless connectivity for optical monitoring of pH. Sens. Actuat. B Chem. 246: 455–460.

Katz, E., Bückmann, A. F., and Willner, I. 2001. Self-powered enzyme-based biosensors. J. Am. Chem. Soc. 123: 10752–10753.

Kianfar, P., Abate, M. T., Trovato, V., Rosace, G., Ferri, A., Bongiovanni, R., and Vitale, A. 2020. Surface Functionalization of Cotton Fabrics by Photo-Grafting for pH Sensing Applications. Front. Mater. 7: 1–9.

Kim, S. J., Song, W., Yi, Y., Min, B. K., Mondal, S., An, K. S., and Choi, C. G. 2018. High durability and waterproofing rGO/SWCNT-fabric-based multifunctional sensors for human-motion detection. ACS Appl. Mater. Interfaces. 10: 3921–3928.

Kinkeldei, T., Christoph Z., Niko M., and Gerhard, T. 2012. An electronic nose on flexible substrates integrated into a smart textile. Sens. Actuat. B Chem. 174: 81–86.

Kong, J., Franklin, N. R., Zhou, C., Chapline, M. G., Peng, S., Cho, K., and Dai, H. 2000. Nanotube molecular wires as chemical sensors. Science. 287: 622–625.

Kruse, C. R., Nuutila, K., Lee, C. C. Y., Kiwanuka, E., Singh, M., Caterson, E. J., Eriksson, E., and Sørensen, J. A. 2015. The external microenvironment of healing skin wounds. Wound Repair Regen. 23: 456–464.

Kutzner, C., Lucklum, R., Torah, R., Beeby, S., and Tudor, J. 2013. Novel screen printed humidity sensor on textiles for smart textile applications, in: 2013 Transducers and Eurosensors XXVII: The 17th International Conference on Solid-State Sensors, Actuators and Microsystems, TRANSDUCERS and EUROSENSORS 2013. pp. 282–285.

Kwon, S., Kim, H., Choi, S., Jeong, E. G., Kim, D., Lee, S., Lee, H. S., Seo, Y. C., and Choi, K. C. 2018. Weavable and highly efficient organic light-emitting fibers for wearable electronics: a scalable, low-temperature process. Nano Lett. 18: 356.

Lai, Y. -C., Deng, J., Zhang, S. L., Niu, S., Guo, H., and Wang, Z. L. 2017. Single-thread-based wearable and highly stretchable triboelectric nanogenerators and their applications in cloth-based self-powered human-interactive and biomedical sensing. Adv. Funct. Mater. 27: 1604462.

Lambers, H., Piessens, S., Bloem, A., Pronk, H., and Finkel, P. 2006. Natural skin surface pH is on average below 5, which is beneficial for its resident flora. Int. J. Cosmet. Sci. 28: 359–370.

Lazarus, G. S., Cooper, D. M., Knighton, D. R., Margolis, D. J., Percoraro, R. E., Rodeheaver, G., and Robson, M. C. 1994. Definitions and guidelines for assessment of wounds and evaluation of healing. Wound Repair Regen. 2: 165–170.

Lee, H., Song, C., Hong, Y. S., Kim, M. S., Cho, H. R., Kang, T., Shin, K., Choi, S. H., Hyeon, T., and Kim, D. -H. H. 2017. Wearable/disposable sweat-based glucose monitoring device with multistage transdermal drug delivery module. Sci. Adv. 3: e1601314.

Leśnikowski, J. 2016. Research on poppers used as electrical connectors in high speed textile transmission lines. Autex Res. J. 16: 228–235.

Li, Q., and Tao, X. M. 2014. Three-dimensionally deformable, highly stretchable, permeable, durable and washable fabric circuit boards. Proc. R. Soc. A Math. Phys. Eng. Sci. 470: 20140472.

Li, Y. H., Zhao, Y. M., Zhu, Y. Q., Rodriguez, J., Morante, J. R., Mendoza, E., Poa, C. H. P., and Silva, S. R. P. 2006. Mechanical and NH_3 sensing properties of long multi-walled carbon nanotube ropes. Carbon. 44: 1821–1825.

Li, Z., and Wang, Z. L. 2011. Air/liquid-pressure and heartbeat-driven flexible fiber nanogenerators as a micro/nano-power source or diagnostic sensor. Adv. Mater. 23: 84–89.

Lieberman, J. 1966. Cyclic fluctuation of sweat electrolytes in women: effect of polythiazide upon sweat electrolytes. JAMA J. Am. Med. Assoc. 195: 629–635.

Liu, N., Ma, W., Tao, J., Zhang, X., Su, J., Li, L., Yang, C., Gao, Y., Golberg, D., and Bando, Y. 2013. Cable-type supercapacitors of three-dimensional cotton thread based multi-grade nanostructures for wearable energy storage. Adv. Mater. 25: 4925–4931.

Liu, S., Hui, K. S., and Hui, K. N. 2016a. Flower-like copper cobaltite nanosheets on graphite paper as high-performance supercapacitor electrodes and enzymeless glucose sensors. ACS Appl. Mater. Interfaces. 8: 3258–3267.

Liu, X., Guo, R., Shi, Y., Deng, L., and Li, Y. 2016b. Durable, washable, and flexible conductive PET fabrics designed by fiber interfacial molecular engineering. Macromol. Mater. Eng. 301: 1383–1389.

Liu, X., and Lillehoj, P. B. 2016c. Embroidered electrochemical sensors for biomolecular detection. Lab Chip. 16: 2093–2098.

Liu, X., and Lillehoj, P. B. 2017. Embroidered electrochemical sensors on gauze for rapid quantification of wound biomarkers. Biosens. Bioelectron. 98: 189–194.

Loffler, M., Schmohl, M., Schneiderhan-Marra, N., and Beckert, S. 2011. Wound Fluid Diagnostics in Diabetic Foot Ulcers. Glob. Perspect. Diabet. Foot Ulcerations. INTECH Open Access Publisher, Croatia.

Lu, L., Hu, Y., and Dai, K. 2017. The advance of fiber-shaped lithium ion batteries. Mater. Today Chem. 5: 24-33.

Lv, J., Jeerapan, I., Tehrani, F., Yin, L., Silva-Lopez, C. A., Jang, J. -H., Joshuia, D., Shah, R., Liang, Y., Xie, L., Soto, F., Chen, C., Karshalev, E., Kong, C., Yang, Z., and Wang, J. 2018. Sweat-based wearable energy harvesting-storage hybrid textile devices †. Energy Environ. Sci. 11: 3431.

Lyu, B., Punjiya, M., Matharu, Z., and Sonkusale, S. 2018. An improved pH mapping bandage with thread-based sensors for chronic wound monitoring. Proc. - IEEE Int. Symp. Circuits Syst. 2018-May, 2018–2021.

Ma, L., Wu, R., Patil, A., Zhu, S., Meng, Z., Meng, H., Hou, C., Zhang, Y., Liu, Q., Yu, R., Wang, J., Lin, N., and Liu, X. Y. 2019. Full-textile wireless flexible humidity sensor for human physiological monitoring. Adv. Funct. Mater. 29: 1904549.

Maity, D., and Kumar, R. T. R. 2018. Polyaniline Anchored MWCNTs on fabric for high performance wearable ammonia sensor. ACS Sens. 3: 1822–1830.

Manjakkal, L., Dang, W., Yogeswaran, N., and Dahiya, R. 2019. Textile-based potentiometric electrochemical pH Sensor for Wearable Applications. Biosensors. 9: 14.

Manjakkal, L., Dervin, S., and Dahiya, R. 2020. Flexible potentiometric pH sensors for wearable systems. RSC Adv. 10: 8594–8617.

Marešová, E., Tomeček, D., Fitl, P., Vlček, J., Novotný, M., Fišer, L., Havlová, Š., Hozák, P., Tudor, A., Glennon, T., Florea, L., Coyle, S., Diamond, D., Skaličan, Z., Hoskovcová, M., and Vrňata, M. 2018. Textile chemiresistors with sensitive layers based on polymer ionic liquids: Applicability for detection of toxic gases and chemical warfare agents. Sens. Actuat. B Chem. 266: 830–840.

Mariani, F., Gualandi, I., Tonelli, D., Decataldo, F., Possanzini, L., Fraboni, B., and Scavetta, E. 2020. Design of an electrochemically gated organic semiconductor for pH sensing. Electrochem. Commun. 116: 106763.

Mattana, G., Cosseddu, P., Fraboni, B., Malliaras, G. G., Hinestroza, J. P., and Bonfiglio, A. 2011. Organic electronics on natural cotton fibres. Org. Electron. physics. Mater. Appl. 12: 2033–2039.

McColl, D., Cartlidge, B., and Connolly, P. 2007. Real-time monitoring of moisture levels in wound dressings *in vitro*: An experimental study. Int. J. Surg. 5: 316–322.

Mccoll, D., Macdougall, M., Watret, L., and Connolly, P. 2009. Monitoring moisture without disturbing the wound dressing. Wounds UK. 5: 2–6.

McLister, A., Phair, J., Cundell, J., and Davis, J. 2014. Electrochemical approaches to the development of smart bandages: A mini-review. Electrochem. Commun. 40: 96–99.

Medbø, J. I., and Sejersted, O. M. 1990. Plasma potassium changes with high intensity exercise. J. Physiol. 421: 105–122.

Mehmood, N., Hariz, A., Fitridge, R., and Voelcker, N. H. 2014. Applications of modern sensors and wireless technology in effective wound management. J. Biomed. Mater. Res. Part B Appl. Biomater. 102: 885–895.

Mendoza, E., Rodriguez, J., Li, Y., Zhu, Y. Q., Poa, C. H. P., Henley, S. J., Romano-Rodriguez, A., Morante, J. R., and Silva, S. R. P. 2007. Effect of the nanostructure and surface chemistry on the gas adsorption properties of macroscopic multiwalled carbon nanotube ropes. Carbon. 45: 83–88.

Menke, N. B., Ward, K. R., Witten, T. M., Bonchev, D. G., and Diegelmann, R. F. 2007. Impaired wound healing. Clin. Dermatol. 25: 19–25.

Mickelsen, O., and Keys, A. 1943. The composition of sweat, with special reference to the vitamins. J. Biol. Chem. 149: 479–490.

Milne, S. D., Seoudi, I., Al Hamad, H., Talal, T. K., Anoop, A. A., Allahverdi, N., Zakaria, Z., Menzies, R., and Connolly, P. 2016. A wearable wound moisture sensor as an indicator for wound dressing change: an observational study of wound moisture and status. Int. Wound J. 13: 1309–1314.

Mohr, G. J., and Müller, H. 2015. Tailoring colour changes of optical sensor materials by combining indicator and inert dyes and their use in sensor layers, textiles and non-wovens. Sens. Actuat. B Chem. 206: 788–793.

Mohr, G. J. 2018. Synthesis of naphthalimide-based indicator dyes with a 2-hydroxyethylsulfonyl function for covalent immobilisation to cellulose. Sens. Actuat. B Chem. 275: 439–445.

Morris, D., Coyle, S., Wu, Y., Lau, K. T., Wallace, G., and Diamond, D. 2009. Bio-sensing textile based patch with integrated optical detection system for sweat monitoring. Sens. Actuat. B Chem. 139: 231–236.

Mousavi, Z., Teter, A., Lewenstam, A., Maj-Zurawska, M., Ivaska, A., and Bobacka, J. 2011. Comparison of multi-walled carbon nanotubes and poly (3-octylthiophene) as ion-to-electron transducers in all-solid-state potassium ion-selective electrodes. Electroanalysis 23: 1352–1358.

Moyer, J., Wilson, D., Finkelshtein, I., Wong, B., and Potts, R. 2012. Correlation between sweat glucose and blood glucose in subjects with diabetes. Diabetes Technol. Ther. 14: 398–402.

Newmark, S. R., and Dluhy, R. G. 1975. Hyperkalemia and Hypokalemia. JAMA J. Am. Med. Assoc. 231: 631–633.

Nishizawa, M. 2009. Carbon Nanotube-Based Enzymatic Biofuel Cells, In: Nanocarbons for Energy Conversion: Supramolecular Approaches. Nanostructure Science and Technology. Springer, Cham.

Nocke, A., Schröter, A., Cherif, C., and Gerlach, G. 2012. Miniaturized textile-based multi-layer pH-sensor for wound monitoring applications. Autex Res. J. 12: 20–22.

Ochoa, M., Rahimi, R., and Ziaie, B. 2014. Flexible sensors for chronic wound management. IEEE Rev. Biomed. Eng. 7: 73–86.

Ozanich, R. 2018. Chem/bio wearable sensors: Current and future direction, in: Pure and Applied Chemistry. De Gruyter, pp. 1605–1613.

Pakolpakçil, A., Osman, B., Özer, E. T., Şahan, Y., Becerir, B., Göktalay, G., and Karaca, E. 2019. Halochromic composite nanofibrous mat for wound healing monitoring. Mater. Res. Express. 6: 1250c3.

Pan, N., Qin, J., Feng, P., Li, Z., and Song, B. 2019. Color-changing smart fibrous materials for naked eye real-time monitoring of wound pH. J. Mater. Chem. B. 7: 2626–2633.

Park, H. J., Kim, W. J., Lee, H. K., Lee, D. S., Shin, J. H., Jun, Y., and Yun, Y. J. 2018. Highly flexible, mechanically stable, and sensitive NO_2 gas sensors based on reduced graphene oxide nanofibrous mesh fabric for flexible electronics. Sens. Actuat. B Chem. 257: 846–852.

Park, J. B., Kim, S. H., and Bae, J. S. 2011. Quinaldine and Indole based pH sensitive Textile chemosensor. Fibers Polym. 12: 696–699.

Parrilla, M., Cánovas, R., Jeerapan, I., Andrade, F. J., and Wang, J. 2016a. A textile-based stretchable multi-ion potentiometric sensor. Adv. Healthc. Mater. 5: 996–1001.

Parrilla, M., Ferré, J., Guinovart, T., and Andrade, F. J. 2016b. Wearable potentiometric sensors based on commercial carbon fibres for monitoring sodium in sweat. Electroanalysis. 28: 1267–1275.

Parvaneh, S., Grewal, G. S., Grewal, E., Menzies, R. A., Talal, T. K., Armstrong, D. G., Sternberg, E., and Najafi, B. 2014. Stressing the dressing: Assessing stress during wound care in real-time using wearable sensors. Wound Med. 4: 21–26.

Patterson, M. J., Galloway, S. D. R., and Nimmo, M. A. 2002. Effect of induced metabolic alkalosis on sweat composition in men. Acta Physiol. Scand. 174: 41–46.

Patterson, M. J., Galloway, S. D. R., and Nimmo, M. A. 2000. Variations in regional sweat composition in normal human males. Exp. Physiol. 85: 869–875.

Peng, Y., Lin, D., Justin Gooding, J., Xue, Y., and Dai, L. 2018. Flexible fiber-shaped non-enzymatic sensors with a graphene-metal heterostructure based on graphene fibres decorated with gold nanosheets. Carbon. 136: 329–336.

Possanzini, L., Tessarolo, M., Mazzocchetti, L., Campari, E. G., and Fraboni, B. 2019. Impact of Fabric Properties on Textile Pressure Sensors Performance. Sensors. 19: 4686.

Possanzini, L., Decataldo, F., Mariani, F., Gualandi, I., Tessarolo, M., Scavetta, E., and Fraboni, B. 2020. Textile sensors platform for the selective and simultaneous detection of chloride ion and pH in sweat. Sci. Rep. 10: 17180.

Power, G., Moore, Z., and O'Connor, T. 2017. Measurement of pH, exudate composition and temperature in wound healing: A systematic review. J. Wound Care. 26: 381–397.

Promphet, N., Rattanawaleedirojn, P., Siralertmukul, K., Soatthiyanon, N., Potiyaraj, P., Thanawattano, C., Hinestroza, J. P., and Rodthongkum, N. 2019. Non-invasive textile based colorimetric sensor for the simultaneous detection of sweat pH and lactate. Talanta. 192: 424–430.

Punjiya, M., Nejad, H. R., Mostafalu, P., and Sonkusale, S. 2017. PH sensing threads with CMOS readout for Smart Bandages. Proc. - IEEE Int. Symp. Circuits Syst. 10–13.

Qi, J., Xu, X., Liu, X., and Lau, K. T. 2014. Fabrication of textile based conductometric polyaniline gas sensor. Sens. Actuat. B Chem. 202: 732–740.

Qin, M., Guo, H., Dai, Z., Yan, X., and Ning, X. 2019. Advances in flexible and wearable pH sensors for wound healing monitoring. J. Semicond. 40: 111607.

Randeniya, L. K., Martin, P. J., Bendavid, A., and McDonnell, J. 2011. Ammonia sensing characteristics of carbon-nanotube yarns decorated with nanocrystalline gold. Carbon. 49: 5265–5270.

Reid, D. O., Smith, R. E., Garcia-Torres, J., Watts, J. F., and Crean, C. 2019. Solvent treatment of wet-spun PEDOT: PSS fibers for fiber-based wearable pH sensing. Sensors. 19: 1–10.

Sato, K., Kang, W. H., Saga, K., and Sato, K. T. 1989. Biology of sweat glands and their disorders. I. Normal sweat gland function. J. Am. Acad. Dermatol. 20: 537–563.

Sato, K., and Sato, F. 1990. Na^+, K^+, H^+, Cl^-, and Ca^{2+} concentrations in cystic fibrosis eccrine sweat in vivo and in vitro. J. Lab. Clin. Med. 115: 504–511.

Sawka M. N., Burke L. M., Eichner E. R., Maughan R. J., Montain, S. J., and Stachenfeld, N. S. 2007. Exercise and Fluid Replacement. Med. Sci. Sport. Exerc. 39: 377–390.

Schaude, C., Fröhlich, E., Meindl, C., Attard, J., Binder, B., and Mohr, G. J. 2017. The development of indicator cotton swabs for the detection of pH in wounds. Sensors. 17: 1365.

Schmid-Wendtner, M. H., and Korting, H. C. 2006. The pH of the skin surface and its impact on the barrier function. Skin Pharmacol. Physiol. 19: 296–302.

Schneider, L. A., Korber, A., Grabbe, S., and Dissemond, J. 2007. Influence of pH on wound-healing: A new perspective for wound-therapy? Arch. Dermatol. Res. 298: 413–420.

Schoolaert, E., Steyaert, I., Vancoillie, G., Geltmeyer, J., Lava, K., Hoogenboom, R., and De Clerck, K. 2016. Blend electrospinning of dye-functionalized chitosan and poly (ε-caprolactone): Towards biocompatible pH-sensors. J. Mater. Chem. B. 4: 4507–4516.

Schreml, S., Szeimies, R. M., Prantl, L., Karrer, S., Landthaler, M., and Babilas, P. 2010. Oxygen in acute and chronic wound healing. Br. J. Dermatol. 163: 257–268.

Schwartz, I. L., and Thaysen, J. H. 1956. Excretion of sodium and potassium in human sweat. J. Clin. Invest. 35: 114–120.

Scott, C., Cameron, S., Cundell, J., Mathur, A., and Davis, J. 2020. Adapting resistive sensors for monitoring moisture in smart wound dressings. Curr. Opin. Electrochem. 23: 31–35.

Sears, M. E., Kerr, K. J., and Bray, R. I. 2012. Arsenic, Cadmium, Lead, and Mercury in Sweat: A Systematic Review. J. Environ. Public Health. 2012.

Seesaard, T., Lorwongtragool, P., and Kerdcharoen, T. 2015. Development of fabric-based chemical gas sensors for use as wearable electronic noses. Sensors. 15: 1885–1902.

Sekar, M., Pandiaraj, M., Bhansali, S., Ponpandian, N., and Viswanathan, C. 2019. Carbon fiber based electrochemical sensor for sweat cortisol measurement. Sci. Rep. 9: 1–14.

Sempionatto, J. R., Raymundo-Pereira, P. A., Azeredo, N. N. F., De Loyola Silva, A. E. A., Angnes, L., and Wang, J. 2020. Enzymatic biofuel cells based on protective hydrophobic carbon paste electrodes: Towards epidermal bioenergy harvesting in the acidic sweat environment. Chem. Commun. 56: 2004–2007.

Shim, B. S., Chen, W., Doty, C., Xu, C., and Kotov, N. A. 2008. Smart electronic yarns and wearable fabrics for human biomonitoring made by carbon nanotube coating with polyelectrolytes. Nano Lett. 8: 4151–4157.

Shukla, V. K., Shukla, D., Tiwary, S. K., Agrawal, S., and Rastogi, A. 2007. Evaluation of pH measurement as a method of wound assessment. J. Wound Care. 16: 291–294.

Siegel, A. C., Phillips, S. T., Dickey, M. D., Lu, N., Suo, Z., and Whitesides, G. M. 2010. Foldable printed circuit boards on paper substrates. Adv. Funct. Mater. 20: 28–35.

Sim, H. J., Choi, C., Kim, S. H., Kim, K. M., Lee, C. J., Kim, Y. T., Lepró, X., Baughman, R. H., and Kim, S. J. 2016. Stretchable Triboelectric Fiber for Self-powered Kinematic Sensing Textile. Sci. Rep. 6: 1–7.

Smith, R. E., Totti, S., Velliou, E., Campagnolo, P., Hingley-Wilson, S. M., Ward, N. I., Varcoe, J. R., and Crean, C. 2019. Development of a novel highly conductive and flexible cotton yarn for wearable pH sensor technology. Sens. Actuat. B Chem. 287: 338–345.

Sonner, Z, Wilder, E., Heikenfeld, J., Kasting, G., Beyette, F., Swaile, D., Sherman, F., Joyce, J., Hagen, J., Kelley-Loughnane, N., and Naik, R. 2015. The microfluidics of the eccrine sweat gland, including biomarker partitioning, transport, and biosensing implications. Biomicrofluidics. 9: 31301.

Stempien, Z., Kozicki, M., Pawlak, R., Korzeniewska, E., Owczarek, G., Poscik, A., and Sajna, D. 2017. Ammonia gas sensors ink-jet printed on textile substrates, in: Proceedings of IEEE Sensors. Institute of Electrical and Electronics Engineers Inc. pp. 1–3.

Steyaert, I., Vancoillie, G., Hoogenboom, R., and De Clerck, K. 2015. Dye immobilization in halochromic nanofibers through blend electrospinning of a dye-containing copolymer and polyamide-6. Polym. Chem. 6: 2685–2694.

Su, P. G., and Liao, Z. H. 2019. Fabrication of a flexible single-yarn NH_3 gas sensor by layer-by-layer self-assembly of graphene oxide. Mater. Chem. Phys. 224: 349–356.

Sun, H., Fu, X., Xie, S., Jiang, Y., and Peng, H. 2016. Electrochemical capacitors with high output voltages that mimic electric eels. Adv. Mater. 28: 2070–2076.

Takamatsu, S., Lonjaret, T., Crisp, D., Badier, J. -M., Malliaras, G. G., and Ismailova, E. 2015. Direct patterning of organic conductors on knitted textiles for long-term electrocardiography. Sci. Rep. 5: 15003.

Takamatsu, S., Lonjaret, T., Ismailova, E., Masuda, A., Itoh, T., and Malliaras, G. G. 2016. Wearable keyboard using conducting polymer electrodes on textiles. Adv. Mater. 28: 4485–4488.

Tamayol, A., Akbari, M., Zilberman, Y., Comotto, M., Lesha, E., Serex, L., Bagherifard, S., Chen, Y., Fu, G., Ameri, S. K., Ruan, W., Miller, E. L., Dokmeci, M. R., Sonkusale, S., and Khademhosseini, A. 2016. Flexible pH-Sensing Hydrogel Fibers for Epidermal Applications. Adv. Healthc. Mater. 5: 711–719.

Tarabella, G., Villani, M., Calestani, D., Mosca, R., Iannotta, S., Zappettini, A., and Coppedè, N. 2012. A single cotton fiber organic electrochemical transistor for liquid electrolyte saline sensing. J. Mater. Chem. 22: 23830.

Tchafa, F. M., and Huang, H. 2018. Microstrip patch antenna for simultaneous strain and temperature sensing. Smart Mater. Struct. 27: 065019.

Tessarolo, M., Possanzini, L., Campari, E. G., Bonfiglioli, R., Violante, F. S., Bonfiglio, A., and Fraboni, B. 2018. Adaptable pressure textile sensors based on a conductive polymer. Flex. Print. Electron. 3: 034001.

Seesaard, T., Khunarak, C., Seaon, S., Lorwongtragool, P., and Kerdcharoen, T. 2014. A Novel Creation of Thread-Based Ammonia Gas Sensors for Wearable Wireless Security System, in: IEEE. pp. 1–4.

Tonnesen, M. G., Feng, X., and Clark, R. A. F. 2000. Angiogenesis in wound healing. J. Investig. Dermatology Symp. Proc. 5: 40–46.

Trengove, N. J., Langton, S. R., and Stacey, M. C. 1996. Biochemical analysis of wound fluid from nonhealing and healing chronic leg ulcers. Wound Repair Regen. 4: 234–239.

Tu, C., Zhang, R. dong, Yan, C., Guo, Y., and Cui, L. 2019. A pH indicating carboxymethyl cellulose/chitosan sponge for visual monitoring of wound healing. Cellulose. 26: 4541–4552.

Tunáková, V., Grégr, J., Tunák, M., and Dohnal, G. 2018. Functional polyester fabric/polypyrrole polymer composites for electromagnetic shielding: Optimization of process parameters. J. Ind. Text. 47: 686–711.

US EPA, O. 1986. Emergency Planning and Community Right-to-Know Act (EPCRA).

Van Der Schueren, L., Mollet, T., Ceylan, Ö., and De Clerck, K. 2010. The development of polyamide 6.6 nanofibres with a pH-sensitive function by electrospinning. Eur. Polym. J. 46: 2229–2239.

van der Schueren, L., and de Clerck, K. 2012. Coloration and application of pH-sensitive dyes on textile materials. Color. Technol. 128: 82–90.

Van Der Schueren, L., De Clerck, K., Brancatelli, G., Rosace, G., Van Damme, E., and De Vos, W. 2012. Novel cellulose and polyamide halochromic textile sensors based on the encapsulation of Methyl Red into a sol-gel matrix. Sens. Actuat. B Chem. 162: 27–34.

Wang, J. 2005. Carbon-nanotube based electrochemical biosensors: A review. Electroanalysis.

Wang, J. 2008. Electrochemical glucose biosensors. Chem. Rev. 108: 814–825.

Wang, L., Fu, X., He, J., Shi, X., Chen, T., Chen, P., Wang, B., and Peng, H. 2020a. Application challenges in fiber and textile electronics. Adv. Mater. 32: 1901971.

Wang, Lie, Wang, Liyuan, Zhang, Y., Pan, J., Li, S., Sun, X., Zhang, B., and Peng, H. 2018. Weaving Sensing Fibers into Electrochemical Fabric for Real-Time Health Monitoring. Adv. Funct. Mater. 28: 1804456.

Wang, R., Zhai, Q., Zhao, Y., An, T., Gong, S., Guo, Z., Shi, Q. Q., Yong, Z., and Cheng, W. 2020b. Stretchable gold fiber-based wearable electrochemical sensor toward pH monitoring. J. Mater. Chem. B. 8: 3655–3660.

Wang, Y., Xu, H., Zhang, J., and Li, G. 2008. Electrochemical Sensors for Clinic Analysis. Sensors 8: 2043–2081.

Wang, Y., Qing, X., Zhou, Q., Zhang, Y., Liu, Q., Liu, K., Wang, W., Li, M., Lu, Z., Chen, Y., and Wang, D. 2017. The woven fiber organic electrochemical transistors based on polypyrrole nanowires/reduced graphene oxide composites for glucose sensing. Biosens. Bioelectron. 95: 138–145.

Weremczuk, J., Tarapata, G., and Jachowicz, R. 2012. Humidity sensor printed on textile with use of ink-jet Technology. In: Procedia Engineering. Elsevier Ltd, pp. 1366–1369.

Windmiller, J. R., and Wang, J. 2013. Wearable electrochemical sensors and biosensors: a review. Electroanalysis. 25: 29–46.

Wu, S., Liu, P., Zhang, Y., Zhang, H., and Qin, X. 2017. Flexible and conductive nanofiber-structured single yarn sensor for smart wearable devices. Sens. Actuat. B Chem. 252: 697–705.

Xiong, X., Waller, G., Ding, D., Chen, D., Rainwater, B., Zhao, B., Wang, Z., and Liu, M. 2015. Controlled synthesis of NiCo2S4 nanostructured arrays on carbon fiber paper for high-performance pseudocapacitors. Nano Energy. 16: 71–80.

Xu, H., Liao, C., Zuo, P., Liu, Z., and Ye, B. C. 2018a. Magnetic-Based Microfluidic Device for On-Chip Isolation and Detection of Tumor-Derived Exosomes. Anal. Chem. 90: 13451–13458.

Xu, Q. B., Ke, X. T., Shen, L. W., Ge, N. Q., Zhang, Y. Y., Fu, F. Y., Liu, and X. D. 2018b. Surface modification by carboxymethy chitosan via pad-dry-cure method for binding Ag NPs onto cotton fabric. Int. J. Biol. Macromol. 111: 796–803.

Xu, W., Lu, J., Huo, W., Li, J., Wang, X., Zhang, C., Gu, X., and Hu, C. 2018. Direct growth of CuCo2S4 nanosheets on carbon fiber textile with enhanced electrochemical pseudocapacitive properties and electrocatalytic properties towards glucose oxidation. Nanoscale. 10: 14304–14313.

Yang, A., Li, Y., Yang, C., Fu, Y., Wang, N., Li, L., and Yan, F. 2018. Fabric organic electrochemical transistors for biosensors. Adv. Mater. 30: 1800051.

Yin, S., Jin, Z., and Miyake, T. 2019. Wearable high-powered biofuel cells using enzyme/carbon nanotube composite fibers on textile cloth. Biosens. Bioelectron. 141: 111471.

Yoon, J. H., Kim, S. M., Park, H. J., Kim, Y. K., Oh, D. X., Cho, H. W., Lee, K. G., Hwang, S. Y., Park, J., and Choi, B. G. 2020. Highly self-healable and flexible cable-type pH sensors for real-time monitoring of human fluids. Biosens. Bioelectron. 150: 111946.

Yun, S. Y., Hong, Y. K., Oh, B. K., Cha, G. S., Nam, H., Lae, S. B., and Jin, J. Il. 1997. Potentiometric properties of ion-selective electrode membranes based on segmented polyether urethane matrices. Anal. Chem. 69: 868–873.

Yun, Y. J., Hong, W. G., Kim, W. J., Jun, Y., and Kim, B. H. 2013. A novel method for applying reduced graphene oxide directly to electronic textiles from yarns to fabrics. Adv. Mater. 25: 5701–5705.

Yun, Y. J., Hong, W. G., Choi, N. J., Kim, B. H., Jun, Y., Lee, and H. K. 2015. Ultrasensitive and highly selective graphene-based single yarn for use in wearable gas sensor. Sci. Rep. 5: 1–7.

Yun, Y. J., Hong, W. G., Kim, D. Y., Kim, H. J., Jun, Y., and Lee, H. K. 2017. E-textile gas sensors composed of molybdenum disulfide and reduced graphene oxide for high response and reliability. Sens. Actuat. B Chem. 248: 829–835.

Yun, Y. J., Kim, D. Y., Hong, W. G., Ha, D. H., Jun, Y., and Lee, H. K. 2018. Highly stretchable, mechanically stable, and weavable reduced graphene oxide yarn with high NO_2 sensitivity for wearable gas sensors. RSC Adv. 8: 7615–7621.

Zamora, M. L., Dominguez, J. M., Trujillo, R. M., Goy, C. B., Sánchez, M. A., and Madrid, R. E. 2018. Potentiometric textile-based pH sensor. Sens. Actuat. B Chem. 260: 601–608.

Zhang, Z., Azizi, M., Lee, M., Davidowsky, P., Lawrence, P., and Abbaspourrad, A. 2019. A versatile, cost-effective, and flexible wearable biosensor for *in situ* and *ex situ* sweat analysis, and personalized nutrition assessment . Lab Chip. 19: 3448–3460.

Zhang, Z., Li, X., Guan, G., Pan, S., Zhu, Z., Ren, D., and Peng, H. 2014. A lightweight polymer solar cell textile that functions when illuminated from either side. Angew. Chemie Int. Ed. 53: 11571–11574.

Zhao, X., Long, Y., Yang, T., Li, J., and Zhu, H. 2017. Simultaneous high sensitivity sensing of temperature and humidity with graphene woven fabrics. ACS Appl. Mater. Interfaces. 9: 30171–30176.

Zhao, Y., Zhai, Q., Dong, D., An, T., Gong, S., Shi, Q., and Cheng, W. 2019. Highly stretchable and strain-insensitive fiber-based wearable electrochemical biosensor to monitor glucose in the sweat. Anal. Chem. 91: 6569–6576.

Zhou, G., Byun, J. H., Oh, Y., Jung, B. M., Cha, H. J., Seong, D. G., Um, M. K., Hyun, S., and Chou, T. W. 2017. Highly sensitive wearable textile-based humidity sensor made of high-strength, single-walled carbon nanotube/poly (vinyl alcohol) Filaments. ACS Appl. Mater. Interfaces. 9: 4788–4797.

Zhu, J., Qin, Y., and Zhang, Y. 2009. Preparation of all solid-state potentiometric ion sensors with polymer-CNT composites. Electrochem. Commun. 11: 1684–1687.

Zhu, Y., Wang, L., and Xu, C. 2011. Carbon nanotubes in biomedicine and biosensing. In: Carbon Nanotubes - Growth and Applications. InTech, pp. 135–162.

Zhu, Y., Ji, X., Wu, Z., and Liu, Y. 2015. $NiCo_2S_4$ hollow microsphere decorated by acetylene black for high-performance asymmetric supercapacitor. Electrochim. Acta. 186: 562–571.

CHAPTER 8

Miscellaneous Materials for Chemical Sensing

Prashant Gupta,[1] *Aastha Dutta*[1] and
Mostafizur Rahaman[2,*]

Introduction

Chemical sensors are devices that convert a chemical/physical property of a specific analyte, component, or chemical moiety into a signal which is quantifiable along with the magnitude of which is proportional to the concentration of the chemical/ material under study (Ohashi and Dai 2006). These are usually based on a single molecule which consists of two vital components: one working as a receptor or an input for selective interaction with the analyte and the other to vary a physicochemical property as read-out reaction to the binding event justifying as input and output, respectively. These molecules have a spacer working as an interlink in the receptor and read-out (Chen et al. 2017). In a nutshell, it comprises of a physical transducer along with a layer with chemical sensitivity. This layer is capable of reacting with the surrounding chemical environment to produce a measurable signal through the transduction principle. Furthermore, its composition is crucial as it has the potential to control a variety of parameters such as selectivity, sensitivity, time of response, and durability among others. The most significant identification quality parameter for a chemical sensor is the signal to noise ratio and quick response which is both sensitive and selective (Regan 2019). The sensors emit response signals upon identifying the presence of certain chemicals for which they are intended. The mechanisms on which sensing may depend include charge carrier concentrations, electroactive properties, chemical reactions (redox), etc. The applications of varying chemical sensors are found in biological, environmental, and industrial monitoring for the presence of gases, vapors, enzymes, proteins, etc., and thus have drawn the research fraternity's

---

[1] Department of Plastic and Polymer Engineering, Maharashtra Institute of Technology, Aurangabad.
[2] Department of Chemistry, King Saud University, Riyadh, 11451, Saudi Arabia.
* Corresponding author: mrahaman1997@gmail.com

attention with regard to the improvement in their effectiveness and efficiencies along with improving the product range (Bhandari 2019).

The chemical sensor market is mainly driven by the rising levels of pollution. The chemical sensor market has been valued at USD 21.39 billion in 2020 with an estimated compounded annual growth rate of 7.5% to reach USD 33 billion by the end of 2026. Their application in the identification of chemical composition in various samples such as blood glucose for old age population, the analyte in chemical industries, etc., along with the advanced ones in defence, healthcare, environmental monitoring, etc., are the driving forces (Mordor Intelligence 2020).

The common materials used in the chemical sensors as reviewed in the literature are graphene (Singh et al. 2017), carbon nanotubes (Schroeder et al. 2018), metal-oxides (Zappa et al. 2018), conducting polymers (Naveen et al. 2017), hydrogels (Hwang et al. 2018), organic-inorganic hybrids (Chen et al. 2019), metal organic frameworks (Jang et al. 2017), biomaterials (Sun et al. 2018), and textiles (Tessarolo et al. 2018), which have been discussed as single chapters each till now.

Miscellaneous sensing materials

Pyrylium

Pyrylium is a cation with the formula $C_5H_5O^+$ which has a six membered ring. It is known to form mono-cyclic and heterocyclic compounds and stable salts with a variety of oppositely charged anions. Pyrylium salts have been successfully employed as sensitizers for photoinduced electron transfer reactions (Che et al. 2006, Banu and Ramamurthy 2009), photocatalysts (Bayarri et al. 2008, Arques et al. 2009) and sensors for volatile organic compounds (El-Roz et al. 2017), anions (Schmitt 2020), amines (Fonseca et al. 2018), amino acids (Beltrán et al. 2020), cyanide (Yahaya and Seferoglu 2018), proteins (Beltrán et al. 2020) and heavy metals (Lazar et al. 2017), thus exhibiting their robustness in various chemical detection methodologies.

Fonseca et al. synthesized a polymeric membrane sensor with an occluded pyrylium salt ([2,6-diphenyl-4 (p-methacryloyloxy) phenylpyryliumtetrafluoroborate]) for the colorimetric determination of trimethylamine (TMA) vapors. A color change of yellow to intense pink was observed with increase in TMA concentration which is an important biogenic amine in food safety applications. The sensory material was reusable (up to 10 cycles) upon HCl washing and exhibited detection and quantification limits of 4.42 and 13.40 ppm, respectively, through the use of UV visible spectrophotometry technique (Fonseca et al. 2018).

Sheng et al. carried out a computational study to understand the role of water molecule and counterion on hydrogen sulphide ion (HS) sensing reactions with the use of para-dimethylamino-2,4,6-triphenyl-pyrylium, a pyrylium derivative. They observed that the HS⁻ sensing mechanism happened via forming an open pyrylium ring complex and subsequent S-C6 bond formation led to the thiopyrylium product (Sheng and Regner 2019).

Chakraborty et al. synthesized a rigid pentacyclic pyrylium fluorescent probe for clinical diagnosis, i.e., measuring pH imbalance during apoptosis by a single step modified Vilsmeier-Haack reaction. The probe exhibited an exceptional "turn on" response for the changes in cell pH values during apoptosis. The center ring of the pyrylium consisting of a strong acceptor linked via two hydroxyl functional groups was observed to be instrumental towards the bright fluorescent dye with superior photophysical characteristics such as photostability, high molar absorptivity and large Stokes shifts. The successful application of this probe to monitor intracellular acidification at physiological pH was further proposed with its tentative use as a chemical tool for screening, drug discovery and dose determination in cancer therapy (Chakraborty et al. 2018).

Hydrazone

The hydrazone functional group is versatile and has easy synthesis routes. Its unique structural characteristics such as modularity and resistance to hydrolysis are useful for various chemical related fields. The functional and structural diversity of the hydrazone group is shown in Figure 1, which shows the versatile characteristics of its structure including nucleophilic imine and amino type nitrogen which are more reactive, imine carbon with electrophilic and nucleophilic character, CQN double bond exhibiting configural isomerism due to its intrinsic nature and an acidic N-H proton. These characteristics are responsible for its physicochemical properties leading to its applications for molecular switches, metal coordination, cation sensing, anion sensing, etc. (Su and Aprahamian 2014). Their hydrazone based compounds such as fluorescent chemosensors are designed with a fluorophore or aromatic ring structures with such a functional group (Wang et al. 2016, Mukherjee et al. 2020).

Ling et al. developed ferrocenyl hydrazone chemosensors for optical and electrochemical sensing of both cations and anions. They used N-tosyl hydrazone for selective and sensitive optical and/or redox sensing of Hg^{2+}, Cu^{2+} and F^- ions, all of whom represent the heavy metal category, that can cause considerable damage to living cells and human body. The redox activity of ferrocene was combined with binding ability of hydrazone and tested towards various cations and anions with the help of electrochemical, spectroscopic, and optical techniques. Two synthesized hydrazones, 2a and 2b, were reported to exhibit high binding affinity and sensitivity towards these cations and anions. The addition of Hg^{2+} into a solution of 2a and 2b changes its color from pale yellow to yellowish green, enabling a naked eye detection. A cathodic shift was observed in Fe/Fe^+ redox couple upon formation of a complex with Cu^{2+} and Hg^{2+}. Also, the addition of F^- ions was indicative in the UV absorption spectra and electrochemical signals due to the binding of synthesized hydrazones with F^- via hydrogen bonding within N-H proton of hydrazone and F^- as determined by ^{1}NMR titration (Ling et al. 2019). The various other chemical ions for which these hydrazone-based sensors are used are listed as Table 1.

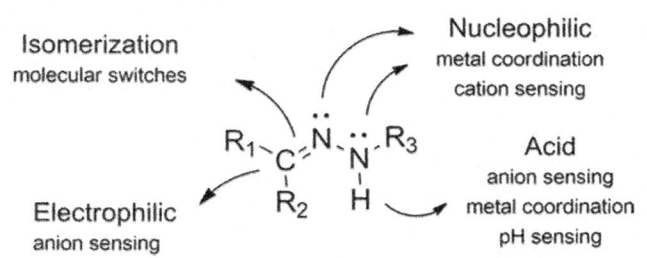

Figure 1. Functional/Structural diversity of hydrazone group. (Reprinted with permission from Su and Aprahamian 2014).

Table 1. List of sensory combinations with hydrazone for sensing of various parameters, cations and anions.

Sr. No.	Hydrazone	Sensory entity	Characteristics	References
1	Rhodamine B-based hydrazone	Cu^{2+}	• Can be used in the fluorescence sensing of Cu^{2+} at the sub micromolar level in buffered aqueous solutions with 25 nM and 3.3 μM as the detection range. • High selectivity against Fe^{3+}, Fe^{2+}, Zn^{2+}, Pb^{2+}, and Hg^{2+}.	(Xiang et al. 2006)
2	Coumarin-hydrazone	Cu^{2+}	• The sensor exhibits a fast and linear response to Cu^{2+} with a 6-fold turn-off signal with a K_a of 0.1 μM in aqueous solution. • It exhibits a highly selective fluorescence response above other transition metal ions to Cu^{2+}.	(Xu et al. 2013)
3	5- (3-hydroxy-4-((2-picolinoylhydrazono) methyl) phenoxy)-N,N,N-trimethylpentan-1-aminium bromide	pH sensing	• The pyridinoyl-hydrazone moiety works as the pH-sensitive fluorophore/chromophore probe. • It shows a noteworthy pH-dependent behavior in both absorption and fluorescence spectra with high sensitivity and a naked eye on-off switch effect at neutral pH.	(Diana et al. 2019)
4	1- ((Z)- (((Z)-1-(5-bromothiophen-2-yl) ethylidene) hydrazono) methyl) naphthalen-2-ol (NAPABTH)	Pb^{2+} and F^-	• Efficient radiometric chemosensors for the selective sensing of Pb^{2+} and F^- ions using thiophene functionalized hydrazone. • The interaction of NAPABTH with Pb^{2+} and F^- ions was visually observed through the development of pink and dark yellow solutions, respectively. • The detection limits were found to be very low for Pb^{2+} and F^- ions as 1.06 ppm and 3.72 nM, respectively.	(Anbu Durai et al. 2021)
5	Various hydrazones derived from pyridoxal and hydrazides of 2-furoic, thiophene-2-carboxylic, pyrazinoic and pyridine-3-carboxylic acids	Co^{2+}, Co^{3+}, Ni^{2+}, and Zn^{2+}	• Changes the solution color, UV-Vis and fluorescent spectra due to the addition of d-metal ions to pyridoxal-derived hydrazones dissolved in buffered DMSO. • High stability constants at neutral pH make PL-derived hydrazones promising agents of heavy metal detoxification	(Gamov et al. 2020)
6	1,8-naphthalimide hydrazide-based hydrazone with substituted furan and thiophene rings	F^- and CN^-	• Fast response, naked eye color change and quenching of fluorescence. • Nitro and methyl furan substitution provided good selectivity to CN^- in THF. • The detection limits of the synthesized sensor molecules were found below 0.3 ppm for both F^- and CN^- detection.	(Saini et al. 2019)

#		Anions		Reference
7	Phosphine substituted hydrazones with or without nitro substituents	F^-, Cl^-, Br^-, I^-, CH_3COO^- and $H_2PO_4^-$	• The binding ability of receptors with anions is reinforced via the introduction of NO_2, a colorimetric group along with electron-withdrawing effect. • Color change observed with F^-, CH_3COO^- and $H_2PO_4^-$ • The binding ability of fluoride with receptor is not affected by the existence of other anions.	(Kumaravel et al. 2017)
8	2-pyridylaldehyde fluorescein hydrazone	Cl^-	• The color change on reaction with hypochlorites was rapid and visible to the naked eye. • Common anions, including F^-, Cl^-, ClO_3^-, NO_2^-, CN^-, S^{2-}, SCN^-, $P_2O_7^{4-}$, AcO^-, CO_3^{2-}, SO_4^{2-}, ClO_4^-, did not interfere with the hypochlorite detection mechanism	(Huo et al. 2012)
9	1,10-phenanthroline-2,9-dicarboxyaldehyde-di-(p-nitrophenylhydrazone)	AcO^-	• Drastic color change from yellow to green in the presence of AcO^- with no changes for other anions • The presence of electron withdrawing groups increases the hydrogen bond donor ability of N-H framework, thus making it favorable for AcO^- sensing	(Qiao et al. 2009)

Black phosphorus

Black phosphorus (BP) is a very recent material which is being used widely in the sensory field. It has two dimensional structure and certain unique properties such as excellent electronic and optical properties, and good biocompatibility (Novoselov et al. 2004, Zhu et al. 2010). It is being used in photoelectric devices, cancer treatment, drug delivery and many other fields. It also has the capacity of carrying large current as well as switching large current ratios. It is non-toxic, thermodynamically stable at room temperature and has a layered-puckered honeycomb structure, in which each phosphorus is bonded by single bond to three phosphorus atoms. BP is used as colorimetric biosensor as it amplifies the signal and enhances the sensitivity of the biosensor. Peng et al. prepared biomarkers using BP and gold hybrid compound (Peng et al. 2017). This biosensor was helpful in detecting carcino-embryonic antigen for colon and breast cancer related samples. However, the disadvantage as reported by Liu et al. was its nature to get degraded in air thereby limiting its usage (Liu et al. 2019). When considered in bulk form, it is seen to behave as a p-type semiconductor. It shows a direct band gap of nearly 0.30 eV and good electrical conductivity (Appalakondaiah et al. 2012).

These BP based sensors have also been reportedly employed in the humidity sensors field due to the properties such as ultra-sensitivity, selective response towards humid air and also the capability of detections at traces level. BP based humidity sensors exhibit configurations of field effective transistors (Korotcenkov 2019). Yasaei et al. developed sensors for humidity sensing applications based on BP. He compared sensors made out of BO, graphene, molybdenum sulphide and concluded that BP based sensors showed the highest sensitivity amongst them (Yasaei et al. 2015). Erande and co-workers developed BP based humidity sensors for conductometric applications (Erande et al. 2016). However, the responses obtained in this case were smaller than that obtained by Yasaei (Nilges et al. 2008). Also, Late prepared BP nano sheets of varying thickness and studied their sensitivity. He concluded that thinner sheets exhibited better response than larger thickness sheets (Late 2016). Yao and co-worker developed quartz crystal microbalance (QDM) based on BP humidity sensors. Two dimensional BP nanosheets were used here and the responses obtained were in form of well-defined logarithmic curve (Yao and Xue 2015). The response was dependent on the number of BP nano sheets used in the microbalance. As the thickness of the BP film increased, the response time increased. However, the recovery time of these sensors was independent of the thickness of the BP sheets. When compared with the humidity sensors of conductometric applications, it was observed that in conductometric based sensors large drift was observed. It was concluded that for higher relative humidity applications, the QDM based BP sensors were more suitable.

Although BP based humidity sensors are finding applications in many fields, certain difficulties arise while using them as sensors such as processing issues regarding the humidity sensitive layers from BP nano sheets. Another drawback is the relative instability of the BP sheets when they were exposed to the open atmospheric conditions and it directly affected its performance capacities (Ling et al. 2015).

Diamond

Diamond has proved to be an excellent material for use as a sensor for studying microscopic structures such as proteins, normally optical microscopes were being used. However, the precision required was not satisfied by using the microscopes and hence the option of diamond as a quantum sensory material was opted for. These diamond-based sensors were produced by two methods, i.e., carbon being subjected to extremely high temperature and pressure to form diamond or its growth on a substrate by the method of deposition of a layer of carbon atoms, more popularly known as "Chemical Vapor Deposition". The second method produces flat samples of diamonds. As the atomic numbers of carbon and nitrogen are very close to each other, therefore, for filling up the spaces of the vacancies of the diamond crystals, nitrogen atoms were introduced. These nitrogen vacancy centers in the diamond crystals make them eligible for being used as ultrasensitive quantum sensor for magnetic fields. These quantum sensors help in studying of protein molecules, bio molecules as well as structure of solids up to the nano level. They can be also used to analyze the structures of the semiconductors. This diamond based quantum sensor is also being used in nuclear magnetic resonance analysis technique (Linares et al. 2009).

Piezo resistive sensors based on boron doped with nano crystalline diamond (NCD) are being commonly used in the medical field (Kulha et al. 2012). Diamond shows strong piezo resistivity along with unique electrical, thermal and mechanical properties. Also, it shows wide band gap property. Due to this piezo resistive property, it can be easily coupled with micro electromechanical systems. The crystalline diamond films are being used as sensors in the medical field. The piezo resistivity of diamond is dependent on the type of film of the diamond crystal. As the grades vary from polycrystalline to nano crystalline thin films, the property changes. The doping level also affects the sensory activity of the material. The grain boundaries as well as the defects arising out of the disordered atoms help in trapping of holes (for p type doping) and create a potential barrier at the boundary and a depletion layer extending inside the grains. In this manner, doping helps in enhancing the sensory activity of crystalline materials, in this case diamond (Kulha et al. 2009, 2012, Venkatanarayanan and Spain 2014).

Electrolyte materials

The need for electrolytic material was realized because of the environmental pollution that was being caused by the conventional materials such as coal and oil-based ones. Normally the electrolytes are either in liquid or in molten state. But some electrolytes are available in solid state and researchers opted for it for being used in sensory applications. The solid electrolytes don't have outstanding properties, which is needed in the field of sensors and thus have limited applications in this field. They are suitable for the applications in which high current densities are not required. One such application is oxygen sensor in which this solid electrolyte is being used. A unique example of it is the oxygen-ion conductor cubic zirconium which is stabilized by using yttrium (Pasierb and Rekas 2009). The researchers are trying to reduce the operation temperature and increase the current density for the zirconium-based electrolytes. Another application of the solid electrolytes is in the field of hydrogen sensors. Generally, these potentiometric hydrogen sensors are made out of a high temperature proton conducting material, i.e., $KTaO_3$ and acceptor-doped strontium or barium cerates. Out of these, the barium cerates exhibit higher conductivity, which comprise of both protonic and ionic ions. The strontium cerates show purely protonic conductivity. These materials were previously being used in humidity sensors and in recent days, they have found applications in hydrogen sensors. The above-mentioned materials become unstable in presence of carbon dioxide gas. Zirconate perovskites compounds also showed active proton conductivity (Kreuer 2003). Also, they have good mechanical and chemical properties. Katahira et al. suggested an efficient hydrogen sensor that is composed of a hydrogen pump and a solid cell with $SrCe_{0.95}Yb_{0.05}O_{3-\alpha}$ as the solid electrolyte (Katharia et al. 2001). This material easily carried out the potentiometric pumping of hydrogen and the response time recorded was less than 60 sec.

The potentiometric sensors are categorized depending upon their modes of operation. In the first type sensors, the gas which is to be detected is converted into a mobile component in the solid electrolyte. This solid electrolyte then works by separating the sensor into a reference and test compartment. In the second type, the gas creates an equilibrium between the components that is different from the dominant mobile species. In third type, "A" type employs sodium carbonate as the solid electrolyte, whereas in "B" type lithium carbonate is being used as the solid electrolyte (Pasierb and Rekas 2009). In continuity, researchers have also been investigating the option of using electrolytes for sulphur dioxide and nitrogen dioxide sensors. The electrodes having poor catalytic activity can be used for absorbing the oxidising components and then they can be electrochemically oxidized with oxide ions with the help of the solid electrolytes. One recent application of solid electrolytes is that of flexible oxide-based neuron transistors that can be operated by using the solid-state electrolytic films. These transistors have been fabricated on flexible plastic substrates and are being used for biochemical sensing applications. These sensors could be operated at very low voltages and also showed superior electrical properties along with easy reproducibility (Liu et al. 2015). As the solid electrolytes are giving results with very short response time, their demand in near future is bound to increase and various other fields of sensors would be explored wherein these electrolytes would be comfortably used (Venkatanarayanan and Spain 2014).

Quantum dots

These are nano sized materials with very unique properties such as fluorescence, photoluminescence, chemiluminescence and electrochemiluminescence. Because of these unique properties, they are finding huge demand in the field of sensors. The quantum dots normally have zero dimensions (Li et al. 2019). The most commonly used quantum dots are made up of carbon or grapheme. The carbon quantum dots (CQD) are also known as carbon dots and are fluorescent in nature and their diameters are less than 10 nm. They have high chemical stability, high conductivity and good broadband optical absorption (Namdari et al. 2017). The graphene quantum dots (GQDs) are chemical derivatives of CQD (Pan et al. 2010). Their properties are similar to graphene or graphene oxide. They appear as small sheets of graphene having diameter less than 10 nm.

CQD was discovered in by Xu et al. in 2004 (Xu et al. 2004). The researchers are constantly striving for various methods that would give a high yield, are less polluted and economic to manufacture these quantum dots. The properties of these quantum dots can be varied by changing the doping material and its amount. The quantum dots can be synthesized by "bottom-up" or "top –down methods" (Bak et al. 2016). In the bottom-up method, the reaction conditions are very rigorous, but they provide products of uniform sizes. The top–down method is simpler and it uses cheaper raw material as compared to the previous method. This method is suitable for producing quantum dots on a large scale.

CQD and GQD have been used as photoluminescence chemo sensors. In this application, they are being used as fluorescent probes which would be used for detecting the heavy metal ions and pesticides (Kozak et al. 2016, Amini et al. 2017), and CQD have been employed to successfully detect the metal ions. Zhang et al. used nitrogen doped CQD for detecting Hg^{2+} and observed that the sensors showed selectivity towards the mercury ions (Zhang and Chen 2014). Qian and co-workers found that a sensor made of graphene oxide and GQD was sensing Pb^{2+} with a fast response time and low detection limit of 0.6 Nm (Qian et al. 2015). Also, this sensor showed good reproducibility. Because of the unique properties of the quantum dots, they are now being widely used in pharmaceutical analysis, environmental pollution detection, etc. The CQD based composites have been used for detecting micro RNA, chlorinated phenols, pentachlorophenol, etc., by using the electro chemiluminescence technique (Zhang and Chen 2014). For detecting hydrogen peroxide which is being secreted by the cancerous cells, Xi et al. developed an electrochemical hydrogen peroxide sensor and it gave results with high sensitivity, low response time as well as low detection limits (Xi et al. 2016).

Chemiluminescence chemo sensors have the advantages of simple instrumentation, high sensitivity, no interference of scattering light and wide linear range of detection (Lin et al. 2013). CQD doped with sulphur and nitrogen have been used for detecting Cu^{2+} in human plasma and water samples. These sensors have also been used for detecting organic species. Shi et al. has used GQDs composites with gold nano particles for synthesizing fluorescence resonance energy transfer biosensor (Shi et al. 2015). This sensor was used for detecting the gene sequence specific to staphylococcus aureus. Also, GQD and CQD have been used for manufacturing immune sensor. It has been reportedly found that such an immune sensor based on CQD was used for detecting carcino embryonic antigens. Yin's group has developed a photochemical biosensor for detecting zeatin (Wang et al. 2018). This sensor showed a linear range from 0.1 to 100 Nm, detection limit of 0.031 Nm and also good selectivity property. Some biosensors based on GQD have also been used for detecting the nucleic acids. Li et al. developed an electrochemical DNA sensor containing GQD (Zhao et al. 2011). The selectivity shown by this sensor in the complex media is still being explored and further studies on this are in progress. So, because of the outstanding properties, these quantum dots are finding more and more applications in today's world.

Fluorine derivatives

Fluorine is the element having the smallest size and outstanding properties such as high temperature resistance, abrasion resistance, good dielectric properties, moisture resistance, chemical resistance, etc. It forms a very stable compound with any other element. An important derivative of graphene being used in sensory applications is fluorographene and its functionalised form fluorographene oxide. It finds application in bio sensing, photovoltaics, medical diagnosis and many more fields. Functional derivatives of graphene formed by introducing hetero atoms, attaching functional groups, adsorbing groups, etc., can abruptly change the electronic properties of graphene as well as remove its low temperature fluctuations. On the other hand, these chemical modifications improve its electrical, and magnetic properties. As graphene is derived from graphite which is available easily, its chemical modifications resulting in some innovative material are easier. In recent days, these graphene derivatives are being preferred in the field of chemical sensors. When fluorine is added to graphite, its hybridization changes from sp^2 to sp^3 and hence its electronic conductivity changes. There are various methods available for synthesizing halogen derivatives of graphene. Fluorinated graphene showed certain unique properties along with the formation of the strongest bond of carbon and fluorine. Fluorographene was first reported as a stable graphene derivative by Sofo and his co-workers (Sofo et al. 2007). The synthesis of fluorinated graphene (FG) was initiated by Robinson et al. by fluorination method (Robinson et al. 2010). It was followed by mechanical exfoliation method of graphite fluoride which was introduced by Nair et al. (Nair et al. 2010). The method introduced later was unsuccessful on a larger scale and hence was not stabilised. But the first method along with variations such as the photochemical fluorination of graphene and fluorographite ionic liquid-based exfoliation has been stabilised. The major factor in this derived compound is the fluorine content.

Fluoro graphene and its various functional derivatives are preferred in field of sensors as it shows high electro activity and electron transfer. Also, due to the presence of carbon fluorine bond, it showed high analytic specificity. The incorporation of fluorine in graphene also modifies its surface properties and can be used as biomarkers (Boopathi et al. 2014). These fluoro graphenes have been used for sensing various bio molecules such as ammonia, dopamine, ascorbic acid, etc. (Urbanová et al. 2016). The C-F bond in fluorographene plays an important role in solvent assisted defluorination and this property is preferable for using these fluoro graphenes in sensors. It was also observed that fluoro graphenes with low content of fluorine exhibit excellent sensing properties. The detection of nicotine amide adenine dinucleotide (NADH) was carried out by involving fluorographene electrodes of varying fluorine concentrations. A neurotransmitter based on ascorbic acid and dopamine causes neurological imbalance in human bodies. For detecting it, sensitive electrodes based on graphene were being used. But it was observed that the electro-oxidation potential of these molecules was dependent on the alignment of the graphene molecules as well as the functional groups attached to it. The best results were obtained when reverse fluorographene oxide electrodes were used. Another bio molecule is ammonia which causes respiratory disorders and various other skin and pulmonary disorders were also sensed with the help of these fluorographene electrodes. The previously used detection methods were metal oxide-based sensors, catalytic ammonia sensors, conducting polymer-based ammonia sensors, etc. But all these methods were costly, had relatively slower detecting times and hence were not preferred. The hydrogen bonding of ammonia with the strongest electronegative element, i.e., fluorine was opted for detecting ammonia at a fast rate. Fluorographene oxide electrodes showed very high sensing for ammonia and it was proved by the high orbital overlapping of ammonia and fluorine orbitals (Tadi et al. 2016). Also, these sensors were reversible in nature. A simple washing of the electrode reverted its impedance to its original value. Recent studies have also shown that ammonia can also be detected by using fluorographene conductometric sensors (Narayanan et al. 2017). Fluorographene has been successfully used for detecting DNA strands by means of impedimetric sensors (Urbanová et al. 2015). Fluorographene is the thinnest 2D insulator and is being used in organic field effect transistors and also as a gate dielectric material (Zhu et al. 2013). Fluorographene has also been used in self-aligned graphene transistors to improve their performance (Ho et al. 2015). Also, they have been used in single atomic layer transistors (Ho et al. 2014). They are being used as probes in the matrix-assisted or surface-enhanced laser ionization mass spectrometry for detecting the trace amounts of chemical compounds. There usage has also been explored for oil-water separations (Chronopoulos et al. 2017).

The sensing of many more bio molecules by using fluorine based compounds is being explored and the future may hold the prospect for the detection of many important bio molecules by these fluorinated sensors.

Meta materials

The metamaterials are artificial materials in which the metallic elements are periodically arranged. Their sizes are much less than the wavelength of the incident electromagnetic waves. The metamaterials fall in the sub wavelength category and are invisible to the incident rays. Because of the periodic arrangement, they show unique responses to the electromagnetic waves. Researchers found metamaterials when they were exploring for materials which would exhibit negative refractive index and unique electromagnetic properties such as backward propagation, reverse Doppler effect, etc. Metamaterials are generally used as sensors for sensing the properties such as displacement, density, strain, material characterization, etc. These sensory activities are controlled by changes in the transmission or the reflection coefficient as a factor of the frequency or the wavelength. Metamaterial sensors are preferred to other analytical techniques such as microwave techniques as they don't need any pre or post treatment of test materials. Also, these materials have very simple topology and low fabrication cost (Vivek et al. 2019).

The metamaterials are being used for sensory applications since 2007. Melik et al. reported of using these materials in nested spring ring resonator (SRR) wherein they were used for sensing the strains developed as a result of plate fitted in long fractured bones (Melik et al. 2009). Wu et al. in 2016 used the above sensors for sensing the displacement activities (Wu et al. 2016). Similarly, complimentary SRR was used for sensing surface crack detection in metals as reported by Albishi and Ramahi (Albishi and Ramahi 2017). When they are being used for material characterization, then depending upon the size, shape and nature of the test samples, they are categorized in 4 classes viz. thin-dielectric layer characterization, bulk-dielectric characterization, micro-fluidic characterization, and bio-molecule characterization.

In thin dielectric layer characterization, the sensing capacity of the sensors towards small quantity of biological and chemical samples is validated by means of thin dielectric films. These thin films are normally made up of silica or polyvinyl alcohol, etc. The thickness of the film is in micron meter range and they are coated over the metamaterials (Vivek et al. 2019). The performance of the sensor is checked by observing the difference in the values of the resonating frequencies with and without sample. The difference in the value of the resonating frequency is dependent on the thickness of dielectric coating, refractive index and permittivity of the film. Huang et al. developed a metamaterial absorber for sensing the permittivity in the mid infrared region (Huang et al. 2018).

In the second technique, the metamaterials are loaded with thick dielectric materials such as Teflon, roger's wood, etc. Here the thickness of the dielectric materials is in mm and above range. This technique is used for checking properties such as moisture, composition, material quality, etc., of the dielectric material (Vivek et al. 2019). When the metamaterials were loaded with these thick dielectrics, their capacitance increased depending on the permittivity and thickness of the dielectrics. The result was observed by a shift in the resonance. Various configurations of SRR such as aligned gap SRR, centred gap SRR, rectangular complimentary SRR were used here in the form of metamaterials (Rusni et al. 2014). Also, researchers prepared metamaterial resonators of various shapes such as S-shaped, H-shaped, X-shaped for employing them in various sensory applications (Sabah et al. 2015).

In the third method, quantification and detection of hazardous chemicals of the environment is carried out. Normally, the microfluidic sensors have a microfluid channel through which the samples are fed. In the metamaterial based microfluidic sensors, the passage for the samples is located in the area where there is availability of high electric field (Vivek et al. 2019). This electric field shows disturbance when the liquid sample flows through the passage. This disturbance depends on the property of the liquid sample and in turn varies the capacitance of the metamaterial, which results in a shift in the resonance frequency. Metamaterial micro fluids having a compact structure are also being used as bio sensors. Withayachumnankul et al. reported the sensing of liquids such as water, ethanol, and their mixture by using microfluidic channel fitted inside PET film over a split gap type SRR (Withayachumnankul et al. 2013). Benkhaoua et al. reported the use of double sided spiral SRR having interdigital gaps (Benkhaoua et al. 2015). These gaps were protected by polyisoprene tube with thin acrylic layer coating. They were used for sensing various liquid samples and their resonance frequency was successfully recorded. Sadeqi et al. reported the use of paper-based metamaterial sensor by using screen printing and wax printing methods. It was a very low cost sensor and wax printing was used on the microfluidic channel made out of paper whereas the resonator cells were designed by using screen printing (Sadeqi et al. 2017).

In the fourth method, the metamaterial sensors were being used in medical field for diagnosing the abnormalities. Here the difference in the resonance peaks of the normal and infected cell or tissue was monitored. SRR was used for sensing various bio molecules (Vivek et al. 2019). For some specific molecules such as biotin and streptavidin, the SRR copper pattern was coated with nickel and gold. Lee et al. used asymmetric SRR for detecting stress biomarkers like cortisol and a-amylase (Lee et al. 2013). Xu et al. developed a flexible biosensor for detecting Bovine Serum Albumin in the visible IR region (Xu et al. 2011). Doležel et al. proposed the use of a metamaterial PMMA-graphene core dielectric medium with gold shell, for detecting damages and tumors in tissues (Doležel et al. 2014). Vafapour reported the use of asymmetric complimentary metamaterial nanostructures as highly efficient bio sensors (Vafapour 2017). So, although metamaterials are being widely used in the sensory field, in days to come, newer materials' exploration in this field is expected.

Ligands

The most common ligand used in sensory applications is chelate ligands which consists of an imine bond formed by the reaction of aldehyde and primary amine group via Schiff's base reaction. Salen ligand has vast applications in various fields such as in molecular magnetism, sensory applications, bioinorganic activities, catalytic field, etc. It has certain outstanding properties such as reusability, recoverability, high efficiency and selectivity. In year 1933, the usage of salen ligands was reported. The word salen is a combination of two chemicals viz. "*sal*" means salicyl aldehyde and "*en*" means ethylenediamine. When salicylaldehyde reacts with ethylenediamine in ratio of 2:1, a tetradentate chelating ligand is obtained which is known as 'salen'. The salen ligand contains the N_2O_2 donor sites which helps the various metal ions to adapt to various geometries (Asatkar et al. 2020).

Salen can be artificially designed suited to the required applications easily. Various properties of salen such as its solubility, chirality, stability, etc., can be easily modified depending on the field of application. In the salen ligand system, there are two reactive sites, i.e., the aromatic ring and the diamine linkage. In the salicylideneimine structure, the reactive positions are 3 and 5. If aromatic ring is substituted, the solubility of ligand improves and if diamine group is substituted, it generates another chiral ligand. The various types of salen ligands are: chiral salen, non-symmetrical salen, conjugated salen, and salen-based metal organic framework.

Metal based salen complexes have excellent sensory properties and have been successfully used for colorimetric and fluorimetric responses. Chan et al. has reported the use of the Pt (II)-salphen based polymeric sensors for detecting various ions such as Pd (II), Cd (II), Hg (II), Zn (II), etc. (Guo et al. 2009). Wezenberg et al. has reported the use of Zn (II)-salphen complexes as metal ion sensors for the demetallation process (Wezenberg et al. 2008, 2010). Song et al. has reported that the chiral salen based fluorescent polymeric sensor can be used for the optical form detection of α-hydroxy carboxylic acids. Also, they have shown that Zn (II) ions can be detected by using salen based fluorescent polymeric sensor (Song et al. 2011, 2012). The salen based chemo sensors have been used for detecting the Al (III) ions by trans metallation process (Cheng et al. 2014). Metal based salen shows many biological activities such as antimicrobial and antioxidant activity and also anticancer propensity and, therefore, they have been used in various therapeutics applications as well as bio sensors.

Crown ethers

Crown ethers are unusually known to form stable complexes with cations. The oxygen atoms align in a planar arrangement surrounding the central cation. This property makes it most viable for being used in the construction of sensory materials (Kakhki and Rakhshanipour 2019). Normally, these crown ethers are connected to molecular scaffolds for sensing the metal ions. Crown ethers are widely used in colorimetric sensors. These sensors have simple instrumentation and in today's analytical field, their demand is increasing by leaps and bounds. Crown ethers were being used in optical transduction techniques in which non-polar chromophores were present. But due to the non-polar nature of the chromophores, their usage in the aqueous medium got restricted. The extinction coefficient of these chromophores affected the detection sensitivity of the sensors (Kakhki and Rakhshanipour 2019). In recent days, these chromophores were substituted by using metallic nano particles. The most commonly used were gold or silver nano particles. These nano particles had very large extinction coefficient as compared to the chromophores and also, they showed a distinct color change when the phase changed from dispersed to aggregated phase.

For the gold nano particles, the color change observed was from red to purple or blue and for silver nano particles, it was from yellow to red or dark green. This color change is seen because of the shifting of the surface plasmon to a longer wavelength range. Gold nano particles have excellent bio stability, bio functionalization property as well as they are easy to prepare and hence they have high demand in the field of sensors (Li et al. 2010). Crown ethers are also being used as multifunctional materials by adding them in supra molecular assemblies (Srivastava et al. 2005). The crown ether-based sensors are being used for detecting the heavy metal ions in the environment. Gold nano particles coupled crown ethers are more preferred in colorimetric sensors because they don't use light sensitive dyes and their instrumentation is very simple. The gold nano particles coupled with aza crown ether were installed in UV spectrophotometer and were used for detecting Pb^{2+} ions successfully. This experiment was carried out by Lin et al. in 2002 (Lin et al. 2002). The lead ion detection was easily highlighted with a distinctive color change. 15-crown-5 functionalized gold nano particles in aqueous medium was used by Lin to detect K^+ ions successfully. The color change observed was from red to blue. Wu et al. took trials of gold nano particles coupled crown ethers for the pre-treatment of complicated urine samples (Wu et al. 2008). For improving the sensitivity of the sensors, efforts were taken to improve the design and synthesis methods of these sensors. Similarly, aza crown ether gold nano particle was used for detecting various other materials such as barium, melamine, etc. When compared with silver nano particle coupled with aza crown ether, it was observed that their selectivity towards metallic ions was poor as compared to the gold nano particles. Bayrakcı et al. reported that aza crown ether coupled with magnetite nano particle was used for studying the human serum albumin (HSA) binding properties (Bayrakcı et al. 2013).

Crown ethers having 18-crown-6 (18C6), dicyclohexyl18-crown-6 (DC18C6) and dibenzo18-crown-6 (DB18C6) have been experimented for sensing of lead ions by using potentiometric sensors. In this application, they were used in PVC membranes based lead ion selective electrodes which were selective towards the lead ions (Zamani et al. 2011). The analysis was carried out by means of potentiometric titrations. This lead selective electrode which consisted of the crown ether was experimented and the selectivity towards mono, di and trivalent ions was studied. The observations were that sensitivity of the crown ethers towards the lead ions was very strong. These potentiometric sensors can be used as indicator electrodes for sensing lead ions from chromate compounds (Zamani et al. 2011).

Porphyrin

Porphyrins are versatile and are capable of signal interacting with the host molecules. They are natural chromophores, possessing an adaptive chelating structure, and have a π-aromatic system. These features are well utilized with a transducer enabling the construction of a variety of chemical sensors based on porphyrinoids. They can bind with gaseous compounds in a reversible manner and are capable of operating photophysical and/or redox chemical processes facilitated by the target analytes. A host of changes can be done to the porphyrin ring via the synthesis route such as coordination metal, skeleton or the peripheral substituent modification. They are well known to mimic their biological functions such as activation of molecular oxygen in order to catalyse the photosynthesis process. These functions further expand the possibilities of sensor designs as they can be extended to different porphyrinoids and they have paved the way for porphyrinoids based chemical sensors for gas and liquid sensing applications (Paolesse et al. 2017). Figure 2 gives an account of a variety of applications within the market of chemical sensing products.

Porphyrins are molecules that can carry out and hold multiple interactions at once. As a result, their selectivity is limited as molecular design can highlight one interaction among others and may give rise to ambiguous signals due to low selectivity. However, if a sample in a compound is known, they can very well be employed for such applications. As informed earlier, they can be utilized for gaseous analytes, wherein the sensor allows the measurement of interaction events between the analyte and sensory layer. The physical quantification of these molecular recognition events can help in determination of four parameters, i.e., mass, electrical impendence, surface potential, and optical characteristics.

Hou et al. reported the development of a three-component pyrene-porphyrin based radiometric fluorescent sensor array for sensing of glycosaminoglycans (GAG). Owing to their complexity and heterogeneity in composition, varying charges, polydispersity, and presence of isobaric stereoisomers, the GAG sample analysis poses considerable challenges to current analytical techniques. They found a structural control over the cross-responsiveness of Hep, HA, Chs, and DS in three porphyrins synthesized using varying pyrene-porphyrin supramolecular complexes. These sensors were potent for discrimination of GAG in PBS as well as 5% serum media. Also, the differentiation of Hep from other biological interferences was reliable along with cent percent accuracy of detection of trace contaminants in Hep (0.1%, wt.%) (Hou et al. 2021).

Colombelli et al. demonstrated the enhanced sensing capabilities of VOCs and oxidizing gas by cobalt bis-porphyrin derivative ($(Co-H)Por_2$) with the help of spectroscopic techniques. A magneto-optical surface plasmon resonance configuration was employed to deposit multilayers of ($(Co-H)Por_2$) onto Au/Co/

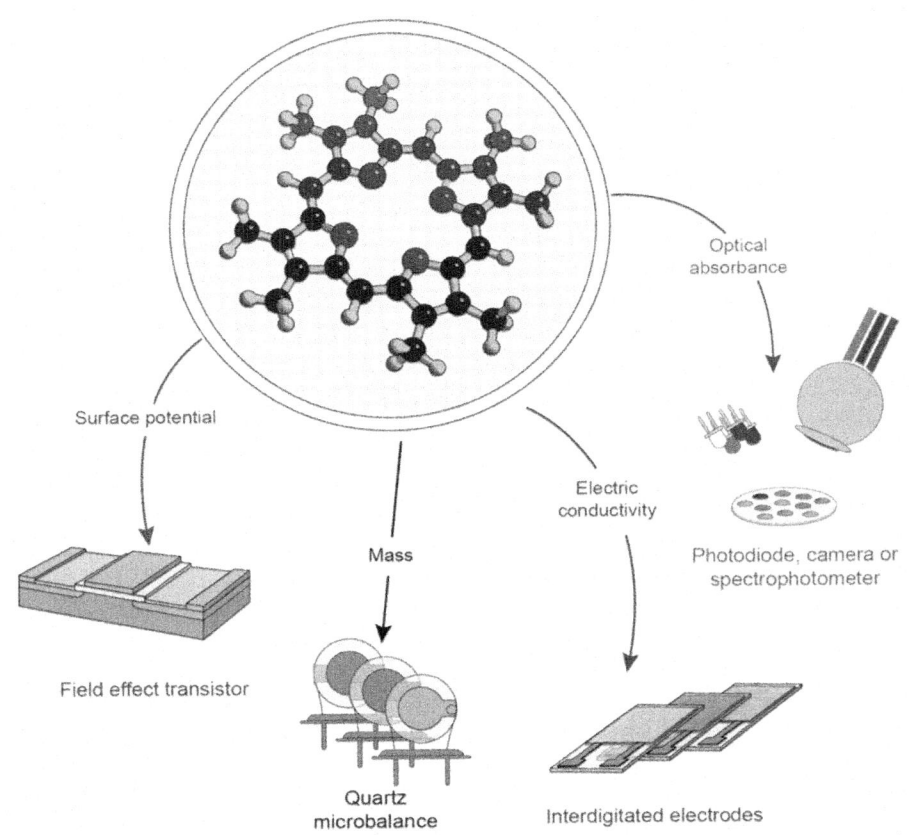

Figure 2. Porphyrin applications as sensing materials. (Reprinted with permission from Paolesse et al. 2017).

Au transducers. The device thus developed showed remarkable sensitivity towards trimethylamine, butanone and NO_2, with an inference that the sensing performance is dependent upon the molecular structure of the alcohols (Colombelli et al. 2017). There have been other reports of NO_2 gas sensing using Meso-5,10,15,20-tetrakis-(4-tertbutyl phenyl) porphyrin (Abudukeremu et al. 2018). The sensing of a variety of other chemicals such as Cd^{2+} ions (Huang et al. 2017), cadmium (Zhao et al. 2020), thiophenol (Chen et al. 2019), diclofenac (Intrieri et al. 2018), and a variety of cations such as Zn^{2+}, Cd^{2+}, Co^{2+}, Fe^{2+}, Ni^{2+}, Cu^{2+}, Ag^+, and Hg^{2+} (Moura et al. 2021), either in solid, liquid, and vapor form, have also been recently reported.

Combination of Multiple Sensory Materials

The combination of various sensory materials can enhance its efficiency and/or functional effect of the sensors in chemical sensing applications. Rahaman et al. developed a polyaniline/graphene/CNT nanocomposite for sensing of 4-aminophenol, an environmentally hazardous chemical. The nanocomposite was prepared by *in situ* polymerization process and was deposited over a flat silver electrode, resulting in the construction of a phenolic sensor with rapid response to 4-aminophenol in a buffer system. The sensitivity and detection limit were observed to be 2.1873 μA cm^{-2} μM^{-1} and 63.4 pM, respectively, with a linear calibration plot for 0.1 nM to 0.01 M concentration range of 4-aminophenol. The SEM micrographs showed significant surface changes due to addition of graphene and CNT in the polymer matrix. Also, investigation of other analytical parameters was done with the help of I-V method at ambient conditions. They revealed the response time, selectivity and reproducibility to be very good and proposed the method to play an important part in selective determination over a wide linear dynamic range with high selectivity and sensitivity, with results having the required levels of satisfaction for use in biomedical and environmental samples (Rahman et al. 2018).

An advanced biosensing application for detection of dopamine was recently reported to consist of a nanocomposite electrode fabricated by polyaniline/N-doped-graphene/DNA-f-CNT (Keteklahijani et al. 2019). The doping of graphene was responsible to increase the catalytic activity of polyaniline. A glassy carbon electrode was coated with a thin layer of DNA-f-CNT and N-doped-graphene. Thereafter, the polyaniline was deposited over the electrode through *in situ* polymerization. This complex structure developed was tested to be a highly sensitive and selective biosensor for dopamine sensing over a linear range of 0.02–1 μM. The detection limit was reportedly 14 nM which certainly draws the pathway for molecular diagnosis of diseases such as Parkinson's disease, a neurological illness.

Matrix Materials

The matrix material in a sensor is majorly chosen on the basis of the applications in which it is intended to be used. In case of a wearable sensor, biocompatibility and flexibility is an obvious characteristic which must be offered by the matrix material. Also, there are other product forms in which the sensors might be required, for example, coatings/thin films, membranes, microneedles, etc. The conductive polymer basket offers good sensing abilities and are employed either alone or in combination with other sensory/matrix materials (Lei et al. 2014, Naveen et al. 2017, Moon et al. 2018, Naseri et al. 2018). For instance, a highly stretchable polymer composite (PMMA/PPy) microtube chemical sensor of microarch shape was reportedly fabricated by the meniscus guided approach. The PMMA/PPy microarches were fabricated on Pt-coated silicon wafers which were embedded on polydimethyl siloxane. As PMMA and PPy components have active interactions with gases of redox nature and inert organic vapors, respectively, they work very well together due to higher surface area and the structural components of the blend (Won et al. 2017). Table 2 gives an account of a variety of polymers being used as matrix materials in chemical sensing applications.

Conclusion

The chapter was focused on providing an overview of miscellaneous materials which can be employed for various chemical sensing applications for biological, environmental, and industrial monitoring of the presence of chemical entities such as gases, vapors, enzymes, proteins, etc. The other chapters of the book were focused on

Table 2. Some recent applications of various polymer matrices in chemical sensing applications.

Sr. No.	Matrix	Sensory material	Applications	References
1	PET	MoS_2 and CNT	Chemical gas flexible sensors	(Kim et al. 2018)
2	PMMA	Zeolitic Imidazolate	H_2 gas sensor	(Kim et al. 2020)
3	PU	FA and HPTS doped silica	pH sensing in acidic, neutral and basic ranges of wastewater	(Duong et al. 2019)
4	PEG	CNT	Selective determination of GSH in spiked rabbit serum	(Rahman et al. 2017)
5	PEO	Nano-CuO and MWCNT	Humidity sensor	(Ahmad et al. 2021)
6	Nylon-6	Graphene	Moisture sensor	(Hossain et al. 2018)
7	PVC	5,5'- (1,4-phenylene) bis (3- (naphthalen-1-yl)-4,5-dihydro-1H-pyrazole-1-carbothioamide)	Chromium (III)-selective potentiometric sensor	(Isildak et al. 2020)
8	PS	Carbon nanoparticles	Biomedical sensing	(Hegarty et al. 2019)
9	PC or PS	Palladium particles	Transdermal sensing	(McConville and Davis 2016)

the major materials employed in sensory use for various chemicals and this chapter in particular exhibits a potential of other materials, for example, fluorescent sensing materials such as pyrylium and hydrazone, black phosphorus, diamond, electrolyte sensors, quantum dots, fluorine derivatives, meta materials, ligand, crown ether, porphyrin, etc. These materials have been well researched and proven effective in a variety of chemical sensing applications.

References

Abudukeremu, H., Kari, N., Zhang, Y., Wang, J., Nizamidin, P., Abliz, S., and Yimit, A. 2018. Highly sensitive free-base-porphyrin-based thin-film optical waveguide sensor for detection of low concentration NO_2 gas at ambient temperature. J. Mater. Sc. 53: 10822–10834.

Ahmad, W., Jabbar, B., Ahmad, I., Mohamed Jan, B., Stylianakis, M. M., Kenanakis, G., and Ikram, R. 2021. Highly sensitive humidity sensors based on polyethylene oxide/CuO/Multi walled carbon nanotubes composite nanofibers. Materials. 14: 1037.

Albishi, A. M., and O. M. Ramahi. 2017. Microwaves-based high sensitivity sensors for crack detection in metallic materials. IEEE Transactions on Microwave Theory and Techniques 65(5): 1864–1872.

Amini, N., Shamsipur, M., Gholivand, M. B., and Barati, A. 2017. A glassy carbon electrode modified with carbon quantum dots and polyalizarin yellow R dyes for enhanced electrocatalytic oxidation and nanomolar detection of L-cysteine. Microchem. J. 131: 9–14.

Anbu Durai, W., Ramu, A., and Dhakshinamoorthy, A. 2021. A visual and ratiometric chemosensor using thiophene functionalized hydrazone for the selective sensing of Pb2+ and F− Ions. J. Fluoresc. 31: 465–474.

Appalakondaiah, S., Vaitheeswaran, G., Lebegue, S., Christensen, N. E., and Svane, A. 2012. Effect of van der Waals interactions on the structural and elastic properties of black phosphorus. Phys. Rev. B. 86: 035105.

Arques, A., Amat, A. M., Santos-Juanes, L., Vercher, R. F., Marín, M. L., and Miranda, M. A. 2009. Abatement of methidathion and carbaryl from aqueous solutions using organic photocatalysts. Catal. Today. 144: 106–111.

Asatkar, A. K., Tripathi, M., and Asatkar, D. 2020. Salen and related ligands. stability and applications of coordination compounds. pp 99. IntechOpen.

Bak, S., Kim, D., and Lee, H. 2016. Graphene quantum dots and their possible energy applications: A review, Curr. Appl. Phys. 16: 1192–1201.

Banu, I. S., and Ramamurthy, P. 2009. Photoinduced electron transfer reactions of pyrylium derivatives with organic sulfides in acetonitrile. J. Photochem. Photobiol. A Chem. 201: 175–182.

Bayarri, B., Carbonell, E., Gimenez, J., Esplugas, S., and Garcia, H. 2008. Higher intrinsic photocatalytic efficiency of 2,4,6-triphenylpyrylium-based photocatalysts compared to TiO_2 P-25 for the degradation of 2,4-dichlorophenol using solar simulated light. Chemosphere. 72: 67–74.

Bayrakcı, M., Maltaş, E., Yiğiter, Ş., and Özmen, M. 2013. Synthesis and application of novel magnetite nanoparticle based azacrown ether for protein recognition. Macromol. Res. 21: 1029–1035.

Beltrán, A., Burguete, M. I., Galindo, F., and Luis, S. V. 2020. Synthesis of new fluorescent pyrylium dyes and study of their interaction with N-protected amino acids. New Journal of Chemistry. 44: 9509–9521.

Benkhaoua, L., Benhabiles, M. T., Mouissat, S., and Riabi, M. L. 2015. Miniaturized quasi-lumped resonator for dielectric characterization of liquid mixtures. IEEE Sens. J. 16: 1603–1610.

Bhandari, S. 2019. Polymer/carbon composites for sensor application. *In*: Rahaman, M., Khastgir, D., and Aldalbahi, A. K. (eds.). Carbon-Containing Polymer Composites, Springer Series on Polymer and Composite Materials. pp. 503–531. Springer Singapore, Singapore.

Boopathi, S., Narayanan, T. N., and Kumar, S. S. 2014. Improved heterogeneous electron transfer kinetics of fluorinated graphene derivatives. Nanoscale. 6: 10140–10146.

Chakraborty, G., Gupta, A., Pugazhenthi, G., and Katiyar, V. 2018. Facile dispersion of exfoliated graphene/PLA nanocomposites via *in situ* polycondensation with a melt extrusion process and its rheological studies. J. Appl. Polym. Sc. 135: 46476.

Che, Y., Ma, W., Ji, H., Zhao, J., and Zang, L. 2006. Visible photooxidation of dibenzothiophenes sensitized by 2-(4-methoxyphenyl)-4, 6-diphenylpyrylium: an electron transfer mechanism without involvement of superoxide. J. Phys. Chem. B. 110: 2942–2948.

Chen, A., Wu, W., and Jones, W. E. 2017. Polymers and molecular wires as chemical sensors. pp. 179–195. *In*: Atwood, J. L. (ed.). Comprehensive Supramolecular Chemistry II. Elsevier, Oxford.

Chen, G., Tang, M., Fu, X., Cheng, F., Long, Y., Li, Y., Jiao, Y., and Zeng, R. 2019. A highly sensitive and selective "off-on" porphyrin-based fluorescent sensor for detection of thiophenol. J. Mol. Struct. 1179: 593–596.

Chen, S., Gao, J., Chang, J., Zhang, Y., and Feng, L. 2019. Organic-inorganic manganese (II) halide hybrids based paper sensor for the fluorometric determination of pesticide ferbam. Sens. Actuat. B: Chem. 297. 126701.

Cheng, J., Ma, X., Zhang, Y., Liu, J., Zhou, X., and Xiang, H. 2014. Optical chemosensors based on transmetalation of salen-based Schiff base complexes. Inorg. Chem. 53: 3210–3219.

Chronopoulos, D. D., Bakandritsos, A., Pykal, M., Zbořil, R., and Otyepka, M. 2017. Chemistry, properties, and applications of fluorographene. Appl. Mater. Today. 9: 60–70.

Colombelli, A., Manera, M. G., Borovkov, V., Giancane, G., Valli, L., and Rella, R. 2017. Enhanced sensing properties of cobalt bis-porphyrin derivative thin films by a magneto-plasmonic-optochemical sensor, Sens. Actuat. B: Chem. 246: 1039–1048.

Diana, R., Panunzi, B., Tuzi, A., Piotto, S., Concilio, S., and Caruso, U. 2019. An Amphiphilic Pyridinoylhydrazone Probe for Colorimetric and Fluorescence pH Sensing, Molecules. 24: 3833.

Doležel, I., Tarparelli, R., Iovine, R., La Spada, L., and Vegni, L. 2014. Surface plasmon resonance of nanoshell particles with PMMA-graphene core. COMPEL - Int. J. Comput. Math. Electr. Electron. Eng.

Duong, H. D., Shin, Y., and Rhee. J. I. 2019. Development of fluorescent pH sensors based on a sol-gel matrix for acidic and neutral pH ranges in a microtiter plate. Microchem. J. 147: 286–295.

El-Roz, M., Awala, H., Thibault-Starzyk, F., and Mintova, S. 2017. Selective response of pyrylium-functionalized nanozeolites in the visible spectrum towards volatile organic compounds. Sens. Actuat. B: Chem. 249: 114–122.

Erande, M. B., Pawar, M. S., and Late, D. J. 2016. Humidity sensing and photodetection behavior of electrochemically exfoliated atomically thin-layered black phosphorus nanosheets. ACS Appl. Mater. Interfaces. 8(18): 11548–11556.

Fonseca, S. E. B., Rivas, B. L., Pérez, J. M. G., Calzada, S. V., and García, F. 2018. Synthesis of a polymeric sensor containing an occluded pyrylium salt and its application in the colorimetric detection of trimethylamine vapors. J. Appl. Polym. Sc. 135: 46185.

Gamov, G. A., Zavalishin, M. N., Petrova, M. V., Khokhlova, A. Y., Gashnikova, A. V., Kiselev, A. N., and Sharnin, V. A. 2020. Interaction of pyridoxal-derived hydrazones with anions and Co^{2+}, Co^{3+}, Ni^{2+}, Zn^{2+} cations, Phys. Chem. Liq. 0: 1–13.

Guo, Z., Tong, W. -L., and Chan, M. C. 2009. Axially rotating (Pt-salphen) 2 phosphorescent coordination frameworks, Chem. Commun. 41: 6189–6191.

Hegarty, C., McKillop, S., McGlynn, R. J., Smith, R. B., Mathur, A., and Davis, J. 2019. Microneedle array sensors based on carbon nanoparticle composites: interfacial chemistry and electroanalytical properties. J. Mater. Sc. 54: 10705–10714.

Ho, K. -I., Liao, J. -H., Huang, C. -H., Hsu, C. -L., Zhang, W., Lu, A. -Y., Li, L. -J., Lai, C. -S., and Su, C. -Y. 2014. One-step formation of a single atomic-layer transistor by the selective fluorination of a graphene film. Small. 10: 989–997.

Ho, K. -I., Boutchich, M., Su, C. -Y., Moreddu, R., Marianathan, E. S. R., Montes, L., and Lai, C. -S. 2015. A self-aligned high-mobility graphene transistor: decoupling the channel with fluorographene to reduce scattering. Adv. Mater. 27: 6519–6525.

Hossain, M. M., Islam, M. A., Shima, H., Kafeela, M., Hasan, M., and Lee, M. 2018. Synergistic effect in moisture sensing of nylon-6 Polymer films through molecular-level interfacial interactions of amide linkages in the presence of graphene. J. Phys. Chem. C. 122: 24672–24683.

Hou, M., Fan, L., Fan, X., Liang, X., Zhang, W., and Ding, Y. 2021. Pyrene-porphyrin based ratiometric fluorescent sensor array for discrimination of glycosaminoglycans. Anal. Chim. Acta. 1141: 214–220.

Huang, H., Xia, H., Xie, W., Guo, Z., Li, H., and Xie, D. 2018. Design of broadband graphene-metamaterial absorbers for permittivity sensing at mid-infrared regions. Sci. Rep. 8: 1–10.

Huang, W. -B., Gu, W., Huang, H. -X., Wang, J. -B., Shen, W. -X., Lv, Y. -Y., and Shen, J. 2017. A porphyrin-based fluorescent probe for optical detection of toxic Cd^{2+} ion in aqueous solution and living cells, Dyes Pigm. 143: 427–435.

Huo, F. -J., Zhang, J. -J., Yang, Y. -T., Chao, J. -B., Yin, C. -X., Zhang, Y. -B., and Chen, T. -G. 2012. A fluorescein-based highly specific colorimetric and fluorescent probe for hypochlorites in aqueous solution and its application in tap water. Sens. Actuat. B: Chem. 166–167: 44–49.

Hwang, Y., Park, J. Y., Kwon, O. S., Joo, S., Lee, C. -S., and Bae, J. 2018. Incorporation of hydrogel as a sensing medium for recycle of sensing material in chemical sensors. Appl. Surf. Sci. 429: 258–263.

Intrieri, D., Damiano, C., Rizzato, S., Paolesse, R., Venanzi, M., Monti, D., Savioli, M., Stefanelli, M., and Gallo, E. 2018. Sensing of diclofenac by a porphyrin-based artificial receptor. New J. Chem. 42: 15778–15783.

Isildak, Ö., Özbek, O., and Gürdere, M. B. 2020. Development of Chromium (III)-selective Potentiometric Sensor by Using Synthesized Pyrazole Derivative as an Ionophore in PVC Matrix and its Applications. J. Anal. Test. 4: 273–280.

Jang, J. -S., Koo, W. -T., Choi, S. -J., and Kim, I. -D. 2017. Metal organic framework-templated chemiresistor: sensing type transition from p-to-n using hollow metal oxide polyhedron via galvanic replacement. J. Am. Chem. Soc. 139: 11868–11876.

Kakhki, R. M., and Rakhshanipour, M. 2019. Application of nanoparticle modified with crown ether in colorimetric determinations. Arab. J. Chem. 12: 3096–3107.

Katahira, K., Matsumoto, H., Iwahara, H., Koide, K., and Iwamoto, T. 2001. A solid electrolyte hydrogen sensor with an electrochemically-supplied hydrogen standard. Sens. Actuat. B Chem. 73(2-3): 130–134.

Keteklahijani, Y. Z., Sharif, F., Roberts, E. P. L., and Sundararaj, U. 2019. Enhanced sensitivity of dopamine biosensors: an electrochemical approach based on nanocomposite electrodes comprising polyaniline, nitrogen-doped graphene, and DNA-Functionalized carbon nanotubes. J. Electrochem. Soc. 166: B1415.

Kim, S., Han, J., Kang, M. -A., Song, W., Myung, S., Kim, S. -W., Lee, S. S., Lim, J., and An, K. -S. 2018. Flexible chemical sensors based on hybrid layer consisting of molybdenum disulphide nanosheets and carbon nanotubes. Carbon. 129: 607–612.

Kim, H., Kim, W., Cho, S., Park, J., and Jung, G. Y. 2020. Molecular Sieve Based on a PMMA/ZIF-8 Bilayer for a CO-Tolerable H2 Sensor with Superior Sensing Performance. ACS Appl. Mater. Interfaces. 12: 28616–28623.

Korotcenkov, G. 2019. Black phosphorus-new nanostructured material for humidity sensors: achievements and limitations. Sensors. 19: 1010.

Kozak, O., Sudolska, M., Pramanik, G., Cigler, P., Otyepka, M., and Zboril, R. 2016. Photoluminescent carbon nanostructures, Chem. Mater. 28: 4085–4128.

Kreuer, K. -D. 2003. Proton-conducting oxides. Annu. Rev. Mater. Res. 33: 333–359.

Kulha, P., Kromka, A., Babchenko, O., Vanecek, M., Husak, M., Williams, O. A., and Haenen, K. 2009. Nanocrystalline diamond piezoresistive sensor. Vacuum. 84: 53–56.

Kulha, P., Babchenko, O., Kromka, A., Husak, M., and Haenen, K. 2012. Design and fabrication of piezoresistive strain gauges based on nanocrystalline diamond layers. Vacuum. 86: 689–692.

Kumaravel, M., Mague, J. T., and Balakrishna, M. J. 2017. Hydrazone derivatives appended to diphenylphosphine oxide as anion sensors. J. Chem. Sci. 129: 471–481.

Late, D. J. 2016. Liquid exfoliation of black phosphorus nanosheets and its application as humidity sensor. Microporous Mesoporous Mater. 225: 494–503.

Lazar, I. G., Diacu, E., Buica, G. O., Ungureanu, E. M., Arnold, G. L., and Birzan, L. 2017. The heavy metals sensing based on 2, 6-Bis (-2- (Thiophen-3-yl) Vinyl)-4- (4, 6, 8-Trimethylazulen-1-yl) Pyrylium Modified Electrodes. Rev. Chim. 68: 2509–2513.

Lee, H. -J., Lee, J. -H., Choi, S., Jang, I. -S., Choi, J. -S., and Jung, H. -I. 2013. Asymmetric split-ring resonator-based biosensor for detection of label-free stress biomarkers. Appl. Phys. Lett. 103: 053702.

Lei, W., Si, W., Xu, Y., Gu, Z., and Hao, Q. 2014. Conducting polymer composites with graphene for use in chemical sensors and biosensors, Microchim. Acta. 181: 707–722.

Li, M., Chen, T., Gooding, J. J., and Liu, J. 2019. Review of carbon and graphene quantum dots for sensing, ACS Sens. 4: 1732–1748.

Li, Y., Schluesener, H. J., and Xu, S. 2010. Gold nanoparticle-based biosensors. Gold Bull. 43: 29–41.

Lin, S. -Y., Liu, S. -W., Lin, C. -M., and Chen, C. 2002. Recognition of potassium ion in water by 15-crown-5 functionalized gold nanoparticles. Anal. Chem. 74: 330–335.

Linares, R., Doering, P., and Linares, B. 2009. Diamond bio electronics. Stud. Health Technol. Inform. 149: 284–296.

Ling, L., Hu, J., and Zhang, H. 2019. Ferrocene containing N-tosyl hydrazones as optical and electrochemical sensors for Hg^{2+}, Cu^{2+} and F$^-$ ions. Tetrahedron. 75: 2472–2481.

Ling, X., Wang, H., Huang, S., Xia, F., and Dresselhaus, M. S. 2015. The renaissance of black phosphorus. Proc. Natl. Acad. Sci. 112: 4523–4530.

Liu, H., Song, H., Su, Y., and Lv, Y. 2019. Recent advances in black phosphorus-based optical sensors. Appl. Spectrosc. Rev. 54: 275–284.

Liu, N., Zhu, L. Q., Feng, P., Wan, C. J., Liu, Y. H., Shi, Y., and Wan, Q. 2015. Flexible sensory platform based on an electrolyte-gated oxide neuron transistor. arXiv preprint arXiv.1505. 04964.

McConville, A., and Davis, J. 2016. Transdermal microneedle sensor arrays based on palladium: Polymer composites. Electrochem. Commun. 72: 162–165.

Melik, R., Unal, E., Perkgoz, N. K., Santoni, B., Kamstock, D., Puttlitz, C., and Demir, H. V. 2009. Nested metamaterials for wireless strain sensing. IEEE J. Sel. Top. Quantum Electron. 16: 450–458.

Moon, J. -M., Thapliyal, N., Hussain, K. K., Goyal, R. N., and Shim, Y. -B. 2018. Conducting polymer-based electrochemical biosensors for neurotransmitters: A review. Biosens. Bioelectron. 102: 540–552.

Mordor Intelligence. 2020. Chemical sensor market - growth, trends, covid-19 impact, and forecasts (2021–2026).

Moura, N. M. M., Valentini, S., Cheptene, V., Pucci, A., Neves, M. G. P. M. S., Capelo, J. L., Lodeiro, C., and Oliveira, E. 2021. Multifunctional Porphyrin-based dyes for cations detection in solution and thermoresponsive low-cost materials. Dyes Pigm. 185: 108897.

Mukherjee, S., Betal, S., and Chattopadhyay, A. P. 2020. A novel turn-on red light emitting chromofluorogenic hydrazone based fluoride sensor: Spectroscopy and DFT studies. J. Photochem. Photobiol. A Chem. 389: 112219.

Nair, R. R., Ren, W., Jalil, R., Riaz, I., Kravets, V. G., Britnell, L., Blake, P., Schedin, F., Mayorov, A. S., and Yuan, S. 2010. Fluorographene: a two-dimensional counterpart of Teflon. Small. 6: 2877–2884.

Namdari, P., Negahdari, B., and Eatemadi, A. 2017. Synthesis, properties and biomedical applications of carbon-based quantum dots. An updated review, Biomedicine and Pharmacotherapy. 87: 209–222.

Naseri, M., Fotouhi, L., and Ehsani, A. 2018. Recent progress in the development of conducting polymer-based nanocomposites for electrochemical biosensors applications: A mini-review. Chem. Rec. 18: 599–618.

Naveen, M. H., Gurudatt, N. G., and Shim, Y. -B. 2017. Applications of conducting polymer composites to electrochemical sensors: A review. Appl. Mater. Today. 9: 419–433.

Nilges, T., Kersting, M., and Pfeifer, T. 2008. A fast low-pressure transport route to large black phosphorus single crystals. Journal of Solid State Chemistry. 181: 1707–1711.

Novoselov, K. S., Geim, A. K., Morozov, S. V., Jiang, D., Zhang, Y., Dubonos, S. V., Grigorieva, I. V., and Firsov, A. A. 2004. Electric field effect in atomically thin carbon films, Science. 306: 666–669.

Ohashi, T., and Dai, L. 2006. C60 and carbon nanotube sensors. pp. 525–575. *In:* Dai, L. (ed.). Carbon Nanotechnology. Elsevier, Amsterdam.

Pan, D., Zhang, J., Li, Z., and Wu, M. 2010. Hydrothermal route for cutting graphene sheets into blue-luminescent graphene quantum dots. Adv. Mater. 22: 734–738.

Paolesse, R., Nardis, S., Monti, D., Stefanelli, M., and Di Natale, C. 2017. Porphyrinoids for chemical sensor applications. Chem. Rev. 117: 2517–2583.

Pasierb, P., and Rekas, M. 2009. Solid-state potentiometric gas sensors—current status and future trends. J. Solid State Electrochem. 13: 3–25.

Peng, J., Lai, Y., Chen, Y., Xu, J., Sun, L., and Weng, J. 2017. Sensitive detection of carcinoembryonic antigen using stability-limited few-layer black phosphorus as an electron donor and a reservoir. Small. 13: 1603589.

Qian, Z. S., Shan, X. Y., Chai, L. J., Chen, J. R., and Feng, H. 2015. A fluorescent nanosensor based on graphene quantum dots–aptamer probe and graphene oxide platform for detection of lead (II) ion. Biosens. Bioelectron. 68: 225–231.

Qiao, Y. -H., Lin, H., Shao, J., and Lin, H. -K. 2009. A highly selective naked-eye colorimetric sensor for acetate ion based on 1,10-phenanthroline-2,9-dicarboxyaldehyde-di-(p-substitutedphenyl-hydrazone). Spectrochim. Acta A Mol. Biomol. Spectrosc. 72: 378–381.

Rahman, M. M., Hussein, M. A., Salam, M. A., and Asiri, A. M. 2017. Fabrication of an L-glutathione sensor based on PEG-conjugated functionalized CNT nanocomposites: a real sample analysis. New J. Chem. 41: 10761–10772.

Rahman, M. M., Hussein, M. A., Alamry, K. A., Al-Shehry, F. M., and Asiri, A. M. 2018. Polyaniline/graphene/carbon nanotubes nanocomposites for sensing environmentally hazardous 4-aminophenol. Nano-Struct. Nano-Objects 15: 63–74.

Regan, F. 2019. Sensors: Overview. pp. 172–178. *In:* Worsfold, P., Poole, C., Townshend, A., and Miró, M. (eds.). Encyclopedia of analytical science (Third Edition). Academic Press, Oxford.

Robinson, J. T., Burgess, J. S., Junkermeier, C. E., Badescu, S. C., Reinecke, T. L., Perkins, F. K., Zalalutdniov, M. K., Baldwin, J. W., Culbertson, J. C., and Sheehan, P. E. 2010. Properties of fluorinated graphene films. Nano Lett. 10: 3001–3005.

Rusni, I. M., Ismail, A., Alhawari, A. R. H., Hamidon, M. N., and Yusof, N. A. 2014. An aligned-gap and centered-gap rectangular multiple split ring resonator for dielectric sensing applications. Sensors. 14: 13134–13148.

Sabah, C., Taygur, M. M., and Zoral, E. Y. 2015. Investigation of microwave metamaterial based on H-shaped resonator in a waveguide configuration and its sensor and absorber applications. J. Electromagn. Waves Appl. 29: 819–831.

Sadeqi, A., Nejad, H. R., and Sonkusale, S. 2017. Low-cost metamaterial-on-paper chemical sensor. Opt. Express. 25: 16092–16100.

Saini, N., Wannasiri, C., Chanmungkalakul, S., Prigyai, N., Ervithayasuporn, V., and Kiatkamjornwong, S. 2019. Furan/thiophene-based fluorescent hydrazones as fluoride and cyanide sensors. J. Photochem. Photobiol. A Chem. 385: 112038.

Schmitt, L. D. 2020. Organic and Organometallic Anion Sensors: Design, Synthesis, and Function – PhD thesis. State University of New York at Binghamton.

Schroeder, V., Savagatrup, S., He, M., Lin, S., and Swager, T. M. 2018. Carbon nanotube chemical sensors. Chem. Rev. 119: 599–663.

Sheng, Y., and Regner, M. 2019. Roles of water molecules and counterion on hs– sensing reaction utilizing a pyrylium derivative: a computational study. J. Phys. Chem. A. 123: 3334–3343.

Shi, J., Chan, C., Pang, Y., Ye, W., Tian, F., Lyu, J., Zhang, Y., and Yang, M. 2015. A fluorescence resonance energy transfer (FRET) biosensor based on graphene quantum dots (GQDs) and gold nanoparticles (AuNPs) for the detection of mecA gene sequence of Staphylococcus aureus. Biosens. Bioelectron. 67: 595–600.

Singh, E., Meyyappan, M., and Nalwa, H. S. 2017. Flexible graphene-based wearable gas and chemical sensors, ACS Appl. Mater. Interfaces. 9: 34544–34586.

Sofo, J. O., Chaudhari, A. S., and Barber, G. D. 2007. Graphane: A two-dimensional hydrocarbon. Phys. Rev. B. 75: 153401.

Song, F., Ma, X., Hou, J., Huang, X., Cheng, Y., and Zhu, C. 2011. (R, R)-salen/salan-based polymer fluorescence sensors for Zn^{2+} detection. Polymer. 52: 6029–6036.

Song, F., Wei, G., Wang, L., Jiao, J., Cheng, Y., and Zhu, C. 2012. Salen-based chiral fluorescence polymer sensor for enantioselective recognition of α-hydroxyl carboxylic acids. J. Org. Chem. 77: 4759–4764.

Srivastava, S., Frankamp, B. L., and Rotello, V. M. 2005. Controlled plasmon resonance of gold nanoparticles self-assembled with PAMAM dendrimers. Chem. Mater. 17: 487–490.

Su, X., and Aprahamian, I. 2014. Hydrazone-based switches, metallo-assemblies and sensors. Chem. Soc. Rev. 43: 1963–1981.

Sun, Q., Qian, B., Uto, K., Chen, J., Liu, X., and Minari, T. 2018. Functional biomaterials towards flexible electronics and sensors. Biosens. Bioelectron. 119: 237–251.

Tadi, K. K., Pal, S., and Narayanan, T. N. 2016. Fluorographene based ultrasensitive ammonia sensor. Sci. Rep. 6: 1–9.

Tessarolo, M., Gualandi, I., and Fraboni, B. 2018. Recent progress in wearable fully textile chemical sensors. Adv. Mater. Technol. 3: 1700310.

Urbanová, V., Holá, K., Bourlinos, A. B., Čépe, K., Ambrosi, A., Loo, A. H., Pumera, M., Karlický, F., Otyepka, M., and Zbořil, R. 2015. Thiofluorographene–hydrophilic graphene derivative with semiconducting and genosensing properties. Adv. Mater. 27: 2305–2310.

Urbanová, V., Karlický, F., Matěj, A., Šembera, F., Janoušek, Z., Perman, J. A., Ranc, V., Čépe, K., Michl, J., and Otyepka, M. 2016. Fluorinated graphenes as advanced biosensors–effect of fluorine coverage on electron transfer properties and adsorption of biomolecules. Nanoscale. 8: 12134–12142.

Vafapour, Z. 2017. Near infrared biosensor based on classical electromagnetically induced reflectance (CI-EIR) in a planar complementary metamaterial. Opt. Commun. 387: 1–11.

Venkatanarayanan, A., and Spain, E. 2014. Review of recent developments in sensing materials. 47: 101.

Vivek, A., Shambavi, K., and Alex, Z. C. 2019. A review: metamaterial sensors for material characterization. Sens. Rev. 39: 417–432.

Wang, L., Chen, X., and Cao, D. 2016. A cyanide-selective colorimetric "naked-eye" and fluorescent chemosensor based on a diketopyrrolopyrrole–hydrazone conjugate and its use for the design of a molecular-scale logic device. RSC Adv. 6: 96676–96685.

Wang, Y., Zhou, Y., Xu, L., Han, Z., Yin, H., and Ai, S. 2018. Photoelectrochemical apta-biosensor for zeatin detection based on graphene quantum dots improved photoactivity of graphite-like carbon nitride and streptavidin induced signal inhibition. Sens. Actuat. B: Chem. 257: 237–244.

Wezenberg, S. J., Escudero-Adán, E. C., Benet-Buchholz, J., and Kleij, A. W. 2008. Colorimetric discrimination between important alkaloid nuclei mediated by a bis-salphen chromophore. Org. Lett. 10: 3311–3314.

Wezenberg, S. J., Anselmo, D., Escudero-Adán, E. C., Benet-Buchholz, J., and Kleij, A. W. 2010. Dimetallic activation of dihydrogen phosphate by Zn (salphen) chromophores. Eur. J. Inorg. Chem. 2010: 4611–4616.

Withayachumnankul, W., Jaruwongrungsee, K., Tuantranont, A., Fumeaux, C., and Abbott, D. 2013. Metamaterial-based microfluidic sensor for dielectric characterization. Sens. Actuat. A Phys. 189: 233–237.

Won, K. H., Weon, B. M., and Je, J. H. 2017. Highly stretchable polymer composite microtube chemical sensors produced by the meniscus-guided approach. Curr. Appl. Phys. 17: 339–342.

Wu, S. -H., Wu, Y. -S., and Chen, C. 2008. Colorimetric sensitivity of gold nanoparticles: minimizing interparticular repulsion as a general approach. Anal. Chem. 80: 6560–6566.

Wu, W., Ren, M., Pi, B., Cai, W., and Xu, J. 2016. Displacement sensor based on plasmonic slot metamaterials. Appl. Phys. Lett. 108: 073106.

Xiang, Y., Tong, A., Jin, P., and Ju, Y. 2006, New fluorescent rhodamine hydrazone chemosensor for Cu (II) with high selectivity and sensitivity. Org. Lett. 8: 2863–2866.

Xu, H., Wang, X., Zhang, C., Wu, Y., and Liu, Z. 2013. Coumarin-hydrazone based high selective fluorescence sensor for copper (II) detection in aqueous solution. Inorg. Chem. Commun. 34: 8–11.

Xu, X., Ray, R., Gu, Y., Ploehn, H. J., Gearheart, L., Raker, K., and Scrivens, W. A. 2004. Electrophoretic analysis and purification of fluorescent single-walled carbon nanotube fragments. J. Am. Chem. Soc. 126: 12736-12737.

Xu, X., Peng, B., Li, D., Zhang, J., Wong, L. M., Zhang, Q., Wang, S., and Xiong, Q. 2011. Flexible visible–infrared metamaterials and their applications in highly sensitive chemical and biological sensing. Nano Lett. 11: 3232–3238.

Yahaya, I., and Seferoglu, Z. 2018. Fluorescence dyes for determination of cyanide. Photochemistry and Photophysics-Fundamentals to Applications. IntechOpen. 179.

Yao, Y., and Xue, Y. 2015. Impedance analysis of quartz crystal microbalance humidity sensors based on nanodiamond/graphene oxide nanocomposite film. Sens. Actuat. B Chem. 211: 52–58.

Yasaei, P., Kumar, B., Foroozan, T., Wang, C., Asadi, M., Tuschel, D., Indacochea, J. E., Klie, R. F., and Salehi-Khojin, A. 2015. High-quality black phosphorus atomic layers by liquid-phase exfoliation. Adv. Mater. 27: 1887–1892.

Zamani, A. A., Khorsihdi, N., Mofidi, Z., and Yaftian, M. R. 2011. Crown ethers bearing 18C6 Unit; sensory molecules for fabricating PVC membrane lead ion-selective electrodes. J. Chin. Chem. Soc. 58: 673–680.

Zappa, D., Galstyan, V., Kaur, N., Arachchige, H. M. M., Sisman, O., and Comini, E. 2018. "Metal oxide-based heterostructures for gas sensors"-A review. Anal. Chim. Acta. 1039: 1–23.

Zhai, Q., Li, J., and Wang, E. 2017. Recent advances based on nanomaterials as electrochemiluminescence probes for the fabrication of sensors. ChemElectroChem. 4: 1639–1650.

Zhang, R., and Chen, W. 2014. Nitrogen-doped carbon quantum dots: facile synthesis and application as a "turn-off" fluorescent probe for detection of Hg^{2+} ions. Biosens. Bioelectron. 55: 83–90.

Zhao, J., Chen, G., Zhu, L., and Li, G. 2011. Graphene quantum dots-based platform for the fabrication of electrochemical biosensors. Electrochem. Commun. 13: 0231–33.

Zhao, L., Zhao, Y., Li, R., Wu, D., Xu, R., Li, S., Zhang, Y., Ye, H., and Xin, Q. 2020. A porphyrin-based optical sensor membrane prepared by electrostatic self-assembled technique for online detection of cadmium (II). Chemosphere. 238: 124552.

Zhu, M., Xie, X., Guo, Y., Chen, P., Ou, X., Yu, G., and Liu, M. 2013. Fluorographene nanosheets with broad solvent dispersibility and their applications as a modified layer in organic field-effect transistors. Phys. Chem. Chem. Phys. 15: 20992–21000.

Zhu, Y., Murali, S., Cai, W., Li, X., Suk, J. W., Potts, J. R., and Ruoff, R. S. 2010. Graphene and graphene oxide: synthesis, properties, and applications. Adv. Mater. 22: 3906–3924.

Index

P

Performance parameters 29, 45
pH detection 180, 187
Polymer dots 138, 147, 151
Pyrylium 208, 224

Q

Quantum dots 214, 215, 224

R

Real-world applications 30

S

Selective recognition 111, 112, 116, 121
Selectivity 19–21, 25–30, 34–36, 38–40, 45, 46
Sensing applications 105, 107, 108, 112, 123, 125
Sensing mechanisms 1, 2, 5, 9

Sensor device architecture 26
Sensors 75–92
Smart dressing 184, 186
Stimuli responsive 79
Sweat analysis 168, 169, 172, 173, 180, 182, 183

T

Textile sensor 164, 165, 167, 174, 177–180, 187, 189–195
Theoretical models 24, 25
Transducer 2, 3
Tunability 125

W

Wearable sensor 163, 164, 167, 169, 173, 174, 180, 181, 195, 196
Wound monitoring 184–188, 196

About the Editors

Subhendu Bhandari

Dr. Subhendu Bhandari is an Assistant Professor in the Department of Plastic and Polymer Engineering at Maharashtra Institute of Technology, Aurangabad, Maharashtra, since 2015. His academic background and research interest encompass diversified fields of polymer science and technology. He pursued Bachelor of Technology in Jute and Fibre Technology from University of Calcutta in 2005 and Master of Technology in Rubber Technology from Rubber Technology Centre at Indian Institute of Technology Kharagpur in 2009. He earned his Ph.D. in the field of nanostructured conducting polymer and its composites for supercapacitor application in the year 2015. His research interest includes conducting polymer, nanocomposites, supercapacitor and fiber reinforced composites. He is also a life member of The Indian Society for Technical Education, Asian Polymer Association and The Society for Polymer Science, India.

Arti Rushi

Dr. Arti Rushi is currently working as an Assistant Professor in Electronics and Telecommunication Department, Maharashtra Institute of Technology, Aurangabad (MS) India. She has received her M.Sc. and Ph.D. degree in 2012 and 2016, respectively, from Department of Physics, Dr. Babasaheb Ambedkar Marathwada University, Aurangabad (MS), India. Her area of interest includes nanotechnology, electrochemical sensors, functionalized materials, macrocyclic compounds, and application of sensors for environment protection. She has worked as reviewer for various journals of Royal Chemical Society.